Regional Economic Development

D0843999

WITHDRAWN
UTSA LIBRARIES

UTSA DT LIBRARY RENEWALS 458-2440

DATE DUE

GAYLORD   PRINTED IN U.S.A.

WITHDRAWN
UTSA LIBRARIES

Robert J. Stimson
Roger R. Stough
Brian H. Roberts

Regional Economic Development

Analysis and Planning Strategy

Second Edition

 Springer

Robert J. Stimson
Professor of Geographical Sciences
and Planning and Convenor
Australian Research Council
Research Network in Spatially Integrated
Social Science
University of Queensland
Brisbane QLD 4072
Australia
r.stimson@uq.edu.au

Professor Roger R. Stough
School of Public Policy
George Mason University
4400 University Dr.
MS 2C9
Fairfax VA 22030
Australia
rstough@gmu.edu

Professor Brian H. Roberts
Centre for Developing Cities
University of Canberra
ACT 2601
Australia
brian.roberts@canberra.edu.au

ISBN-10 3-540-34826-3 Springer Berlin Heidelberg New York
ISBN-13 978-3-540-34826-9 Springer Berlin Heidelberg New York
ISBN 3-540-43731-2 1. Ed. Springer Berlin Heidelberg New York

Library of Congress Control Number: 2006931358

This work is subject to copyright. All rights are reserved, whether the whole or part of the material is concerned, specifically the rights of translation, reprinting, reuse of illustrations, recitation, broadcasting, reproduction on microfilm or in any other way, and storage in data banks. Duplication of this publication or parts thereof is permitted only under the provisions of the German Copyright Law of September 9, 1965, in its current version, and permission for use must always be obtained from Springer-Verlag. Violations are liable for prosecution under the German Copyright Law.

Springer is a part of Springer Science+Business Media

Originally published in the series: Advances in Spatial Science

© Springer-Verlag Berlin Heidelberg 2002, 2006

The use of general descriptive names, registered names, trademarks, etc. in this publication does not imply, even in the absence of a specific statement, that such names are exempt from the relevant protective laws and regulations and therefore free for general use.

Typesetting: Camera ready by author
Production: LE-TEX Jelonek, Schmidt & Vöckler GbR, Leipzig
Cover-design: WMX Design GmbH, Heidelberg

SPIN 11769743   88/3100YL – 5 4 3 2 1 0   Printed on acid-free paper

Foreword

The second edition of this book represents a re-editing and in some cases significant rewriting of the original book published in 2001. Substantial editing occurred and new material has been included in the introductory chapter and in Chap. 10, the concluding chapter. The reason for this was to bring forward to 2006 the original thesis of the book about the need for regions to be prepared to experience increasingly greater shocks and to have increasingly less time to respond in adjusting their economic development to achieve sustainability. Events that have occurred over the past five years - such as the 9/11 terrorist attacks, continuing the rapid advances in technology, the rise of sustained economic growth in China and beginning to unfold in India, the Indonesian Tsunami, Hurricane Katrina devastation in New Orleans, the invasion of Iraq and continued unrest in the Middle East represents on-going and unforeseen risks and new challenges which continue to confront and impact nations around the globe and the regions and localities within them. Regions need to be fast and flexible and agile in order to be not only expeditiously adaptive to change, but also to be proactive in developing strategies to address and shape their futures.

In addition, some significant new material has been added to the other chapters to reflect recent advances in thinking and approaches to regional economic development analysis and planning strategy. For example, one of the most significant additions is the material on leadership and regional economic development in Chap. 8 that reflects considerable new thinking and research on how to conceptualize that leadership and institutions as platforms for sustainable regional economic development. In Chap.6, the discussion on industry clusters and cluster analysis contains new material that reflects an evolving foundation methodology for the initial identification of clusters in an unambiguous and systemic way and the validating mechanisms that may be used. This adds a new and important element to the discussion in the first edition of the book. The section on spatial cluster analysis that was in this chapter in the first edition has been moved to Chap. 9 on decision support systems as it offers a clear application of GIS to a basic analysis issue. Thus, a reader wishing to focus on industry cluster analysis will need to jump forward from Chap. 6 to Chap. 9 to continue the flow from functional cluster analysis to spatial analysis of industry clusters. And in this second edition of the book, in Chap. 7 there is some new material on the nature of regional risk factors and the way regional risk assessment may be assessed and in order that risk management approaches can be incorporated into regional development strategy and plans.

For all the chapters, the methodological components and case studies have been fully and carefully edited to ensure additional clarity for the reader. A number of new case studies have been added in some of the chapters, and old case studies updated where possible or appropriate. In this context, all of the equations and expressions have been reviewed and tested to ensure that they are correct. This should clear up some of the confusion that arose over the expressions used in the first edition, as, for example, for the material on shift-share analysis in Chap. 3. In Chap. 9, in the section dealing with the regional input-output econometric forecasting model for the Northern Virginia and National Capital Region, data has been added to the forecasts so that the methodology can be evaluated in terms of outcomes. This adds a new and interesting dimension to this analysis.

Throughout this second edition of the book, we have taken specific care to more explicitly frame the discussion of theory and applications to reflect the attention being given in contemporary research to the influence of endogenous factors in regional growth and development. In addition, we also more explicitly incorporate discussion of sustainability to reflect the increasingly pervasive concern with principles of sustainable development in regional planning and development. In Chap. 2, case studies on planning approaches incorporating sustainable development have been added.

Finally, we have added many new references to published work that has appeared in the last few years so that readers will be able to pursue both historic and current literature in the field with which this book is concerned.

Acknowledgements

A number of people over the years from the mid 1990s on have worked to help the authors bring this book into a published form for the second time. In this second edition Ms. Emilia Istrate has taken on the task of verifying content, page numbers and eliminating awkwardness in the prose, figures and charts. She has spent many days to make sure that the book is in fine shape. Chunpu Song is due special thanks for reviewing and testing all equations to ensure that they are in a correct form. Donna Sherrard provided keying assistance in making author corrections to the manuscript.

Preface

Regional economic development has attracted the interest of economists, geographers, planners and regional scientists for a long time. And, of course, it is a field that has developed a large practitioner cohort in government and business agencies from the national down to the state and local levels. In planning for cities and regions, both large and small, economic development issues now tend to be integrated into strategic planning processes.

For at least the last 50 years, scholars from various disciplines have theorised about the nature of regional economic development, developing a range of models seeking to explain the process of regional economic development, and why it is that regions vary so much in their economic structure and performance and how these aspects of a region can change dramatically over time. Regional scientists in particular have developed a comprehensive tool-kit of methodologies to measure and monitor regional economic characteristics such as industry sectors, employment, income, value of production, investment, and the like, using both quantitative and qualitative methods of analysis, and focusing on both static and dynamic analysis. The 'father of regional science', Walter Isard, was the first to put together a comprehensive volume on techniques of regional analysis (Isard 1960), and since then a huge literature has emerged, including the many titles in the series published by Springer in which this book is published.

Over time, scholars and practitioners from many fields—including planning, public administration, business, and the management sciences—have also developed various approaches to formulating strategy for regional economic development in a systematic way. This process of regional economic development strategy planning and implementation needs to be informed by regional analysis.

This book is about the analysis of regional economic performance and change, and how analysis integrates with strategies for local and regional economic development policy and planning. Quite deliberately this book is not about the theory of regional economic development, although it provides the reader with an overview of key theoretical and conceptual contexts within which the economic development process takes place. Rather, the deliberate emphasis in this book is to provide the reader—both students and practitioners—with an account of quantitative and qualitative approaches to regional economic analysis and of old and new strategic frameworks for formulating regional economic development planning. This is done within the context of the evolution of society from the industrial to the post-industrial era in which contemporary forces of globalisation and economic restructuring are creating increasing interdependence, rapid change, and high levels of uncertainty and risk for regions at all levels of scale.

The book sets out to provide teachers, students and practitioners in regional economic development with a tool-kit of tried and tested methods for regional economic analysis and strategy planning. But importantly it also introduces the reader to recent innovations and extensions in methodologies for setting about the process of regional economic development planning. This is a 'how-to-do-it' type of book, incorporating many examples of application of tools of analysis and strategic planning processes in local and regional economic development. At the same time it is cast in a story type framework about the evolution of economic development strategy and the ways strategy is being formulated in the post-Fordist era of the 21st century. However, in this context, a considerable part of the book is about the applied research experiences the authors have enjoyed with our collaborating colleagues; but we also draw on the work of many other researchers and practitioners. The bibliography provides the reader with a wide range of theoretical, methodological and applications literature to pursue.

It is the authors' hope that this book will help the reader to better understand the key considerations in regional economic development, to appreciate the value of the tools for regional analysis discussed, and to develop a better appreciation of the importance of good design for the process of regional economic development strategy formulation.

In the complex world of the early 21st century, regions—both large and small—need to be fast and flexible in adapting to the challenges of an increasingly competitive and rapidly changing set of factors that are both exogenous and endogenous to a region. This requires commitment to good practice techniques for analysing regional performance and to the process of regional economic development and strategy planning. It requires commitment to sustained leadership. And it requires the development of comprehensive and integrated information systems to understand and monitor the performance of a region and to help develop and test scenarios for future paths for regional development.

R. J. Stimson
R. R. Stough
B. H. Roberts

Contents

1 Perspectives on Regional Economic Development

1.1 Regions in the New Global Economy

The role of regions in national economies has changed significantly in recent times as a result of globalization and structural adjustment Understanding these processes of change is crucial for undertaking regional economic analysis and in planning for regional development.

In developed economies in the 1950s and 1960s, often industries in the regions of a nation tended to be highly specialized and protected through the operation of tariffs and other government policies that shielded them from international competition. Many regions—especially large urban regions—tended to be characterized by large scale, energy intensive, low labour skill, locally integrated industries producing commodities, manufactured goods and services based on resources and largely local expertise. Many regions exhibited the ultimate outcomes of specialized mass production emanating from the industrial revolution or from roles of centres of administrative control—for example, in the United States, Detroit and its region in the mid west-Great Lakes was the centre of the automobile industry and Pittsburgh was an iron and steel city; on the east coast New York was a financial centre, while Washington was an administrative government centre. Similar examples had existed at an earlier time in Europe but Europe in the 1950s and 1970s was rebuilding from the devastation of World War II.

But in the 1970s dramatic new forces were unleashed generating new processes of change that have reshaped many regions and their economies. There was the oil crisis induced by the Organization of Petroleum Producing Countries (OPEC) which ushered in a new era of increased energy costs, with new technologies re-engineering manufacturing processes and transportation and communications, re-orienting production to become less dependent on the inefficient large scale operations and intensive consumption of energy. New technologies – in particular the invention and development of the semi-conductor silicon chip - set the world on the path towards the information age of knowledge-based industries, with their requirements for new types of highly skilled, flexible labour, management and strategic alliances that are all highly mobile. With the collapse of the Bretton Woods Agreement—which since the end of World War II had seen the world's major nations and their currencies controlled through links to gold and rigid highly immobile exchange rates—began a new environment of floating exchange rates as the world entered a new era of financial deregulation and globalization of capital markets. Progressively, international agreements have been forged breaking down trade barriers as the world moved from an era of protectionism to one of competition in the new global economy.

As a result of these fundamental changes, the developed nations were plunged into an increasingly complex, uncertain and competitive world as they underwent fundamental changes in the structure of their economies. The regional impacts of these changes have been profoundly differential, producing sharp dichotomies between the winners and the losers. A relatively few world cities—such as New York, London, Tokyo, Hong Kong, Sydney—dominate global financial markets. Old industrial regional giants—such as Detroit, Cleveland, and Pittsburgh in the United States, Liverpool in the United Kingdom, and Lille in France—declined. New high technology regions—like the Silicon Valley in the San Francisco Bay area, Route 123 in Boston, the Washington capital region in the United States, Cambridge in the United Kingdom, and Mediterranean France—emerged. Sun belt growth regions—such as in Southern California, Arizona, and Florida in the United States, and the Brisbane-South East Queensland region in Australia—have prospered. Many rural regions have suffered decline as places of agricultural specialization have long been highly protected and thus continue to struggle to be competitive in selling on world markets.

In the 1980s and the 1990s, few regions have not been affected profoundly in some way or other by globalization and structural change, including changes to the international sourcing of goods, materials, services, design, finance, production and marketing. These changes have led to greater inter-regional and international trade, and at the same time resulted in the development of highly specialized agglomerations of new geographic clusters of industries, especially in global cities and mega metro regions that service both national and international markets (Amin and Goddard 1986; Erneste and Meier 1992). These powerful sub-regional economies have Gross Regional Products (GRP) larger than that of many nations (Ohmae 1995). Paradoxically, it is often mega metro regions, and no longer nation states, that are the critical drivers of economic development (Castells and Hall 1994). And in some cases regional economies dominate the national economy from a leading technology or entrepreneurial perspective.

In the past two to three decades, the rise of the post-industrial (sometimes called post-Fordist) orientation of economies across the world has seen the emergence of leading regions in terms of technology and entrepreneurial activity, such as the Third Italy (Italy), West Jutland (Denmark), Bangalore (India), and Silicon Valley and Route 123 (United States). Many studies have identified and analyzed these and other dynamic regions (see, for example, Campagni 1995; Hansen 1992; Scott and Storper 1992; Illeris 1993, Erickson 1994).

The expansion of market boundaries and the reduction of trade barriers have brought new opportunities to regional industries while simultaneously exposing them to increased competition, both domestically and internationally. In nations such as the United States, state and local economic policies have been used to stimulate the vitality and success of firms and to raise local employment and eventually income levels for many decades, and especially to assist regions to develop and adapt different policies to cope with the challenges of restructuring to improve their economies. The contemporary era represents a challenging new context for regional policy formulation. Table 1.1 lists changes in the key attributes of the 'old economy' of the industrial (Fordist) age and the 'new economy' of the post-industrial (post-Fordist) era.

Table 1.1. Attributes of the old and new economies[a]

Economy-Wide Characteristics	Old Economy (Industrial)	New Economy (Post-Industrial)
Organizational form	Vertically integrated	Horizontal networks
Scope of competition	National	Global
Markets	Stable	Volatile
Competition among sub-national	Medium	High
Geographic mobility of business	Low	High
Role of government	Provider	Steer/row/end
Labour and workforce characteristics:		
Labour-Management relations	Adversarial	Collaborative
Skills	Job-specific skills	Global learning skills and cross-training
Requisite education	Task specialization	Lifelong learning and learning by doing
Policy goal	Jobs	Higher wages and incomes (productivity)
Production characteristics:		
Resource orientation	Material resources	Information and knowledge resources
Relation with other firms	Independent ventures	Alliance and collaboration
Source of competitive advantage	Agglomeration economies	Innovation, quality, time to market and cost
Primary source of productivity	Mechanization	Digitization
Growth driver	Capital/labour/land	Innovation, invention and knowledge
Role of research and innovation in the economy	Low moderate	High
Production methodology	Mass production	Flexible production
Role of government	Infrastructure provider	Privatization
Infrastructure characteristics:		
Form	Hard (physical)	Soft (information and organizations)
Transport	Miles of highway	Travel time reduction via application of information technology
Power	Standard generation plant	Linked power grid (co-generation)
Organizational flow	Highly regulated	Deregulation
Telecommunication	Miles of copper wire	Wireless and fibre
Learning	Talking head	Distance learning

[a] For a more extended discussion see Source: Jin and Stough (1998)

A focus on regions and their economies is critically important to the under-standing of the competitiveness of nations in the new era of globalization and structural adjustment (Porter 1990; Dicken 1992). Methods and tools of regional analysis are vital both for research and to inform local and national policy makers and industry leaders in assessing the performance of a region and to formulate strategic planning frameworks to enhance a region to position itself to build and maintain competitive advantage. In addition, over the last decade regional economic development has taken place in the context of increasing concerns over sustainability.

In this book we provide an overview of approaches that have evolved and a re-view of the tools commonly used in regional economic analysis and development strategy formulation and planning. In particular, we emphasize those approaches and tools that are applicable to regions seeking to strategically position themselves to be competitive and to pursue sustainable development in the context of the rapidly changing and increasingly competitive global economy of the 21st century.

1.2 What Is Regional Economic Development?

In perusing much of the literature on regional economic development, it is surprising to find how authors have diversely and often imprecisely defined the term. Regional economic development has been seen as both a product and a process. It is the product of economic development—for example, measured jobs, wealth, investment, standard of living and working conditions, things with which people living, working and investing in regions tend to be most concerned. Generally increases or improvements in these measures are equated with economic development. It is the process—for example, industry support, infrastructure, labour force and market development—with which economists and economic planners tend to be most concerned. Unfortunately, it is often difficult to match the desired outcomes of economic development with the processes used to achieve them. This can be a dilemma for those responsible for managing economic development and for those responsible for developing strategies and plans to achieve some form of congruence between desired product outcomes and appropriate and acceptable economic processes. This dilemma may be compounded further by the unstable and changing nature of economic environments, where 'externalities' (such as exchange rates, new technologies, foreign competition) are playing a greater role in the decision-making processes which influence economic policy and strategy in regions. All too often economists are concerned with seeking to maximize economic yield, rather than looking for new approaches to achieve sustained development and to pursue sustainable development that attempt to generate benefits from economic development for people and business in communities or regions and at a the same time seek to reduce the long-term impacts of excessive environmental consumption.

1.2.1 Neoclassical Theory

Neoclassical economic theory has provided the foundation on which virtually all post World War II economic policies have been grounded. Conventional theories and policies for regional development have tended to focus in one way or another on the capital-labour production function and on responses by the state via a range of economic and non-economic policies. In this framework production (Q) is produced by two inputs capital (K) and labour (L):

$$Q = f(K,L) \tag{1.1}$$

This simple two-factor model can be used to measure productivity of capital and labour output in a region's economy. The model may be expanded to include other functions or factors, such as technology (T) or other variables including learning, to equate as:

$$Q = f(K, L, T...) \tag{1.2}$$

The neoclassical model has provided a useful basis for understanding the implications of labour and capital changes on economic performance of nations and regions (Richardson 1973). However, it does not adequately explain how productivity, performance and other values related to the application of labour, capital and technology affect economic development—especially in regional economies (Malecki 1991, p. 111). Thus neoclassical theories do not adequately identify or explain the behaviour or factors that give definition to regional economic development or economic development processes.

1.2.2 Some Definitions

Blakely (1994, p. xv) defines regional economic development as:

... a process in which local governments or community based organizations are engaged to stimulate or maintain business activity and/or employment. The principal goal of local economic development is to stimulate employment opportunities in sectors that improve the community, using existing human, natural and institutional resources.

This definition introduces a dimension to economic development that makes it concerned with more than labour, capital, prices and production. It is concerned with mobilizing social capital (Coleman 1988). The concept of social and cultural values giving structure to economic development has tended to be ignored traditionally by economists. However, new growth theory is in synchronization with this new perspective. In this context, economic development refers to increases in the quality of life associated with changes but not necessarily increases in the size and composition of population, in the quantity and nature of local jobs, and in the quantity of prices of goods and services produced locally.

Malecki's (1991) definition of regional economic development seeks to encapsulate these concepts. He defines it as:

... a combination of qualitative and quantitative features of a region's economy, which the qualitative or structural [are] the most meaningful...The qualitative attributes include the types of jobs—not only their number—and long-term and structural characteristics, such as the ability to bring about new economic activity and the capacity to maximize the benefits which remains within the region (p. 7).

He goes on to say:

... the standard theory of economic growth and development has concentrated on quantitative changes, despite an increasing awareness that regional growth depends, often critically, on aspects that are understood only in comparison with other regions or nations. The facts of regional development suggest that it is not enough to rely on the concepts of growth without an equivalent concern for the forces which commit growth to take place, or prevented it from occurring. These are the concerns of regional development, whether examined at the national, sub-national or local scale (p. 7).

Economic development thus needs to be seen as having both a quantitative and a qualitative dimension. It is quantitative with respect to the measured benefits it creates through increasing wealth and income levels, the availability of goods and services, improving financial security, and so-on. And it is also qualitative in creating greater social/financial equity, in achieving sustainable development, and in creating a spread in the range of employment and gaining improvements in the quality of life in a region. Economic development also has a product and process dimension. Product is concerned with meeting stated outputs, which might be both qualitative and quantitative. Process is concerned with the policies, strategies, means and resources used to achieve desired outputs including institutional arrangements.

Economic development may be viewed as a matrix of expressed achievable product or service outcomes for a region resulting from acceptable development processes determined by both qualitative and quantitative variables (see Fig. 1.1). This definition suggests we should no longer view regional economic development as something that is concerned primarily with the manipulation of capital, labour and technology to maximize production in response to prices and markets; rather, there are fundamentally new value systems and factors that are beginning to underpin economic systems, many of which we do not yet fully understand.

This multi-dimensional aspect of economic development leads the authors to propose the following definition of regional economic development:

.... Regional economic development is the application of economic processes and resources available to a region that results in the sustainable development of, and desired economic outcomes for a region and that meet the values and expectations of business, of residents and of visitors.

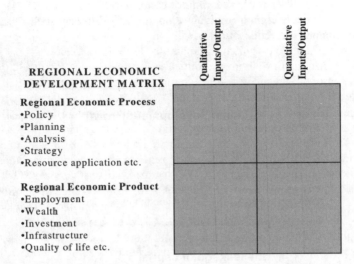

Fig. 1.1. Regional economic development as a matrix of qualitative, quantitative, process and product outcomes

While this definition is far from perfect, it does reflect a shift in economic product and process thinking. The definition also provides a basis for the framework developed in this book for regional economic development strategy and planning, and it assists in the search for factors that underlay the processes, that might support the sustainable development and competitiveness of a region. A balance between the qualitative and quantitative goals of economic development presents a challenge to traditional neoclassical economists and to the emerging breed of economists who recognize that regional economic development can occur on a more sustainable basis. This is not to suggest that neoclassical theory ideas should be dismissed—far from it. Rather, they need to evolve to accommodate those changing values that society holds on expected gains from economic development.

1.3 Understanding How Regional Economies Work and the Challenge that Represents for Development Planning

It is useful to briefly explain how regional economies work and should function under competitive market conditions. Fig. 1.2 depicts how one might conceptualise how open market regional economies 'should' function and what is involved in the development process.

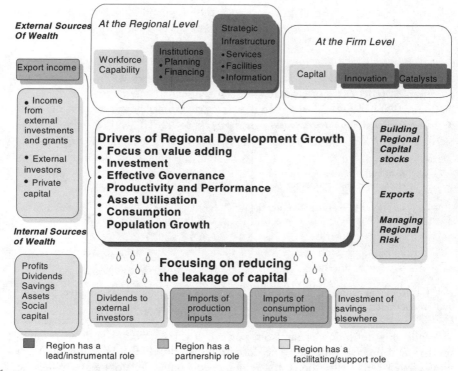

Fig. 1.2. A model of the regional economic development process
Source: Adapted from Lannon 2001

In this model the development of regional market economies is seen to be driven by the need for investment, production, employment, wages and profits and consumption. Investment capital is generated from two primary sources: external capital flows from exports, grants, external investors and repatriated earnings; and internal capital flows from the use of assets, dividends, savings, labour and social capital. The later includes all human non-financial contributions to production or services provided by the community, government and business. In many parts of the world, and in particular in places such as the Philippines, repatriated earnings are a significant external source of capital used to support regional development. And in some regions in the developing world, domestically repatriated funds from family who have migrated to the larger cities are becoming increasingly important in supporting regional development investment and private consumption.

There are several pre-conditions that may be necessary to create a competitive environment to support regional development and investment. These include: a competent workforce, effective institutions, and well-developed infrastructure, finance and logistics systems. Such elements might be loosely referred to as 'strategic infrastructure', which regional governments and institutions need to build to support development and investment. In addition, there is a need for well-developed business networks that are focused on innovation and the commercialization of new products and services. These are activities that primarily occur in the private sector.

An important factor in maintaining the development of regions is to plug the leakage of capital flows that can result from a high dependency on imports into production and consumption, the export of dividends and savings elsewhere, and the loss of human capital. Many Southeast Asian countries and for that matter the Island nations of the Caribbean, for example, have a high level of import dependency to support consumption-driven development. That leads to low regional savings and high capital leakages. In many developed economies, the leakage of pension and insurance fund payments to centralized fund managers is depleting local savings and the capital available for development. To achieve more sustained economic regional development, it is important that income leakages are reduced by creating greater investment opportunities and focusing on the development of import substitution industries.

Regions can approach the development of their economies in many ways – for example, by improving the efficiency and effectiveness of transactions; increasing population; value adding to production, services and logistics systems; closing the waste cycle and converting waste to resources; increasing consumption; increasing exports and reducing imports – and many regional development strategies include a mixture of these. However, for many regions, consumption is the main factor driving the development of the economy. In the long-term, consumption-driven development is not sustainable, especially if accompanied by rising levels of public and private debt. Striking a balance between consumption and value-added growth strategies is difficult, especially for regions that lack competitive strategic infrastructure.

A primary driver of regional development in developed nations - such as the US, Canada and Australia - has been population growth, especially through immigration. Immigration contributes significantly to GDP in Australia (Tian and Shan 1999), especially in regions like Sydney that receive more than two-fifths of Australia's migrants. And within those nations, regions that are the recipient of internal migration streams, such as 'sun belt' regions like Florida and Arizona in the US and the Brisbane-gold coast region in Australia, experience population-led growth. Lesser developed nations also encourage the entry of skilled migrants to meet the growing shortage of skilled labour. The contribution of population to the rapid growth of many coastal regions in China, and most of the capital city regions in South East Asia, is significant. However, large and rapidly growing populations create their own growth management problems. The most heavily populated countries in the world are among the poorest, so population growth does not necessarily lead to sustainable development.

For nations and regions to achieve sustainable development outcomes and growth, there is a need for a balance of policies and strategies to be applied by government and business. These include the following:

(a) a focus on increasing productivity;
(b) competitiveness and reductions in inputs to production;
(c) movement and logistic systems;
(d) the reduction and reuse of wastes;
(e) greater stretch and leveraging of resources; and
(f) the development of demand-driven, export-focused economies.

However, it is not just economic systems that should receive priority to support regional development. There is also a need to focus on:

(a) the development and maintenance of social, cultural and knowledge capital;
(b) risk management; and
(c) improved governance.

These factors are equally important to sustainability.

When considering best practice approaches to regional development, it is important to identify how regions can improve the competitiveness of assets, the inward flow of capital, information to develop knowledge, and linkages with external economies. No two regional economies are the same, so it is important for regions to fully understand where to strategically intervene and to make improvements and decide what practices work best. By adopting best practices, regional economies are able to tap the knowledge and experience of others and adapt different practices to fit local situations.

As we will keep emphasizing throughout this book, one of the difficult challenges facing those responsible for regional development is to know how to develop and sustain competitive performance within the context of regional economic development strategy formulation and planning requires specific skills in regional analysis and evaluation and needs to use best practice processes. It is not unusual for regions and the people in agencies responsible for regional development strategy planning not to have the requisite levels of skills that are necessary in order to both conduct the analyses that are needed and to design the strategy frameworks that are appropriate and which represent best practice approaches and methodologies in order to develop strategy and to formulate plans and implement them in the pursuit of development of a competitive region and development that is sustainable. For example, in those parts of the world that are characterised as more centrally planned economies, such as Vietnam and Indonesia in South East Asia, regional planners, economists and governance specialists often struggle to understand how to restructure provincial and regional economies to operate under open and competitive market conditions. An even more difficult challenge is to identify how to apply a range of best practices to improve the performance of regional economies. Many of these practices are foreign and clash with previous governance structures and institutional cultures. As Hamel and Prahalad (1994) say, there is much 'unlearning' to be done before regional institutions and businesses in those circumstances are able to understand how to adapt and apply best practices to achieve more sustainable regional development outcomes.

1.4 The Changing Paradigms Shaping Economic Development Policy and Strategy

Policy for regional economic development has undergone a number of evolutionary stages since the 1960s, driven by different paradigms of economic thought as shown in Fig. 1.3. Many of those paradigms continue to shape the way communities and people think and plan for the future, but unfortunately much of our thinking on economic development is still embedded in the paradigms of the 1970s, because of an inherent reluctance by many communities to pro-actively embrace change. Subsequently, many regions are not re-equipping themselves fast enough to compete effectively in the global age of business and technology of the post-industrial economy. To compete successfully in the global economy, regional organizations and businesses need to understand the implications of the paradigm shifts occurring in economic policy and strategy, and to build the flexible strategic infrastructure to do so.

1.4.1 Economic Policy Change

Keynesian and monetarist thought are the key policy paradigms that have been most influential in the period 1960–1990. The Keynesian thought that influenced government approaches from post-World War II to the mid 1970s envisaged a role for governments in balancing demand management and the interests of suppliers. This approach provided the rationale for a strong and central role for government. Later, advocates of monetarism were to argue that changes in the level of aggregate money are due essentially to prior money stock changes. Economic activities in the economy can be stimulated or slowed by manipulating the flow of money supply (M1–M6) in response to desired economic outcomes, such as reducing inflation, increasing consumption, and reducing unemployment. Monetarism evolved into economic rationalism, following the lead of the famous Chicago economist, Milton Friedman. However the Asian economic crisis of 1997 led to a shift from rationalism as a focus for economy policy, with a swing back to neo-Keynesian type of policies to boost economic recovery and address the social problems of sudden rises in unemployment—especially in Asian cities. The transfer of traditional public functions to private ownership or management, and the breaking up of public monopolies through full or partial divestiture to private or corporate bodies have accompanied economic rationalism.

These phases in the evolution of economic thought have had a significant impact on paradigms driving economic development policy over the latter part of the 20th century. During the 1950s and 1960s, economic processes were driven by a focus on regulation and by strong government directives and initiatives. That was a time when national governments played a very active role in establishing national industries. For example, national vehicle and steel industries emerged at this time in many nations. Governments played a major role in the provision of infrastructure, planning, industry promotion and marketing systems. As indicated in Fig. 1.3 there was a focus on comparative advantage, by promoting cheap land, utility charges and local tax breaks for new businesses relocating or expanding in a region. These approaches to economic policy were, by and large, positive in developing national and regional economies.

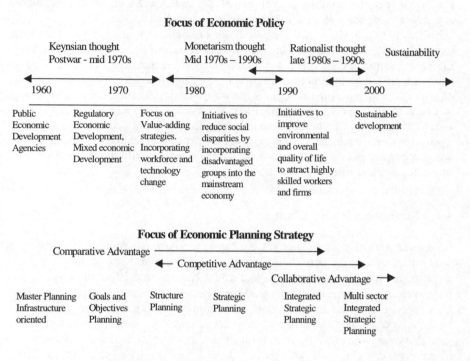

Fig. 1.3. Changing focus for economic development policy and planning strategy

The end of the Bretton-Woods agreement in 1973 removed gold as a standard for currencies and marked the floating of national currencies and an era of globalization, finance and capital emerged. There was a shift in economic thought away from heavy public intervention into industry policy towards a focus on value adding and the application and development of technologies to enhance production processes.

At this time, there was a focus on building technology, science and innovation parks as the catalysts for new age industrial development. Much of the thinking on this in the 1970s was developed by the Stanford Research Institute, where it was recognized there was a close relationship between industry and research establishments in the Silicon Valley (Hall and Markusen 1985; Saxenian 1994) that spawned a large number of technology based industries. Many parts of the world actively pursued policies of technology led development, some of which were successful, but many had something of a 'cargo-cult' philosophy based on sponsorship of technology parks. However, despite this early experience it needs to be stressed how important technology-led regional development has been in the post-industrial era of the information economy.

About this time there was also a shift of focus from comparative advantage to competitive advantage as monetarism began to influence macro economic policy. The older notion of comparative advantage was derived from economic theory on international trade which suggested that a nation or region would specialize in an industry in which it had a comparative advantage related to its particular resource endowment which provided a factor cost advantage in producing a particular good. Later on the emphasis changed towards the notion that regions would need to develop a competitive advantage that necessitated not only a factor cost advantage, particularly related to productivity and quality of goods and services that are traded, but also with respect to other factors that enhance business development and operations, and minimize risk. More recently, economic development and planning has promoted collaborative advantage where firms and regions are encouraged to collaborate in competition for strategic advantage, particularly through partnerships and alliances.

During the late 1980s, social issues concerning disadvantaged groups began to re-emerge as an important element of public sector policy. During this time, equal opportunity, workplace practice safety and other social infrastructure issues had a significant influence on public policy related to regional economic development. At this time many governments were also heavily involved in offering generous incentives for businesses to relocate or set up regional headquarters. Enterprise zones were developed in the United Kingdom and Europe and export processing zones began to emerge in Asia as a means of attracting industries. The 1980s also saw the rapid expansion and development of major new corporations. Significant job losses began to occur in the manufacturing sectors as production by major nationals and multinationals began moving offshore to cheaper sources of labour and sometimes less regulated economies in South America and Asia.

The extravagances of the capital markets splurge in the 1980s, followed by the 1987 stock market crash and a 1989 recession, led to the next paradigm shift with a focus on the principles of sustainable development. In the early 1990s, two parallel events were occurring that would have a significant impact on economic thought and the evolution of best practices. First, globalization was continuing to have a major impact on the economic restructuring of regions, which could be both positive and detrimental to their performance and possible futures. With this change there emerged a new focus on regions rather than just on national economies, as governments placed emphasis on the skill requirement of labour (i.e., labour quality) and on technology driven investment. Second, issues relating to sustainable development and quality of life began to have a significant influence on local economic development and policy. But this was also the age of economic rationalism with the corporatization and privatization of public assets and functions.

However, there was a clash between globalization and increasing community concerns about sustainability issues and quality of life and with how to achieve sustainable development. That has led to the emergence of a new paradigm in economic development thinking. What appears to be emerging is the greater integration of economic, social and environmental factors into decision-making processes about investment and use of limited resources. The United Nations 1992 Rio Declaration and 1997 Tokyo Greenhouse Gas Accord are bringing new dimensions of thought into the economic development processes. The current age is thus one of rapid transition, redefining best practices in economic development in the context of this new paradigm of sustainable development. This poses formidable challenges for regional economic policymakers as they seek to formulate strategy in a new environment of rapid change and uncertainty as well as a concern for achieving sustainable development.

Globalization has also resulted in the emergence of an increasingly borderless society with greater unrestricted movement of information, travel and currency between countries. Greater levels of transparency and standardization are occurring in both processes of government and business. These changes are reducing the importance of the nation state and increasing the focus upon major cities and regions as the centres and engines of economic growth (Knight and Gappert 1989; Ohmae 1995; Salazar and Stough 2006). Regions, particularly some of the world's larger metropolitan regions, are now seen as the dominant focus for the growth of employment, investment decisions and distribution networks in the new global market place. For example, over 85 per cent of employment growth in the United States in the first half of the 1990s had occurred in larger metropolitan cities. The metropolitan region increasingly is the geographic unit of analysis at which competitive economic activities take place, and increasingly this is where prosperity is being generated. In many parts of the world now, regions are not defined by political boundaries but cut across local and international jurisdictions to encompass the broadest definition of contiguous economic activity. Most businesses do not confine their economic activity to a specific jurisdictional boundary, and when the form of their organizational dependence is industry clusters, important economic inter-relationships are even more likely to spread across jurisdictional boundaries.

Given the importance of changes in the world economy as determinants of regional performance it is useful to consider the links between national and regional economies. In terms of policy, Prud'homme (1995) illustrates the absence of a link between national outlook and regional outcomes. He argues how regional policy, through its impacts on regions, has wider economic effects, and that macroeconomic policies, through economic outcomes and their spatial expression, have implications for regional performance. These interrelationships may appear obvious, but the reality of this connection is rarely explicitly accounted for in policy formulation. Regardless of national economic performance and policy intentions, macroeconomic controls and interventions tend not to address spatial inequalities that arise from regional structure and inter-regional patterns. Prud'homme (1995) notes what is lacking in macro-economic planning as it affects regions in these terms:

… The standard macro-economist view of the economy ignores cities, and more generally space. It tries to understand what is produced, how it is produced, for whom it is produced. But it does not care about where it is produced…It considers a country as a geographically undifferentiated whole. As a consequence, macro-economic policies, which are based on this understanding, also ignore cities and regions (p. 731).

The challenge now facing economic development planners is how to formulate economic policy that will respond to global dynamics and sometimes (or often) a national vacuum in macro policy towards regions in many countries. At one time regions were protected from outside competition, and to some extent their economies were able to be manipulated by national governments. But this is no longer the case as the economic rationalism pursued by many national governments has left many regions to fend for them. Regions still continue to look to higher levels of government for support and resources to provide economic direction and investment to stimulate economic development. But unfortunately, many regions fail to understand that globalization has left such governments largely devoid of powers to apply economic and policy mechanisms to enhance the competitiveness of region economies. Thus today it is more and more up to regions to develop and use their own devices to compete internationally in order to survive. To do so regions need first to understand what the factors are that set the dynamics of the emerging new economic age of the 21st century.

1.4.2 Planning Strategy Change

Over time regional economic development planning strategy has played a key role in the implementation of the approaches to economic policies discussed above. The overall thrust of planning strategy has been on achieving economic advantage, and this has undergone three transformations over the post World War II period.

Until the mid 1970s, the focus of planning policy was on comparative advantage. Planning policies primarily were directed towards achieving lowest production costs (labour, materials, energy, taxes and infrastructure) relative to competitors. Comparative advantage was heavily entrenched in supply side economics, where goods and services were produced and surpluses sold (often with the support of subsidies and incentives) in international and domestic markets.

During the 1980s, through the work of people like Harvard management guru Michael Porter (1986), the focus of regional economic planning strategy began to move towards competitive advantage. Competitive advantage has put the focus on 'value factors', including efficiencies, performance and intangibles such as quality of life, human and social capital (Putnam 1993; Fukuyama 1995) rather than just factor cost differentials that define the concept of comparative advantage. But many governments that still continue to promote comparative differences and provide incentives to attract industries to regions have not dismissed the strategy of comparative advantage. Both comparative and competitive advantage strategies are heavily entrenched in a win/lose scenario. With globalization, multi-national or transnational corporations have exploited regional differences created by comparative and competitive advantage strategy, as governments withdrew from protection and interventionist policies. The net result on regional economies, especially in non-metropolitan centres, has generally been catastrophic.

By the early 1990s, the impact of globalization had changed the nature and location of production, resulting in greater specialization or clustering (Dicken 1992). This resulted in a much more competitive environment, with business margins being squeezed as the result of adoption of best practices by government and business relating to management, cleaner production, and quality assurance. The tightening of regulations in response to concerns about the environment and social equity issues have forced business in particular to become more innovative and efficient in production to maintain margins and meet changing community expectations about the environmental friendliness of products. These two factors are bringing a change in business attitude, in which businesses and organizations which might once have considered themselves rivals are now actively seeking alliances, partnerships and other forms of collaboration to explore opportunities for winning and expanding business. There is thus emerging in the search for sustainability and economic growth a win/win strategy to economic development. This strategy is loosely referred to as collaborative advantage. It is a new paradigm emerging in regional economic development strategy and planning, and one that is dependent on greater integration, cooperation and collaboration among business, governments and communities (Huxham 1996). It is this new thrust toward collaborative advantage and how to achieve it that will be a focus of new methods in regional economic development strategy formulation, planning and implementation.

Figure 1.3 lists some of the characteristics of planning strategy under the umbrella of comparative, competitive, and collaborative advantage (Huxham 1996). In the 1950s regional economic planning strategy was guided by master planning and targeted at industry production, infrastructure and market development. Master planning for regions tended to be controlled by government policy agendas designed to address shortages in housing, construction materials, consumer goods, and to create employment. There was a strong focus on the development of industrial estates, many of which were used to support the development of state-owned enterprises as part of a policy of national and regional self-reliance.

From the late 1960s, regional economic planning moved from master planning to a focus on goals and objectives to achieve strategic outcomes. Governments played major roles in setting goals and objectives for regional economic development plans, but involvement and support from industry was also sought in providing expected deliverables. This goal and objectives planning was less deterministic than master planning and was intended to establish direction and targets for economic development. The role of regions in meeting national goals and objectives became important, but the autonomy of regions to shape economic futures was still largely determined by central, national or state government economic policy agendas. Goals and objectives were determined through various analytical approaches and economic visions were set based largely on the future being a continuum of the past. This paradigm was challenged severely following the first OPEC oil shock of 1973.

Structure planning, which was largely concerned with the geography of economic activities, became incorporated into economic development planning in the 1970s and provided a more flexible framework for decision-making. However, in many regional economic plans the thrust was a focus on supporting national government economic policies related to issues of social equity and schemes to encourage greater decentralization of employment and investment, as was the case in many nations, particularly in the developing countries, where five year plans were common in practice in both national and regional development.

By the mid 1970s, strategic planning in business began to influence planning in other sectors of the economy, including planning for regional development. Strategic planning involved the preparation of goals, objectives and strategies for organizations, for businesses and for regions to gain a position of advantage in the context of the environment in which they operated. In most cases, those environments were still considered to be relatively stable, as the effects of the globalization and the opening of national and regional economies to competition were not yet felt. However, strategic planning continued to provide a valuable tool for economic development after the effects of globalization became more noticeable. Strategic planning for economic development began to evolve in the late 1980s to address broader social and environmental issues.

Growing concern about the environment, about social issues, and about economic sustainability led to the emergence of integrated strategic planning for economic development in the 1990s. For example, in New Zealand the Resource Management Act of 1991 provided a new framework that required economic, social and environmental issues to be considered carefully within a framework for sustainable regional development. The States of Queensland and New South Wales in Australia have adopted similar legislation to ensure regional economic development is considered in the context of social, environmental and community values. In the late 1990s, the focus on integrated strategic planning for economic development led to a renewed interest in industry clusters and the role of smart infrastructure in economic development processes. Industry cluster and smart infrastructure development is dependent on multi-sector inputs, which might be best managed under an integrated planning system.

Not surprisingly, the evolution of regional economic development policy and planning strategy has varied significantly between nations, as well as within them. Until the 1990s, socialist nations continued to use master planning as a basis for economic planning at national and regional levels. In developed nations, the shift in the paradigms from 'goals and objectives' planning to 'integrated strategic' planning has been much more rapid—especially in the smaller OECD economies. This is partly the result of the export dependency of many of these economies, and the need for them to substantially enhance their competitiveness following the removal of tariffs and other forms of industry support as the result of economic rationalistic policies. But globalization, combined with growing global concern about environmental and social issues affecting national and regional economic development, is leading to a greater convergence of ideas on economic policy and planning. This has repercussions on the way regions develop strategy to support economic in both basic and non-basic sectors of a regional economy.

1.5 Some Core Theories and Models of Regional Economic Development

1.5.1 Evolution of Approaches

Over time various approaches to theory about economic growth have evolved, and not all have explicitly considered regional growth. Rather, much economic growth theory has been formulated in a non-spatial (or non- geographic) context.

Traditional neo-classical economic growth theory, based largely on the famous Solow model (1956, 2000), has been replaced by a suite of models and arguments that are commonly known as new growth theory. These developments are fundamental as they question the basic assumptions of traditional growth theory—and indeed of neo-classical economic theory in general—which are seen as being inconsistent with the modern economy. However, these developments are welcome for regional economic development analysts because among other things they explicitly introduce a spatial dimension into economic growth theory.

Traditional neo-classical growth theory models focused on the homogeneity of production factors, the price mechanism and the process of capital accumulation, all of which led to convergence thus eliminating inter-regional differences over time. As Maier (2001, p. 115) tells in a review of growth theory, in those models "neither spatial structure nor historical events have any implications of the long term growth path of a region. The latter is determined only by exogenous parameters." He goes on to say that "it is not surprising that regional economists felt quite uneasy about this conclusion" (p. 115), and, as a result, counter arguments to the traditional neo-classical growth theory emerged. These may be grouped under polarization theory as represented early on by the work of Perroux (1950), Myrdal (1957) and Hirschman (1958), and more recently by work on industrial districts and business clusters (see Feser 1998). Advocates of polarization theory argued that production factors are non-homogenous, that markets are imperfect, and that the price mechanism is disturbed by externalities and economies of scale. They argued that deviations from equilibrium are not corrected by counter effects, but rather set off a circular cumulative process of growth or decline, with a complex set of positive and negative feedback loops accumulating to a growth process whose direction is fundamentally undetermined. In a spatial context, these feedback processes generate what are called spread and backwash effects transferring impulses from one region to another. Spatial structure can be an important element in this growth process, generating leading and lagging regions that are highly interdependent. Advocates of polarization theory argue that it is not only economic, but also social, cultural and institutional factors that explain why some regions prosper while others are poor.

New growth theory has "originated in the heartland of economic theory" (Maier 2001, p. 115). Theorists such as Romer (1986, 1990), Barro (1990), Rebelo (1991), Grossman and Helpman (1991) and Arthur (1994) sought to explain technical progress as it generates economic development as an endogenous effect rather than accepting the neo-classical view of long term growth being due to exogenous factors. New growth theory models allow for agglomeration effects (economies of scale and externalities) and for market imperfections, with the price mechanism not necessarily generating an optimal outcome through efficient allocation of resources. Also, the process of capital accumulation and free trade do not necessarily lead to convergence between regions, with positive agglomeration effects concentrating activity in one or a few regions through self-enforcing effects that attract new investment. Thus, new growth theory allows for both concentration and divergence.

What follows is a brief overview of a selection of some of the key theories and models of regional economic development. For an in-depth analysis of these approaches, the reader may pursue the literature cited. The intention here is to make the reader aware of some of the key concepts encapsulated within various theories about regional economic development that are picked up in subsequent chapters.

1.5.2 Economic Base

An important – indeed fundamental – theory of regional economic development is economic base theory (Alexander 1954; Tiebout 1962), which assumes that local regional economies are composed of two parts:

(a)   a non-basic component which exists to serve the needs of the local resident population (local consumption);
(b)   a basic component which produces goods and services for consumption outside the local region (export consumption).

The basic part of the economy is called the export base (or sometimes the economic base).

Work conducted in the 1950s and 1960s (see, for example, Alexander 1954) showed that the basic/non-basic ratio in cities tended to vary between 1:05 and 1:2, with big cities having the higher figure. Ullman and Dacey (1960) proposed a minimum requirements approach to economic base theory, whereby it was possible to predict the non-basic industry employment requirements for cities or towns of various sizes at different levels of an urban hierarchy, with the residual industry employment representing the basic or export component of the sector. This approach was developed further by Czamanski (1964, 1965) in an attempt to link industrial location theory to urban growth. Certainly economic base theory established that there is a degree of regularity in the relationship between the size of a city and certain aspects of its industrial structure (Smith 1971, p. 102), and it has provided an extensively used devise for predicting the impact of new industrial development.

In economic base theory, development is seen to occur through the expansion of the economic base because such development has a multiplying effect. Growth in the export base of a region means that funds flow into the local regional economy from the sale of locally produced goods and services to consumers outside of the region. New local consumption is generated through some of these externally generated funds. This new spending increases the receipts of local suppliers who then spend a part of these new receipts on additional local consumption. The process of cycling and recycling the externally derived receipts continues until the entire above referenced economic base derived receipts leak out of the local regional economy. The initial and subsequent rounds of spending, the so-called indirect and induced economic effects, multiply the effect of the initial increase in the economic base, thereby creating economic development (Stough and Maggio 1994, p. 32)—that is, the growth of jobs, income, output and value added is created by the multiplier effect.

1.5.3 Growth Poles

The French writer Francois Perroux (1950) was influential in introducing growth pole theory into the literature on economic development. This theory argues that economic development strategy should focus investment on a specific sector—that is the growth pole, or sectors, to initiate propulsive development. (For a detailed discussion of Perroux's work see Higgins and Savoie 1988). The growth pole is normally a regional economy's core basic industry. The notion is that as this 'pole' begins to expand, linkages are forged to other sectors as import substitution occurs.

Perroux also proposed that, through appropriate policies, urban centres in a multi-regional context could become growth poles. For example, such centres in developing nations are often the focus of efforts to decentralize economies from a coastal and entry point centred economy—for example, federal government initiatives such as capital city relocation to Canberra in Australia and to Brasilia in Brazil. The term growth pole is often used to refer only to an urban growth node as a consequence of the work of Hirschman (1958), although some researchers, such as Higgins and Savoie (1988), argue that this is due to Hirschman's misinterpretation of Perroux' work, which was written and published in French.

Growth pole theory has been criticized from the associated unevenness of the benefits that stem from successful efforts to implement related strategies. This critique has contributed considerably to the 'balanced vs. unbalanced' growth conflict in the economic literature. With growth pole strategies, benefits accrue initially to the growth pole (sector or region) at the relative expense of other parts of the economy. As a consequence, groups in other sectors or parts of the region or national economy become impatient as the filtering down of the benefits comes later to those sectors or regions. Despite the well documented lag in this 'spread effect', growth pole thinking continues to provide a rationale for targeted development strategies (for example, picking industry winners). Almost all regional economic development strategies have sectoral or regional targeted poles.

Growth pole inspired efforts also have found expression at other levels of scale. For example, for shopping centre development it is often the practice to end-anchor projects with major magnet stores. Such stores are analogous to growth poles that lead to in-fill between the anchors. Further, in more recent approaches to urban development, the concept of a networked or multi-centred polis may be rationalized in terms of growth pole theory in that this approach envisions the development of the metropolis into a network of nodes or poles.

1.5.4 Accumulative Causation

One of the most widely cited theories of regional development is accumulative causation theory (Myrdal 1957), which emphasizes a market focus and the way some places (development nodes) pull in capital, skills and expertise to accumulate competitive advantage over other locations, with backward effects preventing the disadvantaged locations or regions from developing the internal capacity to compete and prosper. The elements and linkages emphasized in Myrdal's theory are set out in Fig. 1.4.

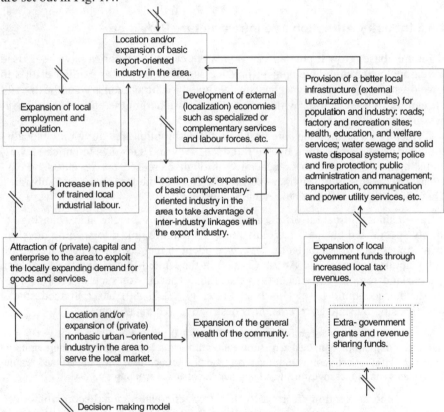

Fig. 1.4. Myrdal's accumulative causation model
Source: Adapted from Myrdal 1957

Galster (1998) demonstrates the processes of accumulative causation in a study of 100 cities showing how changes in the mortgage rate affected housing stock, occupancy, social problems of the neighbourhood, crime and school performance. He demonstrates how economic growth reaches different income groups according to a gearing ratio, reflecting complex patterns of cumulative causation. Krugman (1995) also stresses the importance of accumulative causation and its impact on the economic development processes.

In the new approaches to growth theory, these cumulative processes, which self-reinforce decline or continuing growth, assume a new significance through the explicit recognition of additional dynamics as change processes, including entrepreneurship, learning, education, and acquiring institutional capacity, as well as the migration of firms and households, and the adoption of new technologies and skills (Karlsson et al. 2001, p. 4; see also, de Groot et al. 2004).

1.5.5 Industry Attraction and Infrastructure Provision

Economic base, growth pole and accumulative causation theories have been used in regional economic development practice as justifications for explicit efforts to develop strategies and use mechanisms designed to attract industry and businesses to regions, as well as to provide infrastructure to underpin the business attractiveness of a locality or region.

Typically these strategies are based on the assumption that a region or locality can alter its market position by offering incentives and subsidies to entice new establishments, then growth is generated through increased tax revenues and increased economic output which exceeds the costs of the incentives and subsidies. They also involve measures to retain businesses.

(a) Business attraction efforts range from so-called 'smoke stack chasing' to highly specialized business recruitment efforts that are linked to specific development strategies; for example, the latter is evident in the development strategies for Indianapolis (Indiana) and Austin (Texas) in the United States, and the Far North Queensland region (Queensland) in Australia. The aim of such efforts is to increase the economic base. However, Porter (1990, p. 89) has criticized smoke stack chasing by state governments as destructive to competition:

When states use tax incentives and subsidies to bid against each other for every new plant, the competition is indeed zero-sum. But by investing in specialized training, building cluster-specific infrastructure, and improving the business climate with streamlined regulations, states can attract investment and upgrade the national economy (p. 89).

(b) Business retention efforts also characterized local regional economic development strategies, as seen, for example, in the Mid West of the United States in the 1970s and 1980s, as regions fought to retain economic base manufacturing firms in the face of global economic change to a more services and knowledge based economy. Such efforts seek to retain economic base and related growth potential.

(c) Business creation as an approach has become increasingly important with the rise of the information-intensive or knowledge-based economy. The goal of this approach is to expand a region's economic base by 'growing' new enterprises to take the place of those that 'die', and in so doing add propulsion to the economy.

(d) Import substitution adds another dimension to the business creation approach. Import substitution aims to expand local production of an intermediate good or service that is currently imported, often being adopted as a basic strategy by regions that have well-developed core export base components.

(e) Incentives—such as tax relief, infrastructure augmentation, marketing, and training assistance—have been widely used and are typically an important aspect of regional economic development programs in general, in particular those that focus on business attraction. For example, in the United States the decision to locate a Mercedes Benz plant in Alabama, a BMW plant in South Carolina, and semiconductor plants in Virginia, were all supported by large incentive packages often more than US$ 100 million. Such efforts, while aimed at increasing and even transforming the economic base of the regions to which they are attracted, have been criticized as leading to a general increase in the social cost of such outcomes. Advocates of incentives argue that such efforts are first of all a cost of doing business and second lead to the creation of job and income contributions to the economic base that are sufficient to more than offset the cost of these subsidies. Bartik (1991, 1999), who analyzes the social costs of incentives, concludes that policies in support of incentives often result in net benefits. He shows that job and income benefits often accrue for up to eight years after the incentives are proffered and that these benefits accrue disproportionately to local-low income workers.

What does all of this mean for economic development planning and practice? First, projects that can be rationalized on the basis of their contribution to the economic base are likely to receive stronger support from community groups. For example, instead of trying to attract firms indiscriminately, a program that is designed to attract firms in targeted industries with high local multiplier effects will be more defensible and generally more attractive. Second, linking a group of interrelated projects together in different parts of a region may be used to drive an economic base development program which is often the only way that access to higher level goods in decentralized rural regional economies can be created. Third, many approaches undertake import substitution strategies to achieve improvements in the economic base. In short, economic base theory lies behind much of the strategy and the practice used in many regional economic development plans and programs.

There is, however, much evidence that the costs of such policy instruments are borne by the local workers and employers (Bluestone et al. 1981). Nonetheless, increasingly communities have been packaged and promoted as products, much of which borders on boosterism. But as Blakely (1994, p.58) points out, more recently there has been a change in emphasis from attracting factories to attracting and generating entrepreneurial populations and skills and in improving and promoting amenities and other quality of life attributes of communities. Florida

(2002 and 2005) takes this further arguing that these attributes attract people with high end creative skills that in turn drive and sustain development.

Another important factor involving public intervention in regional development relates to physical infrastructure, which is the foundation upon which rest both the internal geographic interaction (mobility) within a region and the external linkage (trade and innovations) of a region to the outside economy. Thus, investments in infrastructure and their maintenance are seen as being essential to the sustainability and competitiveness of regional economic systems. However, such investments often take time to generate benefits that can be broadly distributed across sectors and throughout a region due to their 'lumpiness' in both time and space. So the infrastructure investment approach shares this shortcoming with the growth pole theory approach.

Infrastructure traditionally has been viewed as physical systems or hard infrastructure, such as roads, sewerage systems, water supply systems, airports and telecommunications hardware. More recently, the concept has broadened to include soft infrastructure (social overhead capital) components, such as education, health, governance, regional leadership, knowledge (for example, production know-how) and amenities that define quality of life suggesting the increasing relative importance of soft infrastructure compared to traditional hard or physical infrastructure in an advanced economy's ability to create and drive technical change. While there are exceptions to this conclusion—such as the need for road improvements to link a remote region to urban centres—soft infrastructure increasingly is becoming the foundation of successful and sustainable economies. In such knowledge-based economies, information, education, amenities, environmental quality, entertainment, venture capital and flexible institutions become the primary attributes of importance and these are now being described as smart infrastructure (see Smilor and Wakelin 1990). Regions with such attributes have high levels of internal networking and related trustful relationships, which serve as central components of their competitiveness.

1.5.6 Location Theory and Central Place Theory

Explicitly spatial models about the nature and functioning of regions and the processes by which firms choose where to locate their activities have been developed by geographers and regional scientists as embodied in *location theory* and *central place theory*.

Location theory addresses the question why economic activities are unevenly distributed across space and addresses the factors that firms consider in selecting a geographic location, both a region and a locality within it. Classic texts on location theory include Hoover (1948), Greenhunt (1956), Beckmann (1968), and Smith (1971). Webber (1984) provides a succinct summary of approaches to industrial location, saying:

... location is a concept that means where something is in relation to other things. So industrial location means a statement not just of the spatial distribution of industry, but also of the relations between that distribution and other phenomena. Industrial location theory explains the spatial distribution of industry by referring to other aspects of society (p. 9).

Location theory has proposed that firms locate so as to minimize costs and seek locations that maximize their opportunities to reach markets and, thus to maximize profits. Much of the focus has been on transport costs, labour costs, other production costs, scale of operation, and agglomeration economics. The evolution of transportation and telecommunications technology has rendered distance (time and costs) less of an impediment, enabling industry to become more footloose, and changing modes of organization of production are emphasizing the importance of business networks, strategic alliances and just-in-time delivery systems. Also, increasingly intangibles such as amenities and business climate are seen as important determinants in the location decision making of firms and managers. Over time, the emphasis in location theory has shifted from least cost/profit maximizing or optimizing type behaviour to satisficing behaviour and the issue of uncertainty as it effects industry costs, efficiency, productivity and profits.

In local and regional economic development strategy formulation and plan implementation, an appreciation of the postulates of location theory is important, particularly in terms of understanding locational advantage or disadvantage of a locality or region for particular types of industries and sizes of firms with respect to their resource input and proximity requirements, labour requirements, market size requirements for local targeted sales, for infrastructure planning, and for assessment of intangible factors in industry attraction and business retention. Location theory provides the rationale for the concentration and dispersal of industries, efficiency of plant size, the impacts of new technologies on things such as transport costs, labour and non-labour costs, and the impacts of externalities such as congestion.

Central place theory derives from the work of Walter Christaller (1933) who posed the question "are there laws which determine the size, number and distribution of central places?" The notion is that urban centres are arranged in a hierarchical pattern of central places of different size and functional complexity, and that there is a systematic spatial arrangement across a large region in the pattern of distribution of central places (urban settlements). The different functions provided by central places reflect a diversity of economic activities that serve their surrounding populations. The focus of central place theory is on the retail and other service functions of a central place rather than manufacturing type activities. The trade area of a specific function is determined by a number of factors, and in particular the inter-relationships between the price of the good or service, the cost of travel for a consumer to gain access to purchase it, the aggregate level of time required to support the business providing it, and the frequency with which it is purchased. Thus, the two prime considerations are the range of a good or service, which sets a spatial limit beyond which people will not travel to access it, and the threshold of the good or service, which refers to the minimum aggregate consumption that the good or service needs to pay for the costs of producing or offering the good or service. The theory goes on to argue that the hierarchical arrangement of central places reflects the functional complexity and order of the goods or services being offered, with levels in the hierarchy of central places being defined by the highest order of good or service being offered. Low order goods or services that are frequently purchased and for which relatively small populations are required within their market areas. Thus those goods or services will be offered frequently and at many locations. Those functions will define the low level centres in the hierarchy of central places, which will be characterized by a few relatively simple economic functions with spatially restricted trade areas that serve small local populations. They include functions such as a local convenience store or an auto filling station. The higher levels of the central place hierarchy will be defined by high order goods and services which are relatively expensive and are consumed by people relatively infrequently, and which require large populations in extensive market areas to generate the high aggregate levels of consumption needed to cover the high costs of providing the good or service. They are functions such as large department stores and large hospitals offering specialist services and facilitations. The trade on market areas of the high order central places, defined by the highest order of good or service they provide, will have within them what is called a nested hierarchy of lower order central places with market areas defined by lower order functions. Christaller proposed a number of theoretical systems of spatial landscapes of nested hierarchies of central places, differentiated by principles relating to either achieving a transport/access optimal efficiency, a market area/size optimal efficiency, or an administratively efficient optimality.

The Christaller central place theory model was based on simple assumptions of an isotropic place (surface) across which the density of people, purchasing power and consumer preferences was homogenous. Of course in reality, these assumptions break down as population densities and the socio-economic characteristics of consumers vary markedly across space. Thus later on Lösch (1940) was to propose an alternative approach to central place theory allowing for variations across space in the density of purchasing power (rich and poor areas), which give rise to different spatial arrangements of market areas for economic functions across space. Even later, others such as Berry and Garrison (1958) and Berry (1967), were to derive a set of threshold values for the population and functions at various levels in a regional central place hierarchy, using a grouping or classification provider to identify hierarchical levels.

Berry and others extended this work much further into detailed analyses of the hierarchical arrangement of central places (shopping and business centres) within large cities, and geographers such as Hagerstrand (1966) used central place theory as a framework to examine the way innovations disperse through a hierarchal urban system from high order places of innovation to lower order places of adoption.

A concise summary of central place theory is available in King (1984), who also shows its link to location theory. Gunnerson (1977) formulated mathematical solutions to show how industries might be optimally distributed among a hierarchy of urban centres, and White (1977) has used a dynamic approach to central place theory to look at growth and decline across a settlement system, linking central place performance to levels of profitability differentials between different places.

King (1984) points out how one of the legacies of Christaller and Lösch has been "the fashioning of regional development plans around central place theory", the idea being that "a well-developed, hierarchal central place system is in some sense an efficient arrangement that is likely to have a beneficial effect upon the economic development of a region in question" (p. 72). Certainly central place theory has been widely used in large city regional planning as it relates to commercial development and centres policy, as well as in planning urban settlement systems in developing nations, as used, for example, in a planning study in Ghana (Grove and Huszar 1964).

In local and regional development strategy planning, central place theory does provide a useful framework for understanding the potentials and limitations of a particular place vis à vis its market situation and functional complexity as influenced by its place in the central place hierarchy. Alternative approaches to central place theory, such as proposed by Curry (1967), introduce the elements of uncertainty and risk for both businesses and consumers into the system, linking central place theory to a communications theory framework.

King (1984) points out how:

... the economic role of an urban center is one that provides goods and services to a surrounding region and to the smaller communities within it is one that is played by virtually all cities and towns. Every urban center is to some degree a central place and is subject to the forecast principles emphasized in central place theory. These economic relationships come sharply into focus when, for example, political questions about metropolitan and regional government reorganization are raised. They surface also in many discussions of local taxation and financial issues (p. 91).

Central place theory can provide useful building blocks for regional development as it provides empirically established general rules about the combination of urban locational patterns, urban-size distributions, and range of functional economic activities.

1.5.7 Agglomeration Effects

A particularly important concept in regional economic development—in particular in explaining why some regions develop large concentrations or clusters of certain types of economic activities—is agglomeration effects. In simple terms, we may think of these effects in numerous ways. First, agglomeration economies are benefits available to individuals and firms in large concentrations of population and economic activity, as found in big cities and in some nodal concentrations of activities within them, such as for producer services in CBDs. Second, another form of agglomeration effects, economies of scale, refers to factors that make it possible for large organizations or regions to produce goods and services more cheaply than smaller ones. Third, economies of scope arise through the opportunities of large concentrations of population and activity provide for diversified activities to occur through linkages among firms of various sizes. Fourth, agglomeration refers to externality effects, which relate to the advantages gained through proximity to diversified business and market opportunities as a result of the concentration of people and activities in particular locations.

Neo-classical economic models do not allow for agglomeration effects because they imply a peculiar spatial structure (Mills 1972), the key assumptions being that of production functions with constant returns to scale and that all input and output markets are competitive. Land and labour are homogenous. Utility functions have the usual properties, and consumers spread themselves over the land at a uniform density, and each consumer has adjacent to them all the industries necessary to satisfy their demands. The assumption of constant returns assumes that production can take place anywhere without loss of efficiency, and all transport costs are voided without the need to agglomerate economic activity. However, if only one industry of this economy had a production function with increasing returns to scale – positive agglomeration factors - then the spatial structure of the economy would differ markedly. Because of positive returns to scale, that industry could produce in a more efficient way when it concentrates production in one or a few locations. Those located closer to the site of this agglomerated industry sector would have an advantage over those further away, and they could either save transport costs or produce larger quantities.

In the traditional model, each consumer provides labour inputs to all the industries at their location, but when one of the industry sectors is agglomerated, it produces at a larger scale and therefore needs more labour input than is available at that location, having to attract additional workers who either commute or migrate to that location. The agglomerated industry sector will have to pay higher wages to compensate costs for those workers commuting long distances or to cover the higher costs of land more densely settled by those living nearby. Thus, inter-regional differences in wages and land prices and in the density of economic activity occur. As a result, "the location of the agglomerated industry turns into a market center" (Maier 2001, p. 188). Specialization of land use occurs as those activities not able to pay the higher wage and land costs at the agglomeration node move away. As shown in the classic location theories such as von Thunen (1826), producers will not locate randomly but according to their bid-rent functions, with those industries whose products are more sensitive to transport locating closer to the market than those whose products can be transported more cheaply.

Maier (2001) notes that with the traditional neo-classical model allowing for only one industry to have a production function with increasing returns to scale, resulting in a spatial structure differentiated in land use, prices and densities will vary with the need to transport products and production factors from one location to another. But, when other industries also are allowed to show increasing returns to scale, or if we add location or urbanization effects, then the results described above will be strengthened. Agglomeration effects in one industry, however, are sufficient for producing spatial structure and differentiation.

Starrett (1978, p. 21) found that in a system with spatial structure, "the usual types of competitive equilibrium will never exist". Whenever there are transport costs in the system and all agents are price takers in competitive markets, all possible allocations are unstable—"that is, for any set of prices and location allocation, some agents will want to move to other locations" (Starrett 1978, p. 25). Thus, economic actors will have an inducement to move closer to their suppliers and/or markets. Maier (2001) concludes that these arguments:

... show that a stable spatial structure, i.e. an allocation of economic activities where the type and amount of economic activity differs between locations and where goods and production factors are exchanged between locations, and agglomeration factors are two sides of the same coin. One requires the other... (p. 119).

Allowing for agglomeration factors will generate many more dynamic processes and a diversified spatial structure in regions, with agglomerated factors constituting a theoretical link between spatial structure of an economy and its growth dynamics.

Agglomeration effects thus play an important part in theories of regional development and concepts of economic growth, particularly for long-term dynamics, which Maier (2001) says has been demonstrated by combining traditional neoclassical growth models with stochastic growth models of innovation, showing convergence towards a distribution where one region has almost all production and future growth as a result of agglomeration economies, while other regions stagnate. New economic growth theory shows that agglomeration effects are essential to understanding the functioning of an economy: "agglomeration effects bring about spatial structure, path dependence of growth processes, 'lock-in' phenomena, and long term implications of historic events" (Maier 2001, p. 132).

A lot of the work on agglomeration economies in fact stems from the concept of the industrial district proposed by economist Alfred Marshall (1920) postulating that benefits derived from proximity among businesses are strongest for small enterprises. Feser (2001, p. 231) describes how:

... with internal economies a function of the shape of the average cost curve and level of production, a small firm enjoying external economies characteristic of industrial districts (or complexes or simply urbanized areas) may face the same average cost curve as the larger firm producing a higher volume of output ... Thus we observe the seeming paradox of large firms that enjoy internal economies of scale co-existing with smaller enterprises that should, by all accounts, be operating below minimum efficient scale (p. 231)

It has been agreed that smaller firms utilize superior flexibility and innovativeness to compete with their larger competitors, that changing market conditions favour vertical disintegration of larger producers and greater use of outsourcing (Scott 1988) resulting in re-agglomeration of economic activity with a shift towards more flexible production modes, with small firms carving out market niches (Patten 1991). This is not to say that only small firms will benefit from local externalities, as large firms gain economies of scope and scale through outsourcing some functions once the market is big enough to support those independent producers. Feser (2001) points out that:

... urban or industry scale might be sufficient proxies for those types of effects, provided demand is localized. Likewise, both large and small firms would be expected to benefit from other types of external economies associated with inter-firm proximity, including a network of suppliers, pools of skilled labor, and knowledge spillovers, though small firms may depend to a greater degree on such advantages (p. 232).

Agglomeration effects are now accepted as an essential element of a modern economy and that it is not possible to understand the functioning of an economy without allowing for agglomeration effects.

1.5.8 Technology Based Explanations of Regional Development

Rees (1979) has proposed that *technology* is a prime driver in regional economic development, and over the last three decades the literature has shown how technology is directly related to traditional concepts of agglomeration, learning and leadership.

Thomas (1975) and later Erickson (1994), among others, have shown how technological change is related to the competitiveness of regions. Norton and Rees (1979) and Erickson and Leinbach (1979) show how the product cycle, when incorporated into a spatial setting, impacts differentially on regions through: (a) an innovation stage, a growth stage; and (b) a standardization stage, where production can shift from the original high cost home region to a lower cost location, often off-shore, which has been hastened through the evolution of the internationalization of the production process. Thus some regions are the innovators, others become the branch plants or recipients of the innovation, and these might even then become innovators via indigenous growth. Markusen (1985) has extended the product cycle theory of regional development by articulating how profit cycles and oligopoly in various types of industrial organization and corporate development can magnify regional development differentials.

The concept of *innovative milieu* was formulated to explain the 'how, when and why' of new technology generation, linking back to the importance of agglomeration economies and localization economies that lead to the development of new industrial spaces (Scott 1988; Porter 1990; Krugman 1991). Some theorists, such as Fukuyama (1995), have suggested that not just economic but also value and cultural factors, including social capital and trust, are important in the rise of technology agglomerations as seen in the Silicon Valley phenomenon, where collaboration among small and medium size enterprises through networks and alliances and links with universities forge a powerful R&D and entrepreneurial business climate. But Castells and Hall (1994), in discussing innovative industrial milieus, note:

… despite all this activity … most of the world's actual high-technology production and innovation still comes from areas that are not usually heralded as innovative milieus … the great metropolitan areas of the industrial world" (p. 11).

However, as Rees (2001) points out, technology based theories of regional economic development need to incorporate the role of entrepreneurship and leadership, particularly as factors in the endogenous growth of regions, and it is the "link between the role of technology change and leadership that can lead to the growth of new industrial regions and to the regeneration of older ones" (p. 107).

1.5.9 From Comparative Advantage to Competitive Advantage: The Emergence of a New Theory

A fundamental principle in regional economic development is that of spatial interaction among regions through the movement of goods, services and people. International trade theory in economics has developed a huge body of literature on furthering our understanding of patterns and characteristics of trade between nations. A fundamental tenant in international trade is the *law of comparative advantage*. It states that a nation (or region) will benefit by exporting a good that it produces at a lower relative cost than other nations (or regions). Conversely a nation (or region) will benefit from importing a good that it might produce at a higher relative cost. Those comparative advantages or disadvantages might be due to factors such as labour cost differentials as well as to resource endowment differences. Often governments of nations have intervened to establish quotas, import duties and tariffs to protect domestic producers who are operating at a comparative disadvantage. Such interventions disrupt the patterns and volume of flows of goods and services among nations that might otherwise occur in a truly open competitive environment of what one may refer to as a 'level playing field'. However, these protectionist policies were widely used by national governments to enhance the development of what were seen as key industry sectors and to promote in part import substitution.

Earlier reference was made to how theorists have also attempted to explain regional economic change, and growth and development, from the perspectives of internal growth, export base, and economies of scale. Rostow's (1960) model of *sequentially staged economic development* exemplifies the internal growth theory approach, attributing economic development to changes occurring within a region, through processes such as the application of technology to a local resource or the rise of purchasing power. It has been shown how the export base theory approach discussed previously views regional economic development as a result of the expansion of exports to other regions, with the income generated from exports driving growth including economic development. An important concept in internal growth theory is that as growth occurs and trade develops, involving a larger number of producers, then certain competitive advantages emerge, as some producers become more efficient than others or their locations will become more advantageous than others. The more competitive producers can sell their products more widely, thus being able to increase their production through employing new technology and driving down the average cost of their product through achieving economies of scale and economies of scope.

The *internal growth theory* approach to regional economic development focuses on the local application of technology, with the evolution of a region over time moving through successive phases of trade in agricultural goods to manufactured goods, and then into the growth of services and information-based economic activities. The export base theory of regional economic development places emphasis on foreign investment and/or outside exploitation, the income raised from exports being the impetus for growth, with regional specialization emerging.

However, emerging *new theory of specialization and trade* emphasizes the role of the functional region rather than the nation. Johansson and Karlsson (2001) propose a simple model incorporating four basic concepts of a new theory of location, trade and regional specialization shown in Fig. 1.5. Here, location specialization and regional growth are organized as being dependent on technology and scale effects together with influences from durable regional characteristics. They point out that until the 1980s, comparative advantages mainly were derived from resource-based models, but that since then a revolutionary change in international economies has witnessed economic specialization to a large extent being dependent on increasing returns and that the differences in resources (factor intensities) explain only parts of trade flow and the location of production (see Krugman 1981, 1991). Johansson and Karlsson (2001) say:

... according to the new theory of trade with its scale-based models, imperfect competition and increasing returns are pervasive features of contemporary industrialized economies. With increasing returns as a basic explanation, trade develops because there exist advantages of specialization also among economies and regions that are very similar to each other as regards resource endowments. If specialization and trade are driven by scale rather than by comparative advantage, the gains from trade that arise because of production costs fall as the scale of output increases (p. 153).

Traditional theory cannot predict what sort of goods will be exchanged between countries (or regions) with similar resource endowments, and neither does it say anything about products that are not traded between nations and regions (that is, the so-called 'non desirables'). However, with increasing returns as a complementary explanation, a much broader assortment of comparative advantages and trade flows exists.

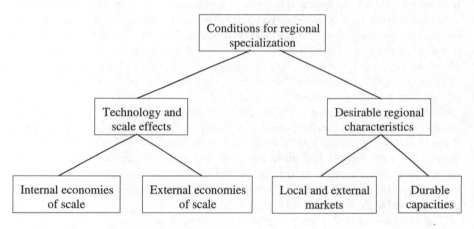

Fig. 1.5. New theory of location, trade and regional specialization
Source: Johansson and Karlsson 2001, p. 153

Johansson and Karlsson (2001) show also how the internal market potential of a functional region is a prime home market which, together with increasing returns to scale, can give rise to processes of endogenous growth (or decline); thus, "resource-based and scale-based mechanisms combine in a dynamic interdependent development process" (p. 154). Regional market size (internal) thus becomes important as well as extends market potential, and when a region has both, its competitive advantage increases, with increased possibilities of the region growing a wide range of industry sectors many of which will be exported to other regions.

There are important further differences in the way regional economic development might be viewed from the perspective of comparative advantage as against competitive advantage.

It has been argued that the most important factor in sustaining economic development is a focus on *competitiveness* which will be the single most important issue facing firms and organizations in the future (Porter 1990; Ohmae 1995). This requires firms and organizations to spend a significant proportion of their time researching and understanding factors that create competitiveness. These include the relative strength of core competencies and skills, entrepreneurship, leadership, innovation, governance, sustainable use of resources and marketing intelligence. It also involves firms and organizations scanning the future horizon to understand the threats and risks in their operational environment, and to anticipate and develop potential new markets that could be created from capitalizing on the factors that give a region its *competitive advantage*.

Competitiveness may be described as involving competition, low enough prices to compare with rival traders, and as having a strong urge to win. According to Porter (1990, p. 1) "competitiveness has become one of the central preoccupations of governments and industry in every nation". Porter first began to explore the linkage of firm competitiveness to national competitiveness through the interplay of several factors, including industry clusters that achieve competitive advantage. But as Porter states:

... there is no accepted definition of competitiveness...Whichever definition of competitiveness is adopted, an even more serious problem has been that there is no generally accepted theory to explain it (p. xii).

This begs the questions: If competitiveness is so important to business and economic development, why do we know so little about how to quantify it? And: Is competitiveness the only factor that drives economic development in regions?

Competitiveness is regarded now as a key element of regional economic development and it is the central thrust of many development strategies and plans. It is embedded in both demand and supply side economics. Markets seek to ensure products and services are made available to consumers and producers at the most competitive prices. The price of products and services are determined by the availability and price of resources, and the efficiencies and effectiveness of infrastructure, technology, skills and capital used to produce them. However, if regional development is primarily concerned with the competitiveness of 'factor costs' of production and the maximization of profits to establish a favourable environment for investment or to compete for trade, economies may evolve along unsustainable development paths, with undesirable social and economic outcomes. There is a need, therefore, to rethink the relationship between competitiveness and sustainable economic development.

For many regions and for many businesses this will involve a significant paradigm shift away from competitiveness viewed solely as a competition in the race to be first. Competitiveness in the global economy is beginning to take on a new meaning as firms, regions and nations, realize the value added by competition through collaboration, strategic alliances, partnership and resource sharing (Moore

1996). This is a new dimension to competitiveness that has as its *raison d'être* in sustainable economic development. This new dimension to both business and economic strategy presents a major challenge for regions to develop more sustainable paths for economic development in the new global economy.

While significant emphasis is now placed upon competitiveness as a thrust in economic strategy in business and in regional economic development, it should be recognized that this is not the only factor that facilitates economic development. Economic development can be achieved through a combination of many factors, including improved leadership, public policy, innovation, population growth and import substitution. However, competitiveness remains a major factor of enhanced economic performance, but it should incorporate elements of the above.

1.5.10 Focusing on Contemporary Perspectives on Regional Economic Development: An Embedded Institutional Approach

In the discussions so far we have outlined some of the basic theoretical approaches used in helping us understand regional economic development. This review has had to be somewhat selective. Certainly the theory part of the discussions could have included other theoretical approaches, such as product cycle theory, profit cycle theory, innovation theory, community development theory, the old and new institutional economics theory, social network theory, and asset or portfolio management theory. To further qualify this discussion, while at the same time opening up additional avenues of thought to the reader, it is important to note that this discussion has drawn from traditional metaphors where economic development is viewed as a growth-inducing, problem-solving business activity. And there are other metaphors that some would offer, such as a vehicle for preserving nature and a sense of place, empowering individuals and groups, exerting leadership, and as a vehicle to achieving equity and distributive goals (see Bingham and Mier 1993, for an expansion of this concept). These non-traditional metaphors offer interesting alternative approaches for expanding the ways economic development theory might be used to rationalize planning and practice.

By way of a general summary of the evolution in approaches to regional economic development over the last three decades or so it helps to further conceptualize the processes involved and to add to explanation on the dynamics of regional economies and their capacity to respond to and to cope with the forces of change, It is evident that there have emerged a number of key themes as to what constitutes regional growth and development and regional competitiveness, Not surprisingly there are differences of views among regional economic development scholars, and some of those differences relate to the relative focus given to the roles of exogenous and endogenous processes and factors. But there does now seem to be an almost universal realization of the *institutional embeddedness* of *endogenous processes and factors* in regional development.

A rather neat overview of this evolution is given by Garlick et al. (2006). They put forth the argument that the key concepts that flow out of the contemporary regional development literature that are particularly important, and which are embedded in the increasingly important discussion of institutions and endogenous regional growth and development, are: structural agglomeration; regional innovation systems; institutional thickness; organization power and control; enterprise segmentation; and social capital.

Garlick et al. (2006) suggest that theoretical and operational approaches in contemporary regional economic development may be differentiated according to the focus that it places on one of the three perspectives.

(a) In the *first perspective*, the notion is that regional competition occurs through business firms within the region or through national policy and programs, and not through the mobilization of the full attributes of the region itself. Successful regional economies are recognized as islands of sustained local accumulation built on superior local productivity (Porter 1998). Competitive regional development results from the existence of a free market critical mass and institutions operating in proximity as rational actors with standard rules as proposed by Porter (1998) and Krugman (1995). The best performing regions might be likely to have a concentration of high-tech firms, knowledge-based industries, and creative human capital as proposed respectively by Saxenian (1994),by Armstrong (2001)and Thrift (2001), and by Florida (1995, 2002), and facilitated by institutional intervention. Here ICT and technology diffusion industry policies, science and technology parks, universities, business agglomeration practices such as clustering and networking (Porter 1998, 2000; Cooke 1996), and an effective institutional presence (Amin 1999), including mechanisms such as advice and information centres, are the structural tools through which businesses and institutions foster regional competitiveness.

This is the Silicon Valley (and other innovative regions such as Austin Texas and the National Capital Region in the U.S. and others) solution for regions in which the emphasis is on the creation and emergence of a regional innovation system (Braczyk et al. 1998).

The presence of those structures and processes are seen to be important to kick-start or catalyze a region's development. This view would suggest that a solution to stimulating regional development comes from the top with institutions playing roles such as facilitating training for new skills development, providing infrastructure and regional information systems, and providing attraction programs and facilitating enterprise and institutional networks. Bottom-up regional processes take such top-down initiatives and try to create synergistic effects out of them.

However, some scholars take the view that an organizational behaviour approach to regional development is important, as promulgated by Amin and Housner (1997). This perspective maintains that the simple availability within the region of institutionally provided structures, programs and their processes alone is not sufficient.

(b) In the *second perspective*, the drivers of regional development are seen to depend on complex processes of integration related to the somewhat 'softer' processes relating to the dynamics of 'regional milieu' that include social capital (Bolton 1992; and Putman 1993) trust (Fukuyama 1999), loyalty and learning regions (Maskell et al. 1998; Keane and Allison 2001), power relations and controls in organizations (Taylor 1982, 1983; Clegg 1989), and Thrift (2001) and organizational culture, norms and rules (Hodgson 1998).

This approach argues that there are deeper organizational management values, cultural issues and norms that exist in local agencies that may determine the extent to which a region's endogenous resources become mobilized through local action. There is an emphasis on the notion of the 'learning region' in which collaboration creates 'institutional thickness' and which bolsters the development process (Amin and Thrift 1994).

(c) In the *third perspective*, it is argued that regions are neither scaled-down versions of national economies nor simple aggregations of competitive firms; nor are they only the result of institutional decision-making. Regions are rather competing with one another on the basis of the broad and full range of their endogenous economic, social, natural, historic, cultural and human attributes (Maillet 1995; Reich 1991; Kanter 1995, Plumber and Taylor 2001; Taylor 2003). Regions are seen to be highly diverse in terms of their attributes, with few of those attributes being used to their full potential. Thus, the crucial issue becomes the 'enterprising capacity' of a region's human capital to "turn creative ideas into results using the region's skills in market identification, risk assessment, persistence, access to development finance, business planning and so on" (Garlick et al. 2006, p. 13). These are ubiquitous but latent abilities that can be activated at a regional scale by facilitating coalitions and bringing resources together to generate convergent regional directions. According to Plumber and Taylor (2001a):

...It is not the setting up of businesses that is the 'enterprise culture'. Rather, the 'culture' is what brings people together in the first place to create, re-create, mould and extend coalitions that seek to exploit business opportunities..... It is not picking winners and subsidizing them. It is about creating forums where potential coalition members might meet and generate ideas – people from the small firms sector, the corporate sector, the public sector, and the local community (p. 12).

The embedded institutional approach referred to here, which seems to be more and more important in contemporary regional development policy, is taken up in more detail in Chap. 8 of this book. It incorporates a discussion of the role of *leadership* in regional development.

1.6 Thinking Global While Acting Local

The contemporary global economy is being driven increasingly by financial and information markets that support production processes that are networked to each other in different parts of the world. Multi-national corporations are playing an increasingly dominant role in orchestrating synchronized production systems spanning multiple regions of the world and that have competitive advantage in the production and distribution of goods and services (Korten 1995). Information technology continues its rapid penetration into every sphere of national and local economies helping to create new products and integrate markets. At the same time production systems are requiring new skills and strategies to cope with dramatically accelerating change motivated by globalization. Thus the new challenge for a region in pursuing strategy for economic development is to 'think global' while 'acting local' (McKinsey and Company 1994).

These changes to the nature of production, trade, information and employment that are impacting firms, regions and nations represent great challenges as to how they can maintain and improve their competitiveness. Economic competitiveness has an increasingly significant impact on investment location decisions, and the current phase of global economic restructuring and competitiveness has changed four important dimensions of economic development:

(a) *Geographic Scale*. Traditionally local government jurisdictions have been the geographic scale at which economic development strategy has focused. However, today the focus is shifting more towards city regions that encompass multiple jurisdictions, and on understanding how industries located throughout a region operate with respect to their inter-sectoral and inter-firm linkages spanning not only the entire region, but also their linkages outside the region.

(b) *Industrial Organization*. Economic strategies have tended to be focused primarily on the company industry sector level, but such strategies fail to recognize the increasing importance of inter-organizational dependencies that exist between industries and companies across industries, regions and nations, particularly in the context of the operation of trans national operations and the strategic linkages and alliances that develop between firms in a global production process.

(c) *Economic Inputs*. The types and location of input factors crucial to the formation, expansion and attraction of industry to a region are changing. Historically material costs, labour, land and taxes were recognized as factors contributing to comparative advantage, but now the focus is on competitive advantage which places a stronger emphasis on value adding factors related to efficiency, technology applications, skills placement and leadership. And as discussed earlier, there is now an emphasis as well on collaborative advantage through strategic alliances and networks.

(d) *Sustainability*. Regional economic performance is no longer being viewed in purely economic outcomes. There is little value to communities if economic development does not improve the quality of life, maintain safe and pleasant living environments, or resolve social disparities within communities. Globalization is bringing about major changes to the flow of information and to governance systems, and the paradigm of sustainable development—the integrating of concerns for economic viability, social equity and cohesion, and ecologically sustainable development—increasingly is being adopted as an underlying principle of regional development strategies and for planning practices.

A consideration of the above represents considerable challenges for governance in regional development. We are witnessing the emergence of a borderless society with nearly unrestricted movement of information, travel and currency between countries. Today there are greater levels of transparency and standardization in government and business. The role of government is also changing.

Governments are having less control and influence over economic development and investment decision-making, and they will thus need to learn how to facilitate and manage development processes that fit the global forces shaping the changes and patterns of investment and production. This situation is often very difficult for governments to accept, especially those transforming from centrally planned to more market orientated economies. Local communities too are demanding greater empowerment, forcing governments to delegate and execute their public responsibilities in a more consultative and efficient manner. New partnerships between government, business and communities are emerging to execute many of the functions and responsibilities traditionally undertaken by government agencies.

Those trends are a paradox for government. On one hand they need to shape public policy and public investment in services and infrastructure to facilitate the growth of competitive business in a global economic context, while on the other hand they need to empower communities and business to mobilize resources to provide and manage much of the infrastructure and services needed to support sustainable economic development. What is emerging is a transformation in governance processes that have been incumbent for a very long time, creating difficult challenges for leadership and the design of new institutions in governance.

Those changes are resulting in the relative decline of the nation state and an increasing focus on regions as the centres of economic growth. Regions – and more particularly larger metropolitan regions – are dominating the growth of employment, investment decisions and distribution networks in the global market place. Increasingly, it is the metropolitan region that is the geographic unit of analysis at which competitive economic activities take place, and increasingly they are the places where prosperity is being generated. Such regions are not necessarily defined by political boundaries, but cut across local – and sometimes international – jurisdictions to encompass the broadest definition of contiguous economic activity. For example, the Shenzhen region of Southern China is an integral part of the Hong Kong economy; the 'Cascadia' region of the Pacific North West of North America—stretching from Portland through Seattle to Vancouver—crosses the international border between the United States and Canada; and Southern California has a continuous urban region stretching from Santa Barbara through Ventura, the Los Angeles metropolis and Orange County to San Diego, then across the United States-Mexico border to Tijuana. Around the globe, mega metro regions are integrating if not already being totally integrated as powerful economic regions.

Most businesses do not confine their economic activity to a specific jurisdiction, and when the form of their organizational dependence is the industry cluster, important economic inter-relationships are even more likely to spread across jurisdictional boundaries. Globalization and restructuring in national economies has resulted in the outsourcing of production and services. As developed economies open up to international competition, national industries are not able to compete on price or achieve economies of scale, and subsequently they restructure or move production offshore. From these restructured industries new hybrid businesses emerge that begin producing and exporting components and services into global markets. This leads to growing networks of suppliers and distributors and the formation of industry clusters. These industry clusters now account for a major proportion of employment in developed economies.

Traditionally, national governments sought to facilitate the creation of sectoral industries as part of a strategy to gain self-sufficiency in demand for domestic goods and services, with surplus production being exported. There was strong horizontal and vertical integration in those nationalized industries. In the past, most nations under policies of economic self-sufficiency supported the development of national steel, automobile, chemical and food processing industries. However, the inter-sectoral activities and thus linkages among national industry sectors were often minimal, resulting in significant duplication of research and resources by sector industries. National restructuring policies subsequently resulted in industries looking at ways to cut cost and improve efficiencies to improve competitiveness. As a result, many industries moved offshore to other locations, which had competitive advantages through factors such as lower labour costs or a more cohesive business environment.

The factors discussed above are substantially changing the way industries develop in regional and metropolitan economies. Governments and businesses that do not respond to these changes may run the risk of losing market position and of missing out on those existing industries and on new economic development opportunities that are created by more open markets. It makes little difference either to global corporations or to financial markets where particular industries and businesses are located, provided they have the capacity to deliver services, commodities and products in a timely, efficient and cost-effective manner, taking into consideration acceptable risk. Thus it is important for regional economic development strategy to be cognizant of the global context of a region and to develop local actions that will facilitate integration with the global economy.

1.7 Sustainable Development

A number of times in this chapter we have made mention of how the issue of *sustainability* and the concept of *sustainable development* are concepts now widely accepted as desirable objectives to guide urban and regional development and planning and they are also going to become more important as principles to consider in regional economic development strategy.

1.7.1 Definitions

The most commonly cited definition of sustainability is the Bruntland Commission Report, *Our Common Future* (Bruntland Commission 1987), which placed the concept of sustainable development and 'futurity' firmly on the international agenda by defining it as:

...development that meets the needs of the present without compromising the ability of future generations to meet their own needs ... [it is] not a fixed state of harmony, but rather a process of change in which the exploitation of resources, the direction of investments, the orientation of technological development and institutional change are made consistent with the future as well as present needs (p. 43).

This definition refers to a specific type of further and future development of society which essentially stresses that development should be such that natural resources are not exploited; rather they are used in a manner that ensures continuous use into the future (van Lier 1994).

The Bruntland Commission statement received some criticism due to a failure to address the critical distinction between development and growth. Often development is equated with growth and these terms have been wrongly used interchangeably, as they are not synonymous. For example, Kozlowski and Hill (1993) argue:

...Development is the realization of specific social and economic goals which may call for a stabilization, increase, reduction, change of quality or even removal of existing uses, buildings or other elements, while simultaneously (but not inevitably) calling for the creation of new uses, buildings or elements. It must be noted that in each case development should lead to progress, expressed primarily by welfare improvements in the communities involved and that it will occur through specific changes (p. 4).

1.7.2 Approaches to Sustainable Development

It is common to identify at least three approaches to sustainable development: the *ecological* (environment); the *economic*; and the *socio-cultural*. At the same time the concept has been applied to other more specific contexts such as transportation. But focusing on the three primary approaches, one can see why sustainability is now being also referred to as a concern for the *triple bottom line*.

The Ecological Approach. This considers sustainable development in the context of stability of physical and biological systems that are essential and critical to the overall ecosystem. Some have argued that sustainable growth is contradictory in ecological terms, as ecosystems do not grow indefinitely. Compared to human ecosystems where resources are used in a once-through linear flow, natural ecosystems use, process, transport and consume resources in a continuous feedback loop. The ecological approach considers that the loss of resilience implies a reduction in the systems' self-organization mechanisms, but not always a loss of productivity. Of course this depends on the extent to which human societies can adapt and function when confronted with stress (Perring 1991; Daly and Cobb 1989).

The Economic Approach. This is based on the concept of the maximum flow of income that can be created, while at least maintaining the renewable stocks or assets that yield these benefits. Underlying this approach is the idea of optimality and economic efficiency in the use of scarce resources. In this context, sustainable development can be simply defined as: "non-declining consumption per capita or per unit of GDP or some alternative agreed welfare indicator" (Turner 1993 p. 6). Thus, according to researchers such as Maler (1990), Turner (1993), and Munasinghe (1993), the primary difference between the economic and ecological approaches is found in the consequences of a loss in ecological resilience. From the discipline of economics we know that under the Hicksian income measure, a society that continues to consume its fixed capital without replacing it is unsustainable.

The Socio-cultural Approach. This aspires to maintain the stability of social and cultural systems, recognizing that people are at the centre of concerns for sustainable development wanting a healthy and productive life in harmony with their environment. The changing form of cities and the nature of the built environment can have dysfunctional and disempowering effects on some groups in society, resulting in social and economic disadvantage. Development objectives might include: poverty eradication; crime and alienation reductions; seeking to minimize women disproportionately bearing most of the social and environmental costs of the built urban environment, which has been largely developed by men and for men; and addressing the multi-faceted urban access issues for a wide range of social groups to avoid further marginalization. Thus, *inter-generational* and *intra-generational* equity is of pivotal importance to this approach. The right to develop needs to be considered in light of meeting the social, developmental and environmental needs of present and future generations in an equitable manner.

1.7.3 The Shift in Paradigm Focus

The emerging concern with *sustainability* has led to the evolution of a new paradigm for viewing growth and development.

The *traditional growth models* were based on premises such as:

(a) the goal of profit maximization;
(b) community production and consumption that was resource intensive and concentrated in large urban-industrial centres;
(c) fossil fuel-based energy using energy consumptive technologies;
(d) large scale production systems that were centralized; and
(e) the assumption that humans dominate the environment which was seen as abundant and limitless; and,
(f) a goal to maximize social benefits.

The *new sustainable development paradigm*, however, is based on premises such as:

(a) the goal of viable, long-term growth;
(b) conserving resources in production through energy efficient technologies and dispersed production centres of lesser scale;
(c) a shift towards alternative energy sources, recycling and conservation of resources; and
(d) the assumption that humans and the environment are mutually interdependent, acknowledgment that resources are exhaustible and often irreplaceable, and that conservation is a principle for long-term viability.

Thus, improvements in regional performance may not necessarily be the outcome of economic growth as typically defined in terms of increasing per capita gross regional/domestic product, and no-growth is not synonymous with no-development. But what is seen as being important by the advocates of sustainable development is the acknowledgment that the need for economic progress goes hand-in-hand with development, although such development should minimize costs (economic, social and environmental) and maximize benefits – a challenging trade-off issue. Jacobs (1991) sees sustainable development as that which delivers basic environmental, social and economic services to all residents of a community without threatening the viability of the natural, built and social systems upon which the delivery of these services depends. Pearce et al. (1989) have suggested sustainability means that the environment should be protected in such a condition and to such a degree that environmental capacities (the ability of the environment to perform its various functions) are maintained over time: at least at levels which give future generations the opportunity to enjoy a similar measure of environmental consumption.

Sustainable development is also often viewed as improving the *quality of life* while living within the carrying capacity of supporting ecosystems, according to the 1992 Maastricht Treaty on European Union (Article 2). The 1992 European Community Fifth Environmental Action Programme of 1993 sees sustainable development as a harmonious and balanced development of economic activities, sustainable and non-inflationary growth respecting the environment, while the International Council of Local Environmental Initiatives (ICLEI) views sustainable development as continued economic and social development without detriment to the natural resources on the quality of which human activity and further development depend.

Overall, *sustainability* may be construed as 'continuing without lessening'; thus a sustainable community or region may be regarded as one that seeks to maintain and improve the economic, environment and social characteristics of an area so that its members can continue to lead healthy, productive and enjoyable lifestyles. Development might be seen as improving or bringing to a more advanced state. Therefore, *sustainable development* may be seen as improving the economy without undermining the environment or society.

1.8 Schools Shaping Thinking About Regional Economic Development Strategy and Planning

So far in this introductory chapter, the discussion has focused on approaches to regional economic development and the evolution of regional economic growth and development theory. The focus now turns explicitly to consider approaches to thinking about strategy for regional economic development which is a core topic of this book.

Mintzberg (1990) has identified ten schools of thought that have shaped much of our current thinking on strategy for regional economic development. These schools have relevance to business and economic development strategy. The schools fall into three broad categories, but they are not mutually exclusive. They are:

(a)   three strategy form schools;
(b)   six strategy process schools; and
(c)   the configurational school.

A summary overview of these is given in the sections that follow.

1.8.1 Strategy Form Schools

The first three schools of strategy relate to how strategy should be formed. These are:

The Design School. This school involves a conceptual process developed in the 1960s primarily by the Harvard Business School. It makes extensive use of methods like SWOT analysis (strength-weaknesses, opportunities– threats) to match a firm's capabilities with its environment. There is a firm belief that strategy and structure must fit in between a firm's capabilities and its customers and markets. The design school approach to strategy has been criticized because of its lack of responsiveness to change, particularly in large organizations. The method seems to work best for simple organizations in stable environments.

The Planning School. This school takes a formal process approach that originated from ideas on corporate strategy (Ansoff 1965) and was used in the 1960–70s. It took the basic approach of the design school, but planners rather than managers became the major players in the process. It makes extensive use of mechanical processes to achieve programs and projects, as opposed to strategic outcomes. There was a strong emphasis on scheduling, programming and budgeting. A major failure of planning school strategy is that it assumes budgets, objectives, strategies and programs form part of a single process. Subsequently, planning has become separated from strategic management. Planners did their job and management and others did theirs without understanding each other. The result has been gaps between strategic and accounting issues (budget/objectives) and action (ad hoc strategies/programs). Another failure is that planners became so dependent on extrapolating demands, outcomes and strategy, and organizational structures became too rigid, that it became almost impossible to manage rapid change.

The Position School. This school involves the use of analytical processes from Porter's early work in the 1980s and the work by the Boston Consulting Group. The approach introduces the ideas of competitive strategy and competitive positioning. Competitive strategy moves away from the concept of strategy as a linear process to achieve specific outcomes to a process of dealing with dynamic economic environments and changing outcomes and positions. The focus of competitive strategy is to look at industry structure and change, and to identify how to position firms in a competitive environment. This approach swept away the design and planning schools that assumed more stable environments. The idea of competitive positioning has its origins deep in history—including Sun Tzu's military strategy (600 BC)—which looked at types of strategy best suited for particular contexts. The criticism of the positioning school is that it only looks at economic issues in the context of the business environment. Until recently, the role of environmental, political and social dynamics as part of the economic environment was dismissed.

1.8.2 Strategy Process Schools

Six schools of strategy are essentially concerned with process. Many elements of these processes have been incorporated into the direction of the schools described above. But the process schools represent a particular emphasis or thrust of strategy.

The Entrepreneurial School. This school builds strongly on the personality of leadership to provide the visionary process and thrust in setting directions to plan and manage the future. The school embraces many qualities of leadership, drawing upon judgment, corporate wisdom, experience, and insight to build a consensus for a sense of direction and vision about the future of an organization and its core business. This school is based firmly on the ideas that entrepreneurs have a key leadership role in developing new combinations of things, doing things differently and taking risks to get things done. The entrepreneurial school recognizes the need to foster internal (firm) and external (industries, regions) entrepreneurship as part of the strategy. The entrepreneurial approach to strategy is most useful for start-ups, turn-around, and small enterprises, but it has limitations in large corporations, organizations and governments where the focus of economic strategy is highly diversified and multidimensional.

The Cognitive School. This school involves a mental process that deals with aspects of cognitive psychology and that contributes to strategy formation. Key concepts in the cognitive school are: the role of perception in dealing with complex issues that contribute to a wide range of economic development and production processes; concept attainment such as mapping and tacit knowledge about an economy, its resources and its development potential; re-conception in changing the mind set of businesses and organizations facing major changes; and cognitive styles involving the way the economic environment is understood (De Waele 1998). The cognitive school is leading to new areas of research into behavioural economics—in particular changes in consumer behaviour—and psychographics, involving the spatial patterns of consumer behaviour. The cognitive school is having a growing influence on economic thinking and strategy, especially as markets become more niche oriented and consumers demand greater choice.

The Learning School. This school, initiated by Lindblom (1965), recognizes that strategic direction approaches to strategy are too simple and severely handicapped by the lack of information needed for sound decision making. Strategies are too complex to develop and implement and do not involve set piece plays on a chessboard, and organizations can only develop strategy as part of an ongoing learning process. The science of muddling through describes the process for dealing with situations, which are too complicated as disjointed incrementalism. Strategy develops from learning over time, rather than through a dependence on analysis.

The Political School. This school is essentially about power, and is deeply entrenched in the ideas of the planning and positioning schools. In a sense all strategies are political, requiring an official stamp of approval before being implemented. However, the different schools of politics—ranging from mandate, consensus building, stakeholder or interest group balancing, to partnership—will have a major impact on strategy development and success. The political school prevails in government as well as a wide range of industries and community organizations. It assumes that the formation of strategy is a power process, based on agreed positions rather than as a proper consideration of alternative points of view. Political ideologies mould strategy and ultimately the organization that will be responsible for its implementation. The implication is that an organization is driven by parochial, rather than common interest. It raises major problem issues about the governance of strategy.

The Culture School. This school is erected on an ideological process based on established patterns of beliefs shared by members of an organization, group or society. These beliefs become the mind that guides the decision-making processes related to the organization, work and production. The cultural school is focused strongly on collective and co-operative effort, with many aspects of culture deeply embedded in work and business practice. Japanese business practice has provided much of the contemporary thinking for the cultural school, many of the ideas having developed from management theory (Ouchi 1981). The cultural school is not new. It dates back centuries and can be traced to early monastic sects, including 19th century Quaker ideas on industrial villages and Zionism. Unlike other strategy schools, the cultural school tends to resist rather than support change. The inflexibility of the Japanese cultural school is one explanation for the failure of the Japanese economy since 1991. The emergence of research on social capital and cultural capital (Coleman 1988; Putnam 1993; Brusco 1995) has added much support to the role of culture in supporting economic development. As the world becomes more globally integrated it appears the re-discovery, re-invention and localization of culture is growing in response to the need to create or maintain identity (Naisbitt 1994). The ideas of the culture school are expected to play a more significant role in economic development strategy in the future.

The Environmental School. This school focuses on those forces outside an organization that have a key role in shaping strategy. The school is primarily concerned with the business environment or context. The approach recognizes the need for a balance between the external environment (customers and markets) and the internal environment (leadership and organizational arrangements and aspirations) in shaping strategy. This school is based on Darwinian theories of natural selection and survival of the fittest. The focus of strategy is on tactics to survive, which usually means getting bigger by consuming others or developing defensive strategies to keep predators at bay. The school fails to consider that defence can be turned into advantage through adaptation or learning. Strategic alliances and networking can improve both the defence and expansion needs of strategy, but only within the context of the known environment. The idea of collaborative competition, which can lead to innovation, is generally hostile to the environmental school. The emergence of environmentally sustainable development (ESD) challenges many of the school's tenets. And the failure of the school to consider the environment in its social, economic and physical context is its major deficiency. Nevertheless, the school has made a major contribution to the recognition that external environments have a significant role in shaping, formation and implementation of strategy.

1.8.3 The Configurational School

The third grouping of schools shaping current regional economic development strategy thinking is the configuration school, which sees the different elements of the strategies described above as stages in organizational history. Mintzberg (1990) proposed strategy as a plan for a course of action that is defined in advance. Such a plan would provide a perspective based on cognitive experience; set patterns of behaviour within the organization and the position an organization has in its operational environment. It can provide a learning process for sensing, analyzing, choosing and acting. The configurational school is eclectic, borrowing from the best elements of schools of strategy according to the situation. Thus strategy is an episodic process, which sees different schools of strategy being configured and used at appropriate times.

1.9 New Tools, New Strategies for Regional Economic Development Planning

Because of the changing role of regional economies within nations and the impacts of globalization, and given the context of contemporary concerns about how to achieve sustainable development, a set of new considerations are now being taken into account in formulating and implementing economic development strategies for regions. While traditional modes of regional analysis remain important and useful as means of measuring regional economic change and for investigating inter- and intra-regional relationships by industry sectors, regional economic development policy and strategy formulation now needs to give attention to factors such as how regions:

(a)   accumulate core competencies;
(b)   develop social capital;
(c)   build strategic leadership;
(d)   manage resources;
(e)   build market intelligence;

(f)   provide strategic infrastructure;
(g)   develop a risk management capability; and
(h)   incorporate principles of sustainability into their regional economic development strategies.

In this chapter it has been noted how understanding regional competitiveness is now a crucial element of the strategic planning and management of economic development of regions. In an age of rapid change and uncertainty, carefully considered strategy planning and implementation for regional economic development is becoming increasingly important for enabling regions to position themselves to build and maintain a competitive advantage.

However, many of the traditional approaches to strategy for regional economic development and the implementation of plans have been deficient or appear to be inadequate to deal with the dynamics of regions having to both compete in the global economy and address issues of sustainability. Thus this book both reviews traditional methods of regional analysis and describes new paradigms for regional analysis. As such it identifies new approaches to the formulation of strategy for regional development, built around the use of both existing and new analytic tools and planning methods, enabling regions to capture and create new opportunities for economic development.

Contemporary thinking is diverse on how to plan for and how to facilitate regional economic development in an environment of global competition, rapid change and the concern over sustainability. Authors such as Imbroscio (1995) advocate strategies of greater self-reliance, while Park (1995), McGee (1995) and Ohmae (1996) advocate the development of regional economic development strategies based on strategic alliances and inter- and intra-regional network structures, including digital networks (Tapscott 1996). Henton (1995), Hall (1995), Waites (1995), Sternburg (1991), and Stough et al. (1995) advocate the need to base regional economic development on the growth of clusters of industries. The concern over sustainability in development increasingly is evident in metropolitan and other regional plans seeking to integrate environmental, economic and social approaches to create urban and regional environments that enhance quality of life, meet environmental quality goals, and achieve economic growth and employment diversification.

Indeed there is no universal model or framework guaranteeing success for regional economic development. Planning for the future is difficult, as the future is unknown and uncertain. However, the future is usually not some kind of accident that will be imposed upon a region by dark chaotic forces. Increasingly in an era of rapid technology development and of better information and improved human knowledge, the future is being seen as something that will emerge from the spirit of the present, forged by attitude and action at least over the short to immediate horizon. It is possible, therefore, for a region to create a path for the future (Rosebury 1994) by laying down the strategic architecture required and appropriate to support a range of economic possibilities, based on the competitiveness of resources, infrastructure, governance, and core competencies.

In arguing the need for this emerging paradigm of regional economic development planning, and to assist regions or communities to undertake the process of strategy development, it is important to identify the key elements for regional economic development strategy building and implementation, and to place these in a process that pulls together resources, infrastructure, social capital and technology to facilitate the economic development of a region in a dynamic globally competitive environment.

Regions and communities within them seeking to develop strategy for regional economic development are able today to take advantage of new methodological frameworks that have been emerging over the past decade or so. While these incorporate well tried traditional tools of analysis of regional economic performance, a number of new tools are now available to enhance the process. The remaining chapters of this book canvass and assess these traditional and emerging new approaches and tools.

Increasingly we might envision the intent of regional economic development strategy being to establish a platform for change to guide the development of a region and to facilitate its competitiveness in a global environment in the pursuit of a sustainable future. This involves mobilizing key actors or facilitators and agents of change, through partnership approaches encompassing business, markets, government and community. Traditional tools of regional economic analysis remain useful to conduct regional economic audits and to measure and evaluate regional economic performance, utilizing existing data by building new integrated data bases, and through tapping existing partnerships, networks and alliances. The framework that we have described proposes a focus on the identification, description, analysis and evaluation of core competencies, resource endowments, infrastructure competitiveness, market intelligence, and regional risk through the combination of qualitative and quantitative methods encompassed in industry cluster analysis (ICA) and multi sector analysis (MSA). Economic possibilities for the future can be identified and evaluated leading to the statement of strategic intent. Alternative development futures or scenarios are evaluated through the participation of stakeholders within the region and, most importantly, external to it to encompass the assessments of key decision-makers controlling capital, trade and other flows to the region. If the assessments of the feasibility of the alternative scenarios by the internal and external stakeholders are incongruent, then there is the potential that inappropriate or unfeasible strategies will be pursued. Thus strategic directions might need to be redefined before formulating an economic development strategy which focuses on industry cluster development and the provision of strategic architecture. But it is important also to emphasize that implementation plans and mechanisms need to be developed and put in place by appropriate agencies in the region. The progress made needs to be monitored, requiring agreement on indicators and benchmarks set to measure and evaluate the performance of the region over time in order to assess the degree of success of the strategy and progress towards achievement of the strategic intent. Inevitably this involves building enhanced regional infrastructure systems along with strengthening existing and building new partnerships, networks and alliances. And in the contemporary era of the global economy, increasingly the pursuit of regional economic development also needs to take place within the context of principles for achieving a sustainable future.

A simple representation of contemporary best practice approach to regional economic development is set out in Fig. 1.6.

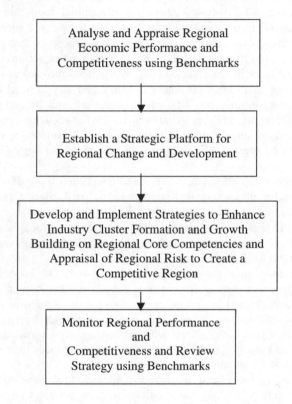

Fig. 1.6. Process for a contemporary approach to regional economic development

1.10 Scope and Structure of this Book

This book is presented in ten chapters that address major components and issues inherent in this new framework for regional economic development discussed above. The intent of the book is as follows:

(a)   To review important long established existing tools of regional economic analysis and strategic approaches to regional economic development.

(b)   To outline recent innovative approaches that have their roots in a wide range of the literature in the economic, behavioural, management, planning, and regional sciences.

(c) To provide a 'how to do it' framework of analytical tools and strategy devel-
     opment approaches to assist regions and practitioners in undertaking regional
     economic development strategy planning and regional performance assess-
     ment processes.
To these ends the book contains a wide range of case studies and examples.

Chapter 2 provides an overview approach of regional economic development
strategy, particularly from a local community perspective, as it emerged during the
1970s and into the 1980s. It focuses on the typical processes that have been involved in developing a strategy and the typical orientation of actions and programs
that have been inherent in the evolution of a widespread movement for communities to plan for and facilitate their future development as practiced in the United
States. In particular, regional profiling and institutional capacity and capability assessment, benchmarking, and organizational issues for regional economic development are discussed and illustrated, including actual examples of strategy processes undertaken by regions. The chapter also discusses strategy for sustainable
development.

Chapters 3 and 4 examine commonly used traditional tools developed by regional scientists, for measuring regional economic performance, how they have
evolved, their limitations and their utility. These chapters provide examples of
their application in regional analysis at different levels of regional scale. The focus
is on measuring the economic base and the use of shift-share analysis in Chap. 3,
and on input-output analysis in Chap. 4.

Chapter 5 reviews and examines the emergence over the last decade or so of
new approaches to path setting for regional economic development, focusing on
strategic intent and the factors and processes for creating a path for the future, including strategic architecture. It presents a new approach to strategy formulation,
derived from ideas that have gained wide acceptance that were developed by a
number of leading management theorists which are being incorporated into the regional economic development literature.

Chapters 6 and 7 discuss and describe two specific tools for assisting regional
economic development strategy formulation; industry cluster analysis (ICA) and
the growth of industry clusters to enhance regional development, and multi sectoral analysis (MSA) for the evaluation of regional competitiveness and assessing
regional risk factors. Examples of applications are provided through case studies.

Chapter 8 turns to the issues involved in capacity building for regional economic development strategy formulation and implementation. It focuses on the
roles of leadership and institutions, and addresses issues such as building social
capital, empowerment, re-engineering institutions and creating organizational
change, the building of strategic alliances and networked partnerships, and the
provision of smart infrastructure and development of supporting information systems. Particular emphasis is placed on the role of leadership in regional economic
development. Examples of such capacity building approaches are provided.

Chapter 9 provides an overview of the way geographic spatial information systems (GIS) are being interfaced with spatial models to build spatial decision support systems (SDSS) that support existing but recently developed tools to inform

decision makers in both understanding impacts in regional analysis and in regional economic development strategy planning.

The concluding chapter speculates on some of the emerging key issues that regions and communities are facing as we enter the 21st century and the challenges they will need to address in strategy and its implementation for regional development in the context of addressing a sustainable future, including regional risk assessment and virtual environments.

2 The Regional Economic Development Movement: The Evolution of Strategy from Early to Contemporary Approaches

2.1 The Changing Context

During the 1950s and 1960s, governments throughout the world became more and more concerned with developing and implementing policies and programs for regional economic development. But the focus on regional growth and development was not restricted to the public sector. The private sector also became heavily involved in the process of regional development strategy and planning and increasingly so did the wider community. In advanced western societies, in what were called the 'long boom' years following World War II—during which time rates of economic growth were high, levels of unemployment were low, and inflation was low—regional and local economies within nations largely were sheltered from external forces during an era of protectionist industry policies and national government controlled exchange rates, and restricted trading arrangements. But regional development policies and programs often took a specific focus, typically being directed towards major infrastructure projects for agricultural regions - as exemplified by the actions of the Tennessee Valley Authority - or development plans for depressed regions - such as the Appalachia Regional Development Program - in the United States. Industry decentralization programs also were used in attempts to relocate industry to regions in decline. For example, in the United Kingdom the government was active in assisting old industrial regions such as Liverpool and Glasgow to attract new industries.

However, as noted in Chap. 1, from the early 1970s the world entered a new era that was characterized by globalization and new policies embracing deregulation and an increasing concern with enhancing national and regional competitiveness. Protective tariff barriers were increasingly dismantled or dramatically reduced; the collapse of the Bretton Woods agreement saw the floating of national currencies; international capital markets began to operate in a deregulated environment; and trade was progressively opened-up. As a result of these changes, widespread and substantial economic restructuring has occurred, differentially impacting the cities, towns and regions within nations. Many traditional industries that were the mainstays of regional economies have declined and in some cases disappeared, while new technologies and changing societal processes have generated new industries. Unemployment rates rose, and rates of inflation rose and fluctuated wildly. Patterns of employment across industry sectors changed dramatically as economic systems moved away from the Fordist era of manufacturing into the post-industrial era of services dominated economy in the information and knowledge age. Female participation in the labour force continued to increase, and in some places cheap immigrant labour dominates some industry sectors. Technology innovation and change now occurs more and more rapidly, and product cycles shorten in time. In sum, during the last few decades we have moved from a world of considerable economic stability and certainty to one of increasing uncertainty and increasing competition. Some regions and localities have been winners, while many others have been losers, and some of those losers were winners in the earlier era of protection and industry subsidy.

As a result of the above, regions as large as major metropolitan cities, and localities as small as neighbourhoods within cities and small rural towns, became more concerned with issues such as how to:

(a) retain existing firms and attract new firms, new industries and investment;
(b) diversify employment and generate job opportunities;
(c) enhance human capital resources;
(d) facilitate development;
(e) improve institutional capacity and capability to grow competitive communities; and
(f) achieve sustainable development.

This period from the 1970s to the 1990s witnessed the emergence of more formal approaches to regional and local economic development, including ways to measure and monitor regional and economic performance, the formulation of strategies to regional or local economic development plans and their implementation, and a focus on developing institutions and organizational arrangements to facilitate the development process, including public-private-community partnerships. By the 1990s there had also emerged an increasing focus in regional development strategy and planning on the issue of *sustainability* and *sustainable development*, which is also referred to as the triple-bottom-line, where there is a concern for environmental, economic and social outcomes in regional growth and development.

This chapter provides an overview of the emergence of this formal and structured approach to planning for regional economic development and implemented by communities, both large and small. The focus we take is on the typical processes involved in undertaking planning for what had been called local economic development that emerged from the late 1960s and early 1970s up to the 1990s and the current decade. Along with the next two chapters — Chaps. 3 and 4—which provide an overview of methods and tools for regional analysis, this chapter provides the background context for the evolution more recently of new methodological approaches to the process for regional economic development strategy planning that has emerged from the early 1990s and which are discussed in the subsequent chapters of this book.

2.2 The Changing Objectives of Local Economic Development Policy

No one theory or set of theories adequately or totally explains regional or local economic development. But it is feasible to suggest, as Blakely (1994, p. 53) has proposed, that it is a sum of a whole range of theories that may be expressed as:

Regional/Local =f(natural resources; labour; capital, investment;
Development   entrepreneurship; transport; communication; industrial
*   composition; technology; size; export market;   interna-*
*   tional economic situation; local institutional capacity;*
*   national, local and state government spending; develop-*
*   ment support schemes)*

All of these factors may be important. However, the economic development practitioner is never certain which factor has the greatest weight in any situation. Thus as foreshadowed in Chap. 1, regional economic development can be viewed as embodying elements of a range of theories, including neo-classical economic theory, economic base theory, location theory, central place theory, cumulative causation theory, and attraction theory. Their relevance is outlined briefly below.

2.2.1 The Evolving Focus

In practice, over the last several decades, regional economic development policies and programs have tended to focus more and more on one or a mixture of the following:

(a) enhancing employment growth and job diversity, and addressing labour costs and productivity;
(b) developing the economic base of communities, particularly through increasing internal institutional linkages;
(c) managing locational assets to enhance location inducing factors; and
(d) developing knowledge-based information-intensive industries.

In this context, Blakely (1994, p. 62) identifies a move from 'old concepts' to 'new concepts' of local economic development in which the quality of a region's human resource base is a major inducement to industry (see Table 2.1). He classifies regional or local economic development policy measures according to their implicit or explicit discrimination between regions, sectors or firms (see Table 2.2).

Many nations have tried variations of these approaches over time, but typically there has been a lack of consistency on which a locality can depend. Because the political-economic framework varies so much among nations (and for that matter, regions), no true prototype of regional or local economic development exists. Blakely (1994) describes it as a 'movement' rather than a strict economic model from which uniform approaches have emerged. Over the years, local economic development has evolved "not merely new rhetoric but a fundamental shift in the actors as well as the activities associated with [it]" (p.49). The next part of the chapter summarizes the evolution of this 'movement' in a number of different broad national geographic contexts.

Table 2.1. Toward a theory of local economic development

Component	Old concept	New concept
Employment	More firms = more jobs	Firms that build quality jobs that fit the local population
Development Base	Building economic sectors	Building new economic institutions
Location Assets	Comparative advantage based on physical assets	Competitive advantage base on quality environment
Knowledge Resource	Available workforce	Knowledge as economic generator

Source: Blakely 1994, p. 62

Table 2.2. Schema for classification of regional policies

Sectoral economic dimension	Spatial dimension of policy		
	None	Implicit	Explicit
No industrial policy	General economic policies	Economic policies with regional implications	Regional policies
Some sectors supported indirectly (e.g. steel, agriculture)	Sectoral policies	Structural policies	Regional development (proper)
Industrial supports			
By sector	Structural policies	Regional structural policies	Local economic development
By firm/plant	Non existent	Local business assistance	Firm-/plant-specific structural policies

Source: Blakely 1994, p. 42

2.2.2 OECD Initiatives

During the 1980s there was a concerted attempt among many OECD economies to link regional and local economic development to a national policy context. This link was encouraged by the OECD (OECD 1986) to moderate the effect of the pace of the economic adjustment on localities and individuals; to cushion the impact of rapid economic change on firms and affected employees; and to revitalize local economies and facilitate adjustment to the economic transformation of the national economy. Five specific objectives of regional or local economic development were proposed:

(a) Strengthening the competitive position of regions and of localities within regions, by developing the potential of otherwise underutilized human and natural resource potential.
(b) Realizing opportunities for indigenous economic growth by recognizing the opportunities available for locally produced products and services.
(c) Improving employment levels and long-term career opportunities for local residents.
(d) Increasing the participation of disadvantaged and minority groups in the local economy.
(e) Improving the physical environment as a necessary component of improving the climate for business development and of enhancing the quality of life of residents.

2.2.3 United States Initiatives

Part of the catalyst for local economic development initiatives in the United States was an increasing interest being shown by policy makers in the economic development issues facing communities. It was reflected in a concern for better management of urban development and services provision and delivery as the imperatives of fiscal constraints required governments to diversify the resource base of cities and regions through enhanced economic performance and increased business activity.

A trend towards this economic development and fiscal diversification orientation was evident by the mid 1970s. For example Meeker (1976), in his study of trends in national and regional growth in the United States, identified a number of broad policy alternatives including:

(a) Growth consequences of environmental regulations.
(b) Choices in the competing adjustment in transportation policy; how the expansion of telecommunications capabilities affected growth.
(c) Changes occurring relating to improving housing and neighbourhoods.
(d) National development and steps towards balanced economic growth.
(e) New directions for guiding and controlling land development; and ways to strengthen the fiscal and management capacity of local government.

That study also noted how there was increasing public concern about the performance of government staff as a factor to assess in considering the public sector's ability to guide growth and development.

In 1978 the White House Conference on Balanced National Growth and Economic Development (Advisory Committee to the White House 1978) also had identified a number of themes, including:

(a) Strengthening local economies, people and jobs.
(b) Improving government budgets.

(c) Improving the management of growth and understanding its varying geographic impacts.

Specific issues discussed were:

(a) Energy, water, environment and growth.
(b) Limited resources and conflicting objectives.
(c) Structural employment; local fiscal plight and regional fiscal policy; growth issues in rural America.
(d) The needs of women and minorities in cities; and federal domestic policies.

Urban strategies for economic development also became more and more common. Warren (1980) shows how ten states set goals and objectives for addressing the needs of local communities. Horizontal strategies were concerned with the coordination or redirection of State government policies and of State agency decision making in assisting local governments, while vertical strategies involving setting policies that guided or changed local government action were addressed. The better targeting of scarce Federal resources to needy communities was identified as a major focus of national urban policy that needed to be better addressed at the State level.

During the 1980s there was widespread activity in cities and in local communities, both large and small, to formulate and implement economic development strategies. A considerable amount of research was published on strategies to undertake this process. It was seen to be a priority in many cities, and places saw it is an imperative.

The management guru Peter Drucker (1991) emphasizes how the factors that were 'uncoupling' national and local economies from one another as many metropolitan areas were thrust into international competition and as the world economy shifted from labour-intensive to knowledge-intensive industries, meant that "practitioners, whether in government or in business, could not wait until there was a new theory. Governments had to act, and their actions were more likely to succeed the more they were based on the new realities of a changed world economy".

By the 1990s, in the United States in particular there was a widespread recognition at both the national and local levels of the renewed importance for regional economic development policies, as cities and regions sought to maintain and improve their competitiveness and to promote and sustain the development process. Many regional economic development policies view the performance of regions as being linked to national economic performance, the thinking being that local economic adjustment and overcoming local barriers to change are vital to the performance of the nation, and that local areas need to be part of the process of national economic growth. However, there also remains a strong attitude of protectionism in some regions, including many agricultural based areas.

2.2.4 The British Experience

Writing on the British experience, Cochrane (1992) tells how the period since the late 1970s has been one in which "the links between business and government have begun to be forged rather more effectively than in the past, as part of the process of moving toward an 'enterprise' state. "The language of welfare has been replaced by the language of growth, regeneration, and public/private partnership, particularly in urban areas" (p. 45).

This enterprise approach has become widespread across the developed economies of the world. Blakely (1994) aptly describes it as:

... essentially a new process by which local governments, along with local corporate firms, join forces and resources to enter into new partnership arrangements with each other, in order to create new jobs and stimulate economic activity in a well-defined economic zone. The central factor in locally based economic development is the emphasis on 'endogenous development' policies using the potential of local business, institutional and physical resources (pp. 49–50).

The importance of endogenous growth is a theme that recurs throughout this book.

2.2.5 The Public/Private and Partnership Approaches

One important theme to emerge in the evolution of the above approaches to regional or local economic development has been what is termed the public/private partnership approach to implementing strategy. Two distinct approaches to how such partnerships are forged have been proposed by Robinson (1989). These are:

(a) A corporate-centre approach, where the emphasis of economic development has been on real estate development and industrial attraction.
(b) An alternative approach, where the concern has been to steer economic development activities towards local disadvantaged sections of the community.

As shown in Table 2.3, these policy approaches are significantly different, although the outcomes might be directed to similar aggregated objectives of growing business activity and new employment opportunities.

Much of the focus in regional economic development theory and of the orientation in regional or local economic development programs up until the 1980s tended to rely heavily on what are known as the 'trickle down' effects of economic growth to improve the conditions of the poor and the disadvantaged.

The alternative approach identified by Robinson focused more on economic development as being process oriented. Blakely (1994) describes it as being concerned with:

... the formation of new institutions, the development of alternative industries, the improvement of the capacity of assisting employers to produce better products, the identification of new markets, the transfer of knowledge, and the nurturing of new forms and enterprises. (p. 50).

Table 2.3. Two economic development policy approaches and their dimensions

Dimension	Corporate-centre approach	Alternative approach
Public and private sector	Primacy of private sector market decisions: private sector lead	Private sector market decisions influenced by public sector interventions: public sector
	Public sector responsible for creating an economic and social climate conducive to private investment	Public sector responsible for guiding private investment decisions so they generate desired economic development outcomes
Public sector planning	Objective favouring growth and tax base expansion	Objectives favouring the creation of direct benefits for low-income and ethnic minority residents
	Planning processes that are relatively inaccessible to low-income and ethnic minority groups	Planning processes that are relatively accessible to low-income and ethnic minority groups
Public sector interventions	Public resources provided as a means of accommodating needs of private industry	Public resources provided conditionally as a means of ensuring specific economic development alternatives
	Intervention in areas likely to generate growth (e.g. attraction of businesses from outside the city)	Intervention in areas likely to produce benefits for low-income and ethnic minority residents (e.g. retraining of displaced workers)
	Targeting of growth sectors (e.g. advanced services, high tech, tourism)	Targeting of growth sectors and sectors able to meet important economic needs
	Targeting of headquarters and branch plants	Targeting of locally owned establishments
	Concentration of projects in central business districts and surrounding areas	Decentralization of project locations
	Emphasis on the creation of jobs for white-collar and highly skilled workers	Emphasis on the range of local labour needs, including those of underemployed, unskilled, and blue-collar workers

Source: Robinson 1989, p. 285

The genesis of this approach was the concern from the mid 1960s to the early 1980s over inequalities in urban areas in some regions of the United States and the interventionalist programs of the Johnson Administration's 'great society' programs, and later of the grass-roots activism that arose as a reaction to the Reaganomics ideology of reduced government funding, privatization, and entrepreneurism (Giloth and Meier 1989, pp. 185–186). The goal was to increase the number and the variety of local jobs through local and neighbourhood action.

2.3 Approaches to Planning for Regional Development in the 1970s and 1980s

2.3.1 Planning Perspectives

Bergman (1981) categorized the approaches that evolved during the 1970s and 1980s to regional economic development planning as:

(a) Approaches responsive to external pressures, which might be pre-active or reactive.
(b) Approaches responsive to local community needs, which might be proactive or interactive.

The planning, policy and development models that evolved within this framework are illustrated in Table 2.4.

Table 2.4. Planning approaches in regional and local economic development

	Responsive perspectives		Planning perspectives	
	Pre-active(I)	Reactive(II)	Proactive(III)	Interactive(IV)
Planning				
Model of practice planning model	Recruitment planning	Impact planning	Strategic planning	Contingency planning
Policy				
Industry	Industrialization	Deindustrialization	New indigenous firms	Building on existing firm base
Enterprise types	Corporate adjustment assistance	Government sponsored	High tech/new tech	Community based
Development				
Intervention model	Industrial inducements	Government program expenditures	Public-initiated development	Community-initiated development

Source: Adapted from Bergman 1981

Much of the actual economic development planning that emerged in practice was reactive. It was government sponsored, and typically it was in response to addressing problems of regional decline as a result of economic restructuring and its deleterious social impacts. Over time, communities increasingly began to embrace more pro-active approaches, with strategic planning becoming widespread by the early to mid 1980s.

However it is important to recognize that regional and local economic development as a field of practice largely remained a "collection of historical activities and reactions to current circumstances" (Blakely 1994, p. 79). In general it was not a well developed, and often was not even a recognized area of government activity. Early on it was associated with chambers of commerce or industrial development or attraction officers, or it was seen as a concern of community-based organizations. As such, much of the emphasis was on community-level institution building, stimulating local ownership, linking employment and economic development through building quality jobs, and public-private venturing. Particularly in the United States, Blakely tells how progressively throughout the 1980s:

... political institutions have forged a new and unusual blend of processes and institutions to create a new concept called 'local economic development': Its key feature is the recognition of the capabilities and resources of local people. It depends on the self-help mentality of the nation (pp. 81–82).

2.3.2 A Model Framework for the Local Economic Development Process

If local economic development is viewed as "a process with a product" (Blakely 1994, p. 64), then if a community is to embark on local economic development planning it is useful to consider a phased approach. In the 1980s, the process followed typically was aimed at developing community capacity to help local institutions to reorient themselves to experience the economic potential of a region or locality. The essential pre-requisites were:

(a) an organization or group of institutions responsible for implementing or coordinating the process; and
(b) clear specification of the economic development area or zone.

The typical tasks involved six phases as set out in Fig. 2.1. But as Blakely (1994, p. 65) notes, too often the actual approach used was ad hoc.

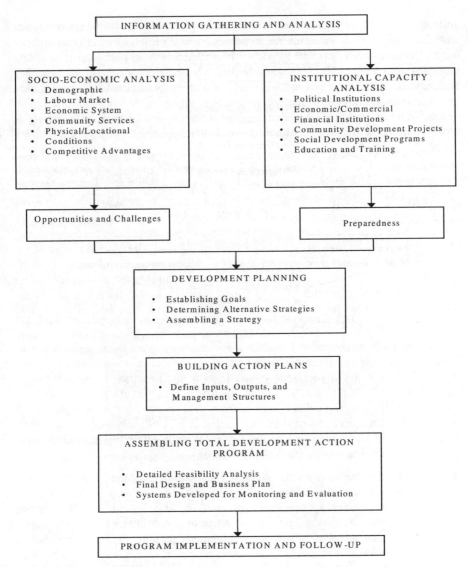

Fig. 2.1. The local development process flow chart
Source: Blakely and Bowman 1986

2.3.3 Profiling a Region: Data Gathering and Analysis in a Regional Audit

Profiling a region to analyze its performance—preferably benchmarked against other regions—is a fundamental phase in the regional economic development process. As shown in Fig. 2.1, this involves both socio-economic analysis and institutional capacity analysis, out of which it is possible to identify opportunities and challenges and to ascertain the preparedness of a region or locality to embark on the economic development process. It is particularly important that this data gathering and analysis be conducted thoroughly and rigorously, for as Bosscher and Voytek (1990) have warned it is only good analysis that enables a community to test its aspirations against the constraints of reality.

Shaffer (1989) outlines six elements that separate this analysis of a region or locality from other types of analysis:

(a) The community is the focal point of the analysis rather than its sub-component firms or households or a residual or small fraction of a larger national economy (although these may also be analyzed).
(b) Internally and externally accessible community leadership and initiatives form an integral part of the analytical framework.
(c) Community resources are important components of the analytical framework.
(d) Citizen participation is the base for developing data; therefore the type of information used and its interpretation are as important as the information itself.
(e) Community analysis is a holistic process that examines the community as an entity but recognizes the different human, physical and economic dimensions.
(f) Citizen attitudes must be explored as both opportunities and constraints in economic development analysis.

The approach to community profiling and analysis in these contexts thus embraces a wide range of dimensions and elements. Analytical tools need to be employed relating to, population, employment, economic structure, community institutional factors, and physical/locational conditions. This involves secondary data analysis, and perhaps the collection of primary data (including surveys). And it involves both quantitative and qualitative data and modes of analysis.

This audit of the region or locality, while varying between communities, sets out to address six basic tasks (Blakely 1994, p.88):

(a) To determine the climate for economic development in the jurisdiction, collect data on economic and political conditions, and decide on goals.
(b) To determine which agencies currently working in economic development are accessible to local government for the purpose of coordination and planning.
(c) To determine whether economic development actually should be conducted by local government alone, or in coordination or consultation with another agency, or by another agency alone.
(d) To identify the barriers to coordination that exist in the jurisdiction, if coordination is desired.
(e) To develop support for the proposed economic development activities from community groups, political groups, and labour unions, among others (e.g., by establishing a formal advisory council with representatives from agencies involved in coordination).
(f) To develop a set of milestones, a means of monitoring progress toward achievement of the milestones, and a plan for revision of activities according to observed results.

Crucial in regional profiling is the requirement to benchmark the region's characteristics and performance against other regions and in particular its neighbours and its competitors. Inevitably this involves many analytical techniques such as location quotients, shift-share analysis, and input-output analysis, which are discussed in Chaps. 3 and 4.

However, as Blakely (1994) points out:

… information on the status of the [regional] economy alone is no indicator of the ability of a community to in fact engage in economic developments. Therefore, in addition to determining the [socio] economic capacity of the area, another analysis of the community institutional profile is required (p.106).

Here the focus typically is on:

(a) community based institutions;
(b) economic structures;
(c) political institutions;
(d) financial institutions;
(e) education and training institutions;
(f) and their roles, skills, organizational structure, and flexibility to cope with and manage change, leadership qualities, and inter-relationships.

And Blakely (1994, p. 108) goes on to show that 'to engage in economic development, a community must have the capacity to perform five development functions through its local institutions', namely:

(a) economic planning;
(b) social and community resource development;
(c) physical and land use planning capacity;
(d) commercial and industrial targeted marketing; and
(e) local finance capacity.

2.3.4 Regional Benchmarking: An Example

By way of example, O'Connor (1996), in his annual *Australian Capital Cities Report*, produces a series of sample graphs that benchmark the performance of the five major metro cities of Australia on a set of statistical indicators relating to population growth and labour force shares by industry sectors, unemployment rates, dwelling commencements, housing price and affordability, international air passengers and international air flights, container traffic, and shares of non-residential construction in industry sectors by zones and sectors within cities. Figs. 2.2 to 2.4 illustrate this benchmarking for a series of housing related indices, such as private dwelling commencements, median house prices, and house price indices. The time series data plots enable the performance of each city to be compared against each other city. Incisive interpretation of the data enables a 'story' to be built up. O'Connor's (1996) interpretation of these four indices is reproduced below.

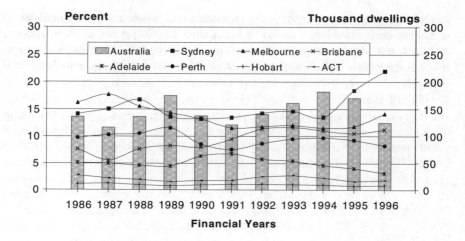

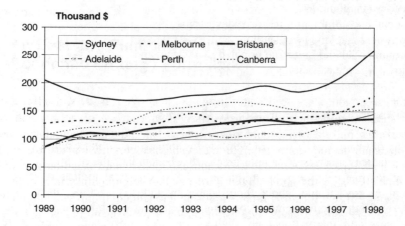

Fig. 2.2. Benchmarking regional housing performance I: Australian capital city shares of national dwelling commencements 1986–1996
Source: O'Connor 1996

Fig. 2.3. Benchmarking regional housing performance II: Australian capital city house prices (at January) 1989-1998
Source: O'Connor 1996

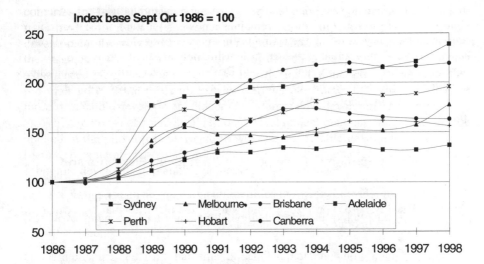

Fig. 2.4. Benchmarking regional housing performance III: Australian capital city house price index, 1986–1998
Source: O'Connor 1996

New house construction in capital cities slowed in 1995, although with more than 95,000 new homes built, activity was at a high level compared to the early 1990s. This activity had important local effects. The share of new houses built in the Sydney market rose sharply to record its highest level since 1986, while the Brisbane market's share began to fall. These outcomes confirmed some of the other trends in the vitality of Sydney, and the local economic problems of Brisbane. The Melbourne and Perth markets recorded small declines in share of total building.

The capital cities recorded a 57 percent share of the nation's new houses built in 1995, maintaining the share recorded since 1986, and in fact a higher level than was the case in 1981. This share was still less than the share of population, but does indicate that the capital city role in the national economy was still strong and was perhaps increasing. The pattern of house prices around the nation changed in only a small way in the previous year. The strength of the Sydney market was retained, with median prices rising steadily after a slump in late 1995. That recovery can be seen in the Melbourne market, although at a lower level.

In contrast, the Brisbane and Perth markets continued to behave erratically with the median price in Brisbane and Perth falling from early 1996. These trends in prices are also shown in the house price index that shows the price of housing relative to its level (set at 100) in 1986. There has been a long term impact of steady demand in Sydney and Brisbane (doubling prices since 1986), substantial gains in value in Perth and Canberra, and the low levels of gain in Melbourne, Adelaide and Hobart. In recent years the relative softening of the Brisbane and Perth markets has weakened, but the Sydney market continued to record steady gains.

The behaviour of house prices suggests that the concentration of high-income high productivity activity in Sydney remained a powerful feature of the Australian economy. Similarly, it shows the restructuring process of the Melbourne economy had yet to yield substantial increases in housing demand by 1996. The data also suggests that the rapid growth of Perth and Brisbane was very sensitive to national economic trends, as their local economies lacked the breadth to cope with short term shifts in some key sectors that were very quickly felt in house price change (pp.18–19).

2.3.5 Regional Economic Development Strategies: Elements and Actions

Teitz and Blakely (1985) propose a framework of the elements that are critical in building a local economic development strategy (see Fig. 2.5).

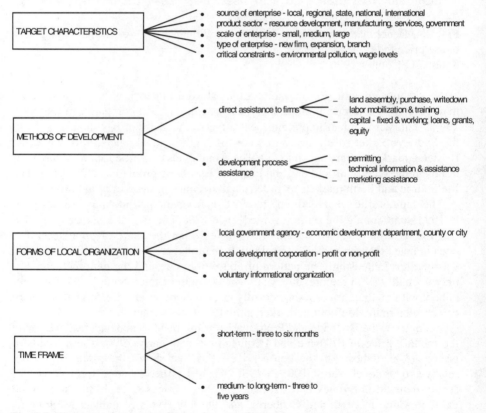

Fig. 2.5. Elements of strategies for local economic development
Source: Teitz and Blakely 1985

These relate to:
(a) What the strategy targets are.
(b) The methods used to accomplish objectives.
(c) The form of economic development organization and institutional arrangements.
(d) The time frame for short-term outcomes and long-term processes and goals.

Strategies need implementation mechanisms and action plans. Typically this requires inputs that Blakely (1994, p. 144) refers to as the '5 Ms' that represent key resources for economic development:

(a) Materials;
(b) Manpower;
(c) Management;
(d) Markets; and
(e) Money.

The institutional arrangements that need to be put into place to manage the strategy typically involve some form of economic development organization. Blakely (1994) tells how experience in the United States in the 1970s and 1980s shows that, particularly at the city and neighbourhood levels, economic development strategies tended to emphasize four strategic approaches:

(a) Locality or physical development, focusing on planning and development controls, townscaping, and household services and housing.
(b) Business development, focusing on small business assistance centres, technology and business parks, venture financing companies, and one-stop business information centres.
(c) Human resource development, focusing on customized training, targeted placement, and local employment programs.
(d) Community-based employment development, focusing on non-profit development organizations, worker owned and managed cooperatives.

Under local economic development strategies, it is inevitable that specific projects need to be financed, which involves identifying sources of capital (usually private but sometimes involving public funding). Also, project planning and implementation procedures need to be put into place. However, in practice there has tended to be:
(a) Over-much reliance on government programs.
(b) A tendency for civic leaders to confuse a particular development tool with a comprehensive strategy plan.
(c) An emphasis on attracting firms from elsewhere rather than building local skills and capacities, a neglect of thorough assessment of local competencies to sustain development.
(d) A lack of long term sustainable commitment to carry out a program or activities.

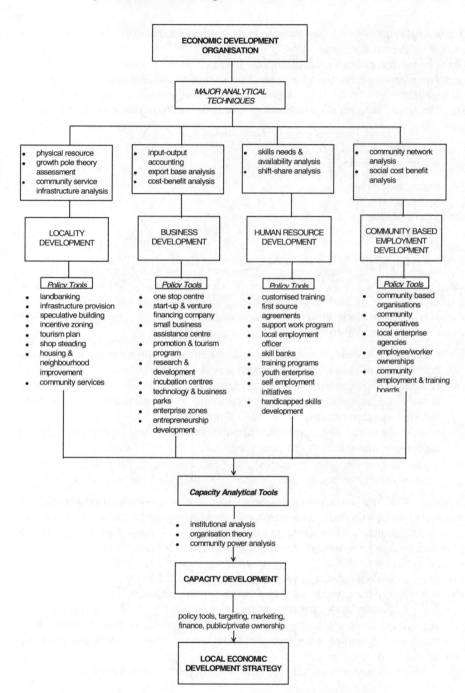

Fig. 2.6. Local economic development analytical and policy framework
Source: Blakely and Bowman 1986

2.3.6 An Overall Framework for Undertaking Economic Development

Throughout the 1980s in the United States, a large number of communities became engaged in the regional or local economic development process discussed above. Blakely and Bowman (1986) proposed a general analytic and policy framework for this process as set out in Fig. 2.6. It encompasses the formation of an Economic Development Organization to undertake the process, and sets out the range of major analytic tools and techniques typically employed to help profile the community and analyze its performance. The four typical strategic processes discussed above are incorporated, along with sets of potential policy implementation tools. Importantly the framework incorporates analysis of institutional capacity and capacity development necessary for local economic development strategy implementation.

2.3.7 An Example of a 1980s Approach to Economic Development Strategy and Organization at the Local Level: Merced County, California

Merced County in California's eastern Central Valley-Sierra region provides a good case study for local economic development. In 1985–86 the County initiated a process through which it sought to develop a strategy that would be feasible (financially, developmentally and politically) and that could be completed within twelve months. The goals were:
(a) jobs for County residents and the creation of an environment of profitability for in-county businesses;
(b) removal or mitigation of specific structural development limitations; and
(c) assisting development of the fiscal and social integrity of the community.

An objective was to ensure that public expenditures made to solve structural problems or take advantage of special opportunities in economic development would provide a 'return on investment' measured by jobs created for County residents. Assessment of past actions indicated that government participation had been a major influence on the economic health of the County, and that economic planning efforts were fragmented and limited in scope. The County worked with a NACO (National Association of Counties) project advisor, through a four-stage planning and action process shown in Table 2.5. A Merced County Office of Economic and Strategic Development was established which, by February 1991, had the organization and functional program shown in Fig. 2.7.

Table 2.5. Four stage process used to develop the Merced County economic development strategy

Stage	
Stage I	Convene the working group
	* Design process
	* Determine membership

Table 2.5. (cont.)

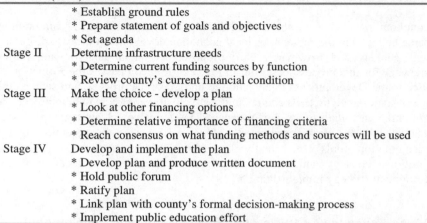

	* Establish ground rules
	* Prepare statement of goals and objectives
	* Set agenda
Stage II	Determine infrastructure needs
	* Determine current funding sources by function
	* Review county's current financial condition
Stage III	Make the choice - develop a plan
	* Look at other financing options
	* Determine relative importance of financing criteria
	* Reach consensus on what funding methods and sources will be used
Stage IV	Develop and implement the plan
	* Develop plan and produce written document
	* Hold public forum
	* Ratify plan
	* Link plan with county's formal decision-making process
	* Implement public education effort

Source: Merced County, Office of Economic and Strategic Development 1991

Fig. 2.7. The organizational and functional program for the Merced County Office of Economic Development. Source: Merced County, Office of Economic and Strategic Development 1991

2.4 Monitoring Outcomes of Economic Development Programs

In the United States, as part of an increasing interest by State and local governments to invest in economic development, regional or local economic development organization managers recognized the need to be able to identify on a regular basis the strengths and weaknesses of individual program initiatives. Thus considerable interest emerged in developing effective means of monitoring and assessing program quality and outcomes.

2.4.1 Designing Performance Monitoring Procedures

In the United States in 1980s the Urban Institute, based in Washington DC, collaborated with the States of Maryland and Minnesota, to design performance monitoring procedures to help economic development officers improve their programs to increase accountability. A range of performance indicators and associated data collection procedures were developed for the following six economic development programs:

(a) business attraction/marketing;
(b) business assistance;
(c) export promotion;
(d) business financing;
(e) community economic development; and
(f) tourism promotion.

These were the concerns of many local economic development agencies at the State and the local levels of scale (Hatry et al. 1990). To a considerable extent, the procedures developed relied upon client-based assessment of performance to generate data largely by incorporating it into the normal operations of a regional economical development agency's mechanisms that could be used to provide regular periodic reports to identify trends in performance over time.

Twelve key determinants were adopted as criteria for developing measurement procedures. These were (Hatry et al. 1990, pp. 4–6):

(a) The performance monitoring system should focus on service outcomes and quality.
(b) The performance monitoring system should focus on helping program managers improve their operations.
(c) The procedures should provide frequent and timely performance information.
(d) The performance monitoring procedures for individual programs should focus on the outcomes accruing to clients of program services (business and communities).
(e) Multiple performance indicators are needed to assess service quality and outcomes.

(f) Non-traditional data sources, such as client surveys and unemployment insurance data are needed and should be used to help assess service quality and outcomes.
(g) Performance indicators should include both 'intermediate' and 'end' outcomes.
(h) The procedures should include indicators that attempt to show the extent of the contribution of agency assistance to the outcome(s) reported by clients.
(i) The system should provide breakouts that array service quality and outcome indicators by client characteristics.
(j) The system should provide comparisons of performance for previous years, for target levels, and across categories of clients.
(k) The system should include explanatory factors as well as performance data.
(l) The data collection and management procedures should be designed to be as inexpensive as possible and to keep demands on personal time to a minimum.

2.4.2 An Example of Performance Indicators: Assessing Community Economic Development Assistance Programs in the United States

During the 1980s, most States in the United States implemented programs that sought to improve the capacity of local government economic development efforts. Typically the intermediate objectives of these programs were to increase the capacity of local governments to perform certain economic development functions and to improve assistance received by business clients of local government economic development programs and with the ultimate aim of enhancing employment and economic conditions of the state. The performance indicators for community economic development assistance programs proposed by The Urban Institute are listed in Table 2.6. It is worth noting that sets of indicators such as these need to be tailored to the specific activities undertaken by a State to assist local communities, which vary somewhat between States and across programs. The indicators PI-1 to PI-4 measuring 'service quality' require data collected through surveys of clients' assessments of attributes of the program or service. PI-5 provides information on assessments of the agency's overall contribution to a community's economic development during the reporting period (to give an intermediate outcome measure), while the end (long-term) outcome indicator PI-6 seeks information on the additional jobs in a community resulting from the assistance program. Regular surveys need to be conducted to generate these performance data.

Most performance indicators relate to specific programs or services, but in the field of regional economic development it is important for success or progress to be measured against a set of 'global indicators' of the type listed in Table 2.7. The data required to measure these global indicators typically is available through secondary sources, such as data series generated by the US Department of Commerce or the Bureau of the Census. But often these data produce somewhat crude measures. For example, indicators 2 and 3 in Table 2.7 provide information on the number of new business starts or business failures, but not on how many jobs were added or lost as a result. However these global indicators are important in that most of them may be disaggregated by industry and region permitting comparisons over time, and they are derived from standardized data collection procedures, thus producing reliability if not validity of measures.

Table 2.6. Performance indicators for community economic development assistance programs

Performance Indicators

Indicators of service quality:

    PI-1 Number and percentage of respondents who rated each service they received as excellent or good.

    PI-2 Number and percentage of respondents who rated the cooperativeness of state personnel as excellent or good.

    PI-3 Number and percentage of respondents who rated the timeliness of the state's services as excellent or good.

    PI-4 Number and percentage of respondents that rated the adequacy of communication with the state about state plans and policies as excellent or good.

Intermediate outcome indicators:

    PI-5 Number and percentage of respondents who reported that the state economic development agency had contributed at least somewhat to their communities' own economic development in the past 12 months.

End (long-term) outcome indicators:

    PI-6 Number and percentage of respondents who reported that additional jobs had resulted over the past 12 months because of the program.

Note: The information in parentheses refers to the source of data for each indicator
Source: Hatry et al. 1990, p. 126

Table 2.7. Global indicators

Global Indicators

1. State (local) employment (both in total and broken out by region/county and industry)

2. Number of new business starts

3. Number of business failures

4. Unemployment rate (both in total and broken out by region/county)

5. Average jurisdiction wage rate

6. Amount of tax revenue receipts

7. Percentage of businesses that rate the jurisdiction as a good place in which to do business

8. Ranking on various national 'business climate' indices

Source: Hatry et al. 1990, p. 137

2.5 The Strategic Planning Approach

2.5.1 The Policy Context

Borrowing from the experience of management theory and its applications in the business world, from the late 1970s regional economic development planning also began to develop a new perspective that was based on principles and methods in strategic planning. Bryson and Einsweiler (1988, p. xi) show how, in the United States during the 1980s strategic planning for cities became a 'hot topic—the subject of an increasing number of professional conference sessions, journal articles, and practitioner-oriented texts'. For example, the President's National Urban Policy Report and the National Urban Policy Report on Strategies for Cities, and a US Department of Urban Development Report, Strategic Planning Guide in 1984, strongly urged the adoption of strategic planning, outlining the benefits that could accrue to communities. This interest in strategic planning in the public sector surged for two reasons:

(a) First, leaders of governments, public agencies and communities faced various difficulties and challenges as their environments become increasingly turbulent (Emergy and Trist 1965) and interconnected (Luke 1988) because changes were occurring more frequently and unpredictably and changes in one area dramatically affect other areas (Bryson and Einsweiler 1988).

(b) Second, organizations, including governments, were being held more accountable for their organizations and communities, despite the increasing difficulty of controlling their environments, and they needed more help to manage the complexity of change (Bryson and Einsweiler 1988).

Strategic planning, as applied in a regional economic development context, represented a more systematic methodological development of the process framework for local economic development outlined by Blakely (1994) and discussed earlier in this chapter.

2.5.2 Strategic Planning: The SWOT Model

The SWOT (Strengths, Weaknesses, Opportunities and Threats) model developed initially for appraising and evaluating business development strategy was one of the commonly used analytical frameworks for strategic planning for regional economic development. Strategic planning provided a systematic way to help cities and communities manage change and to create a better future. It provides the 'creative process for identifying and accomplishing the most important actions in view of strengths, weaknesses, threats and opportunities' (Sorkin et al. undated).

Figure 2.8 illustrates what is typically involved in strategic planning. It may be regarded as a holistic approach involving the consideration of both the external environment (i.e. the community, the city, the region and beyond) and the internal environment (that is, the local authority as an institution). It requires the identification of all the actors, plus an analysis of strengths, weaknesses, opportunities, and threats or constraints—known as SWOT analysis—evident in the external and the internal environment:

(a) The external environment is assessed according to its evolution, the current situation, and the likely future trends.
(b) It covers things such as demography, the economy, social issues, environmental concerns, and technological trends.
(c) It involves the conduct of a comprehensive audit of all aspects of the environment and characteristics of the community, city or region.
(d) It involves climate assessments and the development of future scenarios.
(e) It involves the widespread consultation with the public and businesses in the community to find out their aspirations, desires, expectations, opinions and preferences.
(f) It involves an appraisal of the capacity, capability and preparedness of the institutions and the region to manage change.
(g) It involves the development of strategic aims and objectives, and the formulation of policies and action plans to address the strategic issues that are identified through these processes.

Bryson and Einsweiler (1988) demonstrate how strategic planning has helped public sector decision-makers—including both elected representatives in city government and senior managers of city administrations—to do the following:

(a) think strategically and develop effective strategies, clarify future direction;
(b) establish priorities;
(c) make today's decisions in light of their future consequences;
(d) develop a coherent and defensible basis for decision-making;
(e) service maximum discretion in the areas under organizational control;
(f) make decisions across levels and functions;
(g) solve major organizational problems; improve organizational performance;
(h) deal effectively with rapidly changing circumstances; and
(i) build teamwork and expertise.

Operationally there are typically seven steps in strategic planning for a city (or a community) using the SWOT model (Sorkin et al. undated). These are:

(a) scan the environment;
(b) select the issues;
(c) set mission statements or broad goals;
(d) undertake external and internal analysis;
(e) develop goals, objectives and strategies with respect to each issue;
(f) develop an implementation plan and carry out strategic citations; and
(g) monitor, update and scan.

The process needs to be interactive, beginning anew and continuing over time.

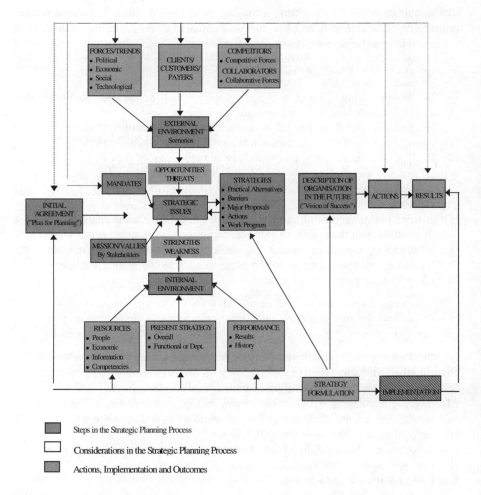

Fig. 2.8. Strategic planning framework
Source: Bryson and Einsweiler 1988

There is an absolute requirement for both public and private involvement in the process of strategic planning. It is a process that creates the opportunity for the development of powerful and lasting public-private partnerships. It can produce better coordination and concerted action within a community as it enables many different groups and programs to act on the same information and to arrive at the same conclusions, that is, it produces a 'shared vision' linking individual efforts to a consensual goal.

Strategic planning typically incorporates forecasting techniques, including expert opinion, extrapolation of trends and modelling.

Expert Opinion or judgmental forecasts are not sophisticated, are qualitative, but nevertheless are valuable. This is so particularly when past behaviour is not a good indicator of future behaviour, when information is difficult to quantify, when quantitative data is not available, and when the forecast is for a relatively short period. Methods such as the Delphi technique or the *Nominal Group Technique* are useful for achieving a consensus in groups. Addressing critical issues once selected may require more formal approaches to achieve more in depth analysis.

Extrapolation of trends is familiar techniques to project events of the past into the future, but these have difficulties where conditions are changing.

Modelling is a more sophisticated approach to forecasting. Models can be extremely complex, but they are particularly useful for making sense of multiple sets of interrelated data.

Strategic planning techniques were thus seen as being particularly useful as a means to address the competitive position of a corporation or a region and to help in evaluating performance against benchmarks and to changing circumstances, building on strengths and addressing weaknesses. But in the end, implementation is the key to strategic planning for regional development. This requires institutions of government to focus on the allocation of scarce resources to critical issues. Relatively few places have established a continuing strategic planning process. Instead the development of urban and regional development strategies have tended to be one-off projects that are often considered complete after initial implementation and without monitoring. Also, all too often governments do not even proceed into the implementation phase as the strategic plans sit on the shelf. In part this is because, by the very nature of the processes involved, strategic planning will bring into the open issues and possible situations that involve potentially difficult political action and unpopular decisions in the eyes of some constituents.

2.5.3 An Example: A Strategic Plan for the Economic Future of Dade County, Florida

Dade County, incorporating Miami is in Southern Florida in the United States. By the early 1980s there was mounting concern throughout the community for its economic well-being. Its position in the competitive market place was slipping. Projections for future growth in population and business indicated a shortfall of more than 116,000 jobs by the year 2000.

In early 1984 several local businesses and government leaders initiated a process to develop a strategic plan for the community. Some 450 individuals came forward from every sector of the community to volunteer their time and effort. The Metro-Dade Board of County Commissioners engaged Arthur Anderson and Co. to work with an executive committee of community leaders. As a result, what became known as the 'Beacon Council' emerged to formulate and implement a strategic plan for Dade County through a public-private partnership. The community embarked on developing the process for a strategy through a four-phase process (Fig. 2.9). What follows is a summary of the process and outcomes involved in this strategic planning exercise (Beacon Council 1985).

Phase 1 involved an 'environmental scan' which examined six subject areas:

(a) demographics;
(b) income and taxes;
(c) business and employment;
(d) education;
(e) public services; and
(f) housing and living conditions.

Major findings in each were outlined. This enabled the identification of those issues that were of greatest importance to the community's future. These were:

(a) organization for economic development;
(b) image and quality of life;
(c) infrastructure and regulation;
(d) education;
(e) international business and commerce;
(f) health care services;
(g) visitor services and tourism; and
(h) industry targeting and domestic marketing.

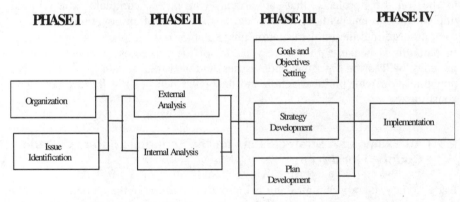

Fig. 2.9. Phases in the Dade County economic development strategy
Source: Beacon Council 1985, p. 7

Phase II, conducted between November 1984 and January 1985, was a broad-based consensus-building process undertaken through 'task forces' comprising representatives of the public and private sectors to examine the issues or factors that could affect Dade's future. Each issue was examined for internal and external factors that could impact the future, then each issue was ranked by the likelihood of its occurrence and the impact it could have on the County. Based on careful analysis of the high-priority risk factors, the 'task forces' defined a series of assumptions that formed the basis for setting Dade's strategic goals. The key findings were:

(a) *Organization for Economic Development*:
   (i)   Five areas of potential overlap were identified: planning and policy development; program and project support; financial packaging; market research; and development of promotional material.
   (ii)   Six areas of significant gaps were identified: retention and expansion of business in the local market; reconstruct and prospecting in the national and international markets; and advertising and promotion in the local, national and international markets.
   (iii) The most significant gaps between Dade and its competitors existed in national advertising and promotion.
   (iv)   In recruiting and prospecting in the national market, all competitor areas outspent Dade.

(b) *Image and Quality of Life*:
   (i)   The overall external image of Dade as a place to live and work is more negative than positive.
   (ii)   Immigration from the Caribbean and Latin America will continue to increase.
   (iii) There is a significant disparity of income among Dade's ethnic groups.
   (iv)   Dade has an abundance of recreational opportunities.
   (v)   The vulnerability of the fragile source of drinking water could limit economic development.
   (vi)   Visual pollution is spreading.

(c) *Infrastructure and Regulation*:
   (i)   Dade's public services and facilities are sound relative to comparative areas.
   (ii)   The development permitting process is confusing and time-consuming.
   (iii) The development of regional input process is cumbersome, costly and lengthy.
   (iv)   Tax and location incentives are not competitive.
   (v)   Funding future infrastructure will require innovative methods.

(d) *Education*:
   (i)   Despite its strengths, public school education in Dade is perceived to be of poorer quality than its counterparts elsewhere in the country.
   (ii)   Dade's educational system has a number of strengths.
   (iii) There is a local need for trained technicians and professionals.
   (iv)   Traditional patterns of funding education are changing.
   (v)   Corporate support of local post-secondary education is relatively low.

(e) *Internal Business and Commerce*:
   (i)   Dade has the major share of international trade in Florida.
   (ii)   Dade's range of trading partners is limited.

(iii)   Political instability will continue to disrupt the economies of some of Dade's major trading partners.
(iv)   International resources are fragmented.
(v)   The county's unique cultural and ethnic mix is conducive to international trade.
(vi)   There is a particular need to foster professionalism and ethics among some local traders.

(f)   *Health Care Services*:
   (i)   Dade already has a relatively high capacity to provide health care services.
   (ii)   The current national organization of the health care indirectly will provide opportunities for development.
   (iii) Dade is strategically positioned to provide health care services to international markets.

(g)   *Visitor Services and Tourism*:
   (i)   Mass transit to convention facilities and tourism attractions is inadequate.
   (ii)   Service employee skills require improvement.
   (iii) The number of hotels and hotel rooms in Dade exceeds that of any tourism competitor.
   (iv)   Marketing, promotion and administrative efforts are duplicated.

(h)   *Industry Targeting and Domestic Marketing*:
   (i)   The size of the existing employment base within industry categories relative to total United States employment in that category; its durability, based on projected job growth rates, and the number and quality of jobs it would provide, and its compatibility with Dade's infrastructure and environment; and its multiplier potential for attracting new business and influencing the number and quality of future jobs was assessed.
   (ii)   As a result, 27 targeted industry categories were identified.

It is important to note that the process in Phase II identified a range of strengths, weaknesses, opportunities and threats facing Dade in these issues, using the SWOT analysis methodology.

Phase III ran from mid January to April 1985 and involved each task force developing overall goals and quantitative objectives for each issue. Over 200 recommendations were proposed as means of meeting the goals, 14 were selected by the Executive Committee as strategies. These were:

(a)   *Economic Development, Marketing and Coordination:*
   (i)   Create a new sales and marketing organization by: Coordinating the marketing of the community to target businesses, and developing an Image-Watch/Image-Building Program; advocating and providing leadership influencing public policies that create a positive business climate; and follow, measure and report on program of the Beacon Council strategies.
(b)   *Community Agenda*: Actions to be taken by existing organizations and by entities other than the Beacon Council marketing organizations;

(ii)   Create an International Business Center.

(iii)   Crete a Health Industry Council.

(iv)   Emphasize Dade's educational assets and provide educational programs and curricula that meet the needs of business.

(v)   Develop existing graduate programs or start new ones in engineering, computer service and related high-tech fields, financial services, communications, agribusiness, and health care administration and organization.

(vi)   Recruit and increase the presence and prominence of black executives and professionals in the community.

(vii)   Develop high-tech research and development park(s).

(viii) Develop computer-assisted business development permitting and procedures manual.

(ix)   Refocus and streamline the Development of Regional Impact process and other regulatory processes affecting economic development.

(x)   Endorse and support anti-crime activities and initiatives.

(xi)   Expand the Miami Beach Convention Center and exhibition facilities.

(xii)   Create a countywide authority to coordinate and manage all convention, event and exhibition facilities.

(xiii) Consolidate tourism agencies and resources.

(xiv)   Expand the 'Welcome to Greater Miami and the Beaches' campaign.

Note that these strategic objectives and actions were a mixture of initiatives oriented towards enhancing the competitiveness of Dade and its industries, targeting potential industries to enhance its performance and job creation, enhancing capital development, and advising institutional and regulatory reform.

Phase IV of the strategic planning process called for a new Beacon Council organization to implement and monitor, and as necessary, review the strategies to meet Dade's future economic development needs. It recognized that strategic planning is an on-going process, with recommendations:

(a) focusing on a limited number of high-priority issues and strategies

(b) recognizing the interrelationship of these issues and the implications for public policy decisions

(c) providing an organized, analytical framework for monitoring changes in Dade's business environment

(d) promoting joint public-private sector participation in addressing the community's long-term needs.

The Beacon Council Organization was structured as shown in Fig. 2.10.

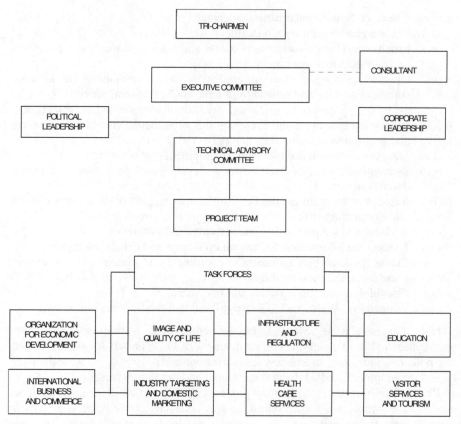

Fig. 2.10. The Beacon Council Organization to oversee Dade County's economic development strategy
Source: Beacon Council 1985, p. 29

2.6 The Concern with Sustainable Development

By the late 1980s there had emerged an increasing concern both in planning and in regional economic development with sustainability issues and how to incorporate the principles of sustainability as objectives in regional development strategies and plans. In some ways sustainable development has become a 'catch cry' which represents many different things for the diverse actor interests in local communities and regions. As we discussed in Chap. 1, while there is some confusion about what sustainability might mean and lack of clarity and specificity about the principles of sustainable development, nonetheless there can be no doubt about the pervasiveness of the concept 'sustainable development' in contemporary approaches to regional development and planning as a commonly stated goal, with sustainability criteria being specified objectives in regional development plans. And there may well be considerable benefits to be derived through the application of the notion that environmental and social multipliers may enhance economic multipliers in achieving local and regional development.

2.6.1 Guiding Principles for Sustainable Regional Development

All societies and cultures are governed by commonly held principles, rules and values etc, which guide community and personal decisions about the collective and personal use of resources and assets. These same principles are shaped, reinforced or changed by custom, past experience and/or events elsewhere. There are many examples of principles developed to support sustainable development. Some of these have been formulated by international agreements, such as the Rio Earth Summit and Kyoto protocols.

It is useful to group principles of sustainability under four broad headings as shown in Fig. 2.11. The inter-relationship between these sets of principles potentially leads to economic, social and business efficiencies. Governance is the primary mechanism for achieving these efficiencies. Sustainability governance principles reflect values and rules for the use of all other forms of capital resources at a society and individual/firm/agency operational level.

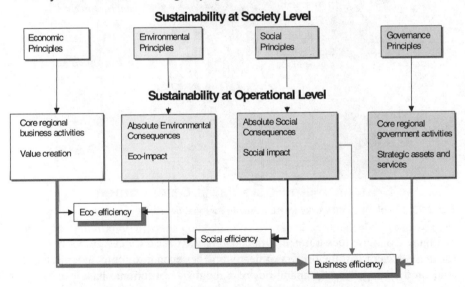

Fig. 2.11. Principles and outcomes of sustainable development

2.6.2 The Required Drivers

There are at least four key drivers for the regional development process which have a significant influence upon the nature and rate of development and how sustainable it might be. These drivers are: institutions and governance, capital investment in strategic infrastructure, catalytic processes such as networks, and business enterprises and their developments and market penetration. There are also many factors that shape the nature of these drivers and the way they influence environmental, social and economic development outcomes. The goal of regional development should be to focus these drivers in a way that will achieve more sustainable development outcomes. For regions to develop strong and competitive economies there is need to constantly review and reshape these drivers in response to environmental changes, opportunities identified, and risks that have the potential to affect a regions competitive advantage.

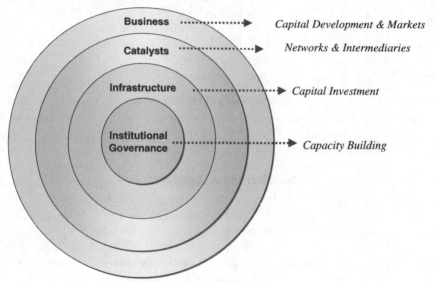

Fig. 2.12. Enabling frameworks for sustainable regional development

Figure 2.12 provides a framework showing the necessary categories of support for these drivers. All the drivers of development depend, to a greater or lesser extent, on the capacity and capabilities of a region's institutional base, the level of investment in strategic infrastructure in the broadest sense, and on the creation and attraction of 'catalysts'. Catalysts are organizations or individuals and in some cases institutions such as leadership that pull together resources, factors of production and finance to invest in projects and services that add value to regional economies. Catalysts introduce a new 'linking' function heretofore not emphasized in development economics. They are analogous to the effective organization and use of capital and entrepreneurship in an enterprise.

The next step in fostering regional economic development is to ask the question: 'What elements of support for these drivers need to be put in place to make development happen?' Capacity building is needed to improve institutional governance. Investment finance is needed to build strategic infrastructure and 'networks' are needed to foster the development of catalysts. Again the concept of networks is new. Catalysts undertake a linking function – networks assist them, with information or systems, to undertake this linkage function more efficiently.

2.6.3 A Framework for Applying Best Practices to Achieve Sustainable Development

There are literally hundreds of best practices that might support regional development, and there are many ways that these can be grouped and/or categorized under different functions or processes. Figure 2.13 shows an overall framework which enables various applications of regional development best practices to be grouped or sorted into the driver processes described above. These groups of processes support the elements of regional development that relate to the elements of sustainability that are represented by the *triple-bottom-line*, i.e., a concern with environmental, economic, and social –outcomes, and in addition and institutional arrangements and governance development. Not surprisingly, applications of best practice supporting sustainable regional development will apply across a range of governance frameworks.

The framework shown in Fig. 2.13 enables best practice applications to be developed and implemented for processes directed towards institutional capacity building; investment in strategic infrastructure and development of catalysts and networks ranging from a national to a community level and across the public, business and community sectors. The framework establishes a mechanism for integrating and coordinating related development processes; for example, policies developed under institutional governance at a regional level should align and link closely with national development policies, while regional policy should inform and where expedient introduce best practice at a national level. It is important that the application of best practice at different governance levels provide for a learning process gained from the exchange of information, ideas and experiences across all levels of government and with the private and community sectors.

The framework described is conceptual, but it does provide a basis for simplifying our thinking about, and the management of, many complex sets of factors and systems that are involved in regional economic development. Governance decision-making processes will become increasing complex in the future in the light of stronger demands from communities for greater transparency and accountability in public institutions and the growing impact of external environmental factors on regional development. In reality, regional institutions and business are already responding to greater complexity by juggling work programs and budgets to respond to change, albeit not in a systematic or optimistic way. Regional planning, development and management decision making processes need to be able to deal with increasingly complex matters ranging from responding to the impacts of changes in foreign exchange rates and commodity price on the competitiveness of regional industries, to improvements to environmental quality and security matters and to maintain the reputation of regions as an attractive place to invest.

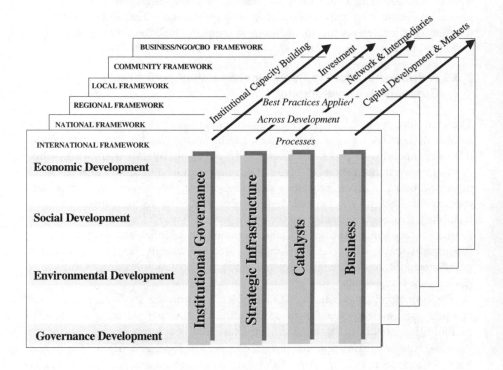

Fig. 2.13. Overall framework for applying best practice to regional development

The framework presented in Fig. 2.13 also helps organizations responsible for regional development to understand the importance of strategic fit in aligning processes supporting regional development - such as through project funding, approvals, etc - with other governance frameworks. For example, regional financial, information, planning budgeting and monitoring processes often might need to be better aligned with national systems. However, in many countries that have recently undergone decentralization, they are not. It is important that regions be encouraged to develop innovative practices and processes that contribute to sustainability that can be passed on as best practice to central government and other regions. This should be promoted as part of a 'smart region learning process'. Most innovative best practices originate at a local level, and encouraging their development and replication more widely – even nationally - is important in the pursuit of sustainable development.

2.6.4 Increasing Multipliers as a Key to Sustainability

An aim of sustainable development practice might well be to take full advantage of the multipliers that result from exchanges of capital to support development determined by the three value systems. This is achieved by maximizing the opportunities to leverage resources between industry or business sectors and stretching resources more effectively within sectors and industry clusters. If organizations can be trained to 'opportunity seek' and 'resource share', then the net result is likely to be an expansion of social, environmental and economic multipliers and outcomes.

There is growing interest by economist Krugman (1995) and others (including Steinfels 1999; Galster 1998; Martin and Sunley 1996; Skott and Auerbach 1995) on the inter-relation of social capital, environmental capital and cultural capital and their roles in supporting development. This suggests that sustainable development should be concerned with actions that result in net positive cumulative causation effects, such as projects and programs which set in motion processes that generate increasing returns on capital investment. The generation of net positive cumulative causation cycles, whereby social and environmental multipliers are used to enhance economic as well as social and environmental outcomes, is central to the sustainable development debate.

The notion that greater investment in natural capital and social capital projects can add value to economic capital introduces a new approach to sustainable development. However, there is scepticism by some economists about the linkage between social capital and natural capital investment and economic multiplier benefits. This is caused partly because of the difficulties that exist in translating social and environmental value into economic terms.

As discussed in Chap.8, there is some evidence to suggest that investment in projects involving the development of social capital (Putnam 1993; Cohen and Fields 1999) and in the restoration of natural environments (Constanza 1994), may lead to new opportunities for economic development, such as eco-tourism and eco-waste materials reprocessing involving industrial ecology. For example, many cities have discovered that investment in improving the quality of urban waterways and rivers provides a catalyst for investment, which also results in public health and well being multipliers that in turn generate other social and economic benefits. Cities such Singapore, Melbourne in Australia, and London, have invested heavily in river clean-up and waterside redevelopment projects that have in turn resulted in billions of dollars of investment for comparatively small public capital outlays.

One way that can help to think creatively about using capital to increase project multipliers that will add to the net value of a region's capital base is shown in Fig. 2.14. To calculate the value added to regions of a project or program we apply input-output multipliers in a matrix table. In the model shown, we consider the use of three types of capital (environmental, social and economic) to support development projects and programs.

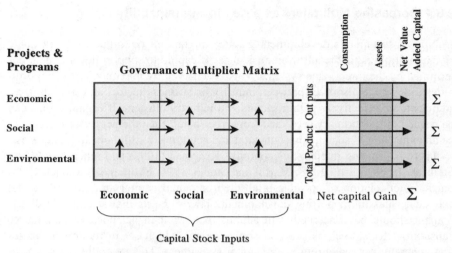

Fig. 2.14. Leveraging three forms of capital to increase the net value of the capital stock of a region

If we want to assess the net economic value of the project we can use econometric, environmental and social accounting techniques to quantify this. This leads to 'triple bottom line' net value-added capital, once consumption and other costs associated with the capital used for an activity are netted out (right column of the Fig. 2.14). The process is similar to conducting a full benefit cost analysis (BCA). Changing the capital input variables can result in different outcomes in the net value-added capital created by a project.

The trouble with using BCA to measure capital formation is that it overlooks other opportunities to add value to projects and programs, although a BCA variant called the 'benefit balance sheet' addresses this shortcoming of the pure BCA approach (Lichfield 1996). If the intent of sustainable development is to maximize total capital multipliers, then we should be investigating additional means to add value to an economic development project by incorporating into the design complementary environmental and social capital building projects. This is the concept of 'project leveraging' proposed by Hamel and Prahalad (1993). This is done for many development projects by including environmental improvements or human resource development training components. However, these components are often short term and are considered to be a cost. As a result, the potential to substantially improve project multipliers may be lost (Roberts and Cohen 2002).

To maximize potential project and program capital multipliers, it is necessary to design projects or programs from a total value-added capital perspective; that is, to look at the project from an economic, social and environmental perspective. The net effect of this approach leads to a matrix of increased multiplier opportunities. If we sum the added value of the combined economic, and supplementary social and environmental projects and these are net capital positive, then the outcome is more sustainable. There is a need to identify if specific initiatives generated from this process could be secured, or are merely speculative. The matrix structure is useful in the sense that it provides a governance framework to make choices about which capital multipliers are important in maximizing the potential for more sustainable project outcomes.

The governance framework in a region also plays an important role in defining the matrix multipliers, as it sets the rules for the way we are permitted to use capital. These 'rules' are largely determined by the values and principles we employ to make choices and decisions about the permitted use of existing capital stocks for development and consumption, and what is used to create new forms of capital that replenish or contribute to an improved balance in total capital accounts.

The conceptual framework described above moves planning for sustainable development beyond BCA and the 'triple bottom line accounting' concepts .This approach provides important indicators of change, but we often have no idea of the value of the capital stock measures. The 'ecological footprint' concept enables measurement of regional demand and consumption patterns, but it does not provide a mechanism to define how best to use capital to add value to the capital stock. BCA does not allow us to maximize potential capital multipliers. What needs to be achieved through sustainable regional development is to create governance structures that attempt to maximize the potential multiplier effects of different combinations of capital utilized for projects and programs to service a range of societal and other environmental needs and to minimize the consumption of all forms of capital used in the process.

2.6.5 Adding Value to the Project Cycle

The framework for sustainable development discussed above helps us to visualize what may need to be done to ensure that regional development projects and programs are sustainable and generate net positive gains in all three capital accounts. Few projects achieve this outcome, and the multiplier potential to leverage projects to enhance the environmental capital and social capital accounts are often overlooked. It is common for the environmental account to be deficit.

The World Bank began to investigate this by using a more holistic approach to project design in its work in developing countries the late 1990s. That experience indicates that during the formulation, assessment, development and implementation phases of a project cycle, the highest concentration of an agency's non-capital work expenditure occurs during project appraisal. Failure to examine the consequences of multi-sector issues during the initial stages of feasibility often results in expensive measures being taken during the appraisal process to address the cross-sector effects of major development projects. The first step in formulating a major project should involve an identification phase, where multi-sector agency issues and opportunities to enhance economic, social, and environmental multipliers are discussed. Initiatives like this would likely yield significant internal cost savings and enhance the overall performance of government agencies.

Figure 2.15 shows conceptually the staff weeks of involvement in a project cycle using a multi sector and a traditional approach to project design and development. On the vertical scale are the staff week inputs. The horizontal scale represents steps in the project cycle beginning at identification and progressing to preparation, appraisal and implementation. This scale varies depending on the size and nature of the project. The area under the curve represents the total staff week inputs. The highest point on the curve using the traditional approach to the project cycle is project appraisal. This is because significant multi-sector issues are often realized at this stage in the project cycle requiring further research or possible re-design of the project.

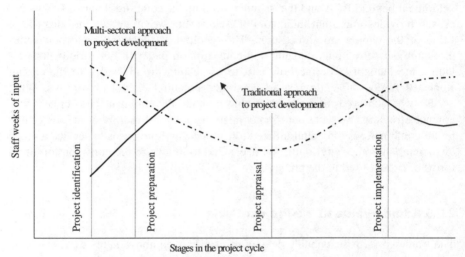

Fig. 2.15. Staff week inputs in the project cycle showing a multi-sector approach versus a traditional approach to project development
Source: Cohen and Roberts 2002

2.6.6 Case Study in Sustainable Regional Development I: The Cairns Regional Economic Development Strategy

The Cairns Region, located in the northern part of the state of Queensland in Australia, covers an area of approximately 50,000 km2. The region contains two world heritage-listed sites: the Great Barrier Reef and the Wet Tropics, which is one of the world's most ancient rainforests. In the 1960s, Cairns was a backward agricultural region. In the early 1980s, the region experienced a remarkable transformation, which was brought about by the influence of national economic reforms, environmental policies, and the vision of a few community leaders to build an international airport. The opening of the international airport in 1984 started a development process which has led to Cairns becoming one of the most internationalized and rapidly growing regional economies in Australia.

In 1990 the economy of the region collapsed due to a national airline pilots strike, a loss of Japanese foreign investment in the tourism sector, and a global recession. In 1992, the Queensland government, in response to concerns about the economy and the environment, supported initiatives to re-engineer the development of the region. The objective was to develop a strategy for sustainable development. A partnership between government, local business and the community was established to prepare an Integrated Regional Development Plan to guide the growth of the region to 2010 and beyond. One of the key initiatives under the plan was the Cairns Regional Economic Development Strategy (CREDS) and the establishment of the Cairns Regional Economic Development Corporation (CREDC), which provides the mechanism for the implementation of the strategy.

Since the mid 1990s, the region has diversified and restructured its economy. With a population of 250,000 and an economy of US$5 million, it has achieved an annual regional growth rate of more than 6 percent. It has done this by implementing a strategy for regional development that is focused strongly on sustainability, competitiveness and quality of life. Protection and restoration of the natural environment has been central to the development of its strong economy, with more than 20 percent of the economy revolving around the tourism industry, particularly eco and business tourism. In excess of 45 percent of the economic activity involves national and international trade, one of the highest for any region in Australia.

The region has focused on developing world-class strategic infrastructure to support the development of tourism and other sectors of the economy. The Cairns Port Authority, a public corporation, has been involved in the foreshore redevelopment, and built its airport and port (Cooke and Morgan 1998). In addition, it constructed a convention/exhibition center to diversify into business tourism; this facility is the most intensely used in Australia. The local governments have cooperated to build key regional facilities to support investment in tourism and new industries.

The CREDS prepared under the integrated regional growth management framework has been the key driving force behind the development of the economy. The strategy recognized the prime importance of developing a secure, diverse regional economy and strong export-oriented service industries. The strategy acknowledged there are critical issues that need to be addressed if regional economic growth is to have a more sustainable outcome in future. These were:

(a)   the growing trade imbalance;
(b)   reduction of capital out-flows;
(c)   retention of profits in the region;
(d)   lack of corporate identity; and
(e)   reversing the decline in foreign investment.

The CREDS recognized the need to capitalize on the benefits that tourism had introduced to the region. This was to be achieved by enabling cross-industry collaboration with other industry sectors. Additionally, the region needed to increase business co-operation and networking, improve resource management, and identify changing international trading structures, particularly in Asia, to enhance the region's export potential. While the wider Far North Queensland (FNQ) regional economy showed every sign of prospering, these key development issues needed to be addressed if the region was to realize its full potential and maintain its standing as a highly competitive international trading economy.

One of the key success factors responsible for the economy's recovery has been the focus on establishing industry clusters. The region has developed an effective approach to industry clustering (Roberts and Dean 2001). Based on a philosophy of regional businesses competing collaboratively in the global marketplace, the system has proven successful in assisting the growth and development of a diverse range of industry sectors. Twelve industry cluster organizations have been established. The process of cluster building has taken many years. Each industry cluster operates autonomously whilst remaining under the umbrella of CREDC. This allows for industry-driven economic development with tangible and holistic support from other clusters and economic and management expertise from CREDC.

Good governance has been an important element in building a sustainable regional economy. The regional and local government planning process involves extensive engagement with the community and stakeholder groups – especially absentee land and property holders. This has been achieved through well developed regional information services established by CREDC and other organizations involved with promoting the development of the region.

Integrated regional strategic planning has also been important. Strong partnerships have developed between government, business and community organizations to implement regional development projects. A consultative rather than a regulatory process is used to drive the integrated regional planning process, under the umbrella of the Regional Organization of Councils (ROC) comprising the City of Cairns and seven other local governments. Generally, the collaborative process has worked well, leading to mutual cooperation between public agencies and a unique ability to lobby both State and Federal governments for development funds and special project grants. The structure of that regional planning framework is set out in Fig. 2.16.

The CREDS and the broader Integrated Regional Plan summarized above relate strongly to key best practice attributes of sustainable development as summarized in Fig. 2.17.

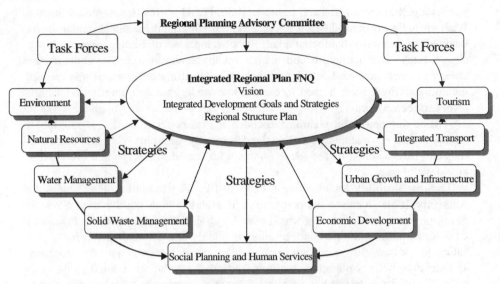

Fig. 2.16. The Integrated regional planning framework for Cairns-FNQ
Source: Derived by the authors from the CREDS

Case Study Best Practice Attributes	
Stewardship of Capital	✓
Good governance	✓
Strategic infrastructure	✓
Engaged communities	✓
Responsive economic policies	✓
Proactive planning	✓
Enabling mechanisms	✓
Value multipliers	✓
Quality of life	✓

Fig. 2.17. Sustainable best practice attributes evident in the Cairns region case study

2.6.7 Case Study in Sustainable Regional Development II: Singapore

Singapore is an island economy and a city-state with a population of just over 4 million, located at the southern end of the Malacca Straits. It has been a trading center for many centuries. In 1960, Singapore had a GDP of less than US$1,650 per capita. In a period of just over 40 years, it had developed into one of the most dynamic economies in Asia, with a GDP of US$ 27,800 per capita). Now Singapore is a global hub of finance, transportation, IT, education, trade and commerce. With virtually no natural resources and very limited land for development, it developed its economy by focusing on the development of human capital, world-class strategic infrastructure and a high quality environment. Singaporeans are among the best educated in the world, with outstanding universities and training institutions which have helped to build a knowledgeable community and a high-level service economy.

A strong emphasis on planning has been a primary factor in developing a high quality living environment for the island economy, and part of Singapore's success can be attributed to the focus on best practices designed to achieve sustainable development.

The responsibility for planning Singapore lies with the Urban Redevelopment Authority. It has overseen the preparation of strategic plans that have involved the development of several new towns, linked by a rapid integrated road and rail transit system enabling the population to move quickly and conveniently from one location to another. Singapore has applied strict zoning controls over development, and strongly polices planning, building and other regulations related to development. Good urban design has been encouraged by incentives to developers to provide open space and civic amenities, in return for being able to construct additional floor space granted through special dispensations.

In Fig. 2.18 it is shown how Singapore has initiated an integrated approach to development planning. Governance has played an important role in Singapore's growth.

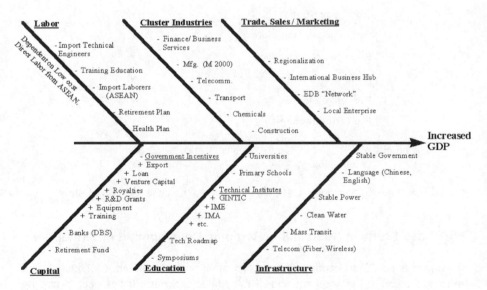

Fig. 2.18. Singapore's integrated approach to regional development planning
Source: Boulton et al. 1997

It is a self-governing country, with an elected parliament. It has had a long period of stable government which has developed a platform of policies enabling it to cultivate competitive advantage in several industry sectors. The rule of law has been strictly upheld, providing confidence within the community, the business sector and investors. Singapore has been criticized for not having a more open and democratic system; however there is a high level of accountability in all government processes.

Singapore's airport and seaport facilities are among the best in the world. Singapore airport is a major hub for the Asian region. It is managed by the Singapore Airport Authority, which is run as a corporate entity. The Port of Singapore is a major transit center for all the world's major shipping lines. It is strategically linked to some regional ports in Malaysia and Indonesia. The island has a well developed network of freeways and distributor roads. It has a policy of road pricing designed to encourage greater use of public transport by imposing a toll electronically on vehicles that enter into the city centre after 7 am. There is a high level of infrastructure maintenance, which ensures that its road network continues to operate at above average levels of efficiency.

Singapore is one of the few places in the world to have effectively solved its housing problem. During the 1960s the government established the Singapore Housing Authority which had the responsibility of providing housing to all income groups. Housing was funded from pension funds, by requiring that 25 percent of all pension fund payments be invested in the housing sector. While private developers now provide much of the recent housing for upper and middle-income groups, pension funds are still used for the construction of replacement substandard housing.

Singapore has placed a strong emphasis on the conservation of its heritage, cultural and natural environment. During the 1960s, much of its traditional housing was destroyed and replaced by high-rise apartments. It was also criticized that Singapore had lost much of its cultural charm as it pursued the quest to modernize. In response to community concerns about the loss of traditional housing, a strong focus on conservation and construction using traditional styles has emerged.

The economic development for the island has been guided by policies to transform Singapore into Asia's leading knowledge economy. In the 1990s, the Singapore government placed a strong emphasis on building a knowledge economy. It has constructed high-quality academic institutions to attract world-class academics and scientists, and encouraged the setting up of foreign universities as a means of promoting competition in the education sector. These initiatives, combined with increasing levels of expenditure on research and development have resulted in Singapore establishing a reputation as a global center of excellence in education, information management, research and development. Singapore offers very attractive incentives (Government of Singapore 2006) to entice regional headquarters of global corporation to be located in recognition of the role these play in developing industry clusters (Boulton et al. 1997).

Singapore is now a world class financial center. It has one of the largest futures markets in Asia, and provides a wide range of banking and other financial facilities for the business sector. It is a major supplier of capital for foreign direct investment into other Asian countries and Australia. It is also an important center of insurance, which is becoming an increasingly large industry in Asia. The principal reason for its success as a financial center is a well-regulated financial market; a high level of transparency and accountability is required from companies that operate from Singapore.

In the 1980s, Singapore realized it was losing competitive advantage in its manufacturing sector. Production costs were rising due to higher wages, higher land costs and stricter environmental conditions being imposed on new industries locating in the country. Singapore businesses realized there was competitive advantage, if they relocated industries to the nearby regions of Johore Bahru and Riau on the islands of Batam and Bintan in Indonesia. This led to the development of new manufacturing industries in these neighboring countries, which use Singapore as their hub for providing finance, business, technology and transportation services. Singapore was, therefore, able to develop a high level service economy, but at the same time ensure that it did not lose the value of manufacturing, even though this is now conducted offshore.

While the Singapore model of development cannot be replicated easily in other Southeast Asian countries, Singapore has many best practices, which if applied in other Asian countries, might result in more sustainable development outcomes in those countries.

2.7 An Enabling Planning Framework for Regional Economic Strategy Formulation

As we will see in the discussion at numerous times throughout this book, increasingly the concern in undertaking regional planning in general and regional economic development planning in particular is to address issues to do with sustainability and sustained growth. Today few regional development plans do not include a goal or statements about sustainability. The emerging approach to sustainability is one of communities and regions taking greater responsibility for formulating policies and implementing development programs. Emerging approaches to regional development involve communities and other stakeholders (including catalysts, civic entrepreneurs, investors) having a greater say in the development and implementation of regional development plans, and new governance structures are emerging to replace old regional development models.

The days of governments totally controlling the agenda on regional development planning are disappearing, even in centrally planned economies in the developing world, such as Vietnam. Nevertheless, central governments will continue to have an important role in defining the rules and frameworks that guide or direct regions to adopt sustainable development. Under decentralization, a number of countries in Asia have been developing regional and local planning and development models giving regions greater autonomy in the organization and control of development. However, many of these planning models are not responding well to the changes occurring in national and global economies. Some of the problems experienced with their implementation involve institutional change (an issue that is addressed in detail in Chap. 8), and working out frameworks that will lead to more sustainable development outcomes and improved development practices.

While there is not a universal regional development planning model that can be applied in the pursuit of sustained growth and sustainable development, there do seem to be a number of common features that have been emerging within regional economic development planning processes. These include:

(a) Preparing and engaging with communities in regional development processes.

(b) Developing good information gathering, analytical, policy and plan formulation systems.

(c) Establishing better mechanisms for detailed planning and implementation.

2.7.1 Stages in the Framework

Figure 2.19 outlines an enabling framework for institutions to support a wide range of activities associated with regional economic development. That involves addressing the issues through the steps outlined below.

Community preparedness. The first step in this process involves preparing communities to participate in a wide range of planning and development processes. That involves a learning process. Unless communities understand and are engaged in the design and ongoing development of planning processes, it is highly unlikely that they will buy into projects and programs with which they have little sense of ownership. There is a range of techniques that can be used to prepare and mobilize community support and engagement in the regional development process. Some of these have already been discussed earlier in the chapter.

Information gathering and analysis. The second stage of a regional development planning process involves the gathering of information and analysis to identify trends, issues, problems, risks and opportunities. The information process from the analysis will not only be used for supporting forward-planning and estimating and implementing development projects and programs, but also for monitoring and evaluation of policies and strategies developed as part of the planning process.

Policy formulation and strategic planning. An important element of regional development is the preparation of policies that describe the desired future directions of development in a region.

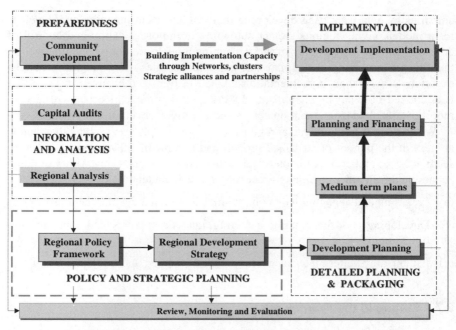

Fig. 2.19. An enabling framework for regional economic development

Policies provide the basis for the formulation of a range of strategies shaping development programs for such things as infrastructure, transportation, housing, economic development, education, recreation, health and community services. It is common practice for regions to develop a strategic plan with objectives that are focused on targeted development outcomes. Regional policies are used for a wide range of planning, budgeting and investigation purposes. In some country contexts, regional policies will be shaped by national policies, decrees, orders and priority development programs; however, local input into the formulation of regional development policies is critical to gaining community support to implement regional development plans that will be developed to support these.

Detailed planning and packaging. Policies and strategic plans provide a long-term perspective for regional development. There is however a need to undertake more detailed planning to investigate the feasibility and acceptability of many projects and programs that might be outlined in a needs development plan prepared as part of a strategic planning process. It is common in many developing countries, for example, for regional and local agencies responsible for regional development to formulate medium-term development and financial plans. These need to be linked closely with medium and long-term budgets. It is good practice to undertake an annual review of all medium and long-term development and financial plans, as circumstances often require adjustments to the time frame for many projects and programs. A key feature of a financial plan is a description of the mechanisms proposed to fund projects and programs that are scheduled for implementation under an annual or medium-term development plan.

Operationalizing. The final stage of the regional economic development planning process involves project implementation. As discussed earlier in this chapter, there is a range of mechanisms regional organizations and institutions can employ to implement projects and programs outlined in a development plan.

2.7.2 Moving from a Cyclical to a Dynamic Planning Process

It is common practice to depict regional development planning process as being 'cyclical.' Figure 2.20 shows a traditional eight stage process beginning with pre-planning and following on to evaluation and review. There are many variations of this models, the concept of a cyclical and sequenced process of activities is the same in most models. The cyclical framework provides a sound framework for strategic and detailed planning processes; however, it is not particularly responsive to changes or shocks. Modern planning systems should be flexible enough to respond to changes in circumstances. Change may be imposed upon regions by external factors or may be initiated by regions to enhance institutional competitiveness. While regional development planning is cyclical, a new more dynamic process has emerged which involves constant interaction between the various stages in a regional development planning cycle process (Fig. 2.21).

For regions to operate under more open and competitive market conditions, planning processes need to adopt a more 'interactive' and 'dynamic' approach to regional economic development. As shown in Fig. 2.21, operationalising a project or program requires constant attention to strategy, engaging with key stakeholders, attention to financing and monitoring and evaluation review. In some cases it is necessary to backtrack in the planning process to consider alternative options. The necessity to constantly interact with forward and previous stages in the planning cycle adds to the complexity of the regional development strategy and planning process. It also calls for very high levels of skills development, which many regions lack.

The change from a straight forward traditional planning cycle model to a dynamic model requires a significant change of thinking within regional development organizations. Business corporations are continually being forced to review and evaluate changes in demand for products and services, and to respond to opportunities created from new information gathered from intelligence sources. This necessitates constant changes within institutional governance and planning systems.

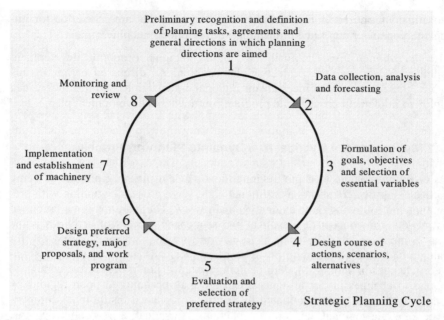

Fig. 2.20. Traditional planning cycle

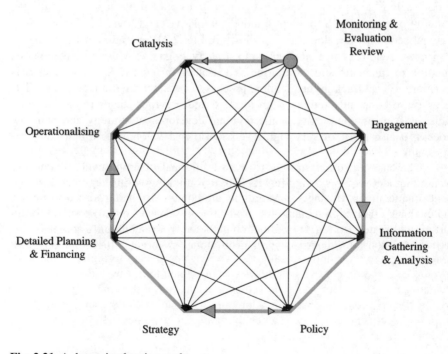

Fig. 2.21. A dynamic planning cycle

Institutions involved with regional planning need to do the same if they are to provide the necessary support to encourage development and investment. This requires a fundamental shift from traditional supply-side approaches to regional planning and development to a more demand-driven approach.

Old planning models, that involved central or regional development agencies determining planning outcomes and funding and managing the regional development process, are no longer capable of responding to demand-driven needs of economies. Many regional institutions do not know how to change - or find it difficult to do so - and are often unwilling or reluctant to adopt more dynamic institutional and regional development planning models. The experience of some countries in the developing world to the introduction of dynamic planning models has, in some cases, been outright hostility.

The challenge to replace the traditional planning cycle model with a dynamic model approach for regional economic development planning is thus considerable. There is often a fear of change, as in some nations and regions there exists limited experience, unwilling leadership, and a limited capacity within public institutions to make the necessary changes. However, without a substantial transformation of the governance arrangements and planning systems used to guide regional development, most regions - especially in some of the developing nations but also in many regions in developed nations - will find it difficult to capture opportunities or develop the strategic infrastructure needed to support regional development. Programs to introduce change management within regional development institutions become crucial considerations if regions are to develop the capacity and planning systems to support sustainable development.

2.8 Summary

In this chapter the discussion has been focused on the evolution of systematic approaches to regional economic development planning and strategy formulation during the 1970s and into the end of the pre 2001 millennium, drawing specific attention to the stages involved in the process of information gathering and analysis through conducting a regional socio economic audit and an institutional capacity and capability assessment, creating a development strategy and implementing plans through actions and programs, and the evaluation of outcomes. The use of SWOT analysis as a methodological framework to assist this process was discussed, as were the importance of benchmarking and the identification of appropriate policy tools for implementation of various approaches to regional and local development. A number of case studies was presented to illustrate the applications of such approaches. This was followed by a discussion of the emerging focus during the 1990s on the concern with achieving sustainable development. Finally, we proposed a conceptualization of an enabling planning framework for approaching regional economic strategy formulation.

Reference was made also to a number of tools of analysis that may be used to assist the conduct of an audit of the performance of a region and to benchmark performance against that of other competing regions. Some of these quantitative analytical tools are now discussed in detail in Chaps. 3 and 4.

3 Traditional Tools for Measuring and Evaluating Regional Economic Performance I: Economic Base and Shift-Share Analysis

3.1 The Need to Measure Regional Economic Performance

Regional economic development policy is basically about the allocation or reallocation of resources to enhance the economic performance of industries. Planners and policy makers need to be able to measure and evaluate that performance. Thus it is necessary to:

(a) Measure the degree to which economic activity and employment in a region is related to serving local demand as against serving demand external to the region (i.e. exports).
(b) Assess a region's overall performance relative to that of other regions.
(c) Assess which industry sectors are performing better in the region.
(d) Assess a given sector's efficiency relative to other industry sectors' performance in the region.

Tools are required to both measure and analyze the performance of a region over time with respect to its industry sector performance on a variety of parameters or indicators. This is important because a given industry sector may be seen as a 'good fit' for a region even if it is inefficient. For example, in the short-term an industrial sector in a region may be 'leading' in terms of volume of activity, value added, or its share of labour demand, but it may not be operating efficiently when compared to the same sector in another region. Thus, a less dominant sector might be a better candidate for growth and development because of its competitive or relative efficiency.

Numerous techniques have been developed by regional scientists to address these issues, and this chapter and in Chap. 4 some of the commonly used methodologies, what they measure, and their advantages and limitations are discussed. This chapter begins with an overview of economic base theory, then goes on to consider shift-share analysis and data envelopment analysis. Chapter 4 discusses input-output analysis. These are all techniques commonly used to measure and analyze regional economic performance. They are quantitative tools of analysis using secondary data, such as those obtained from the census.

3.2 Economic Base Theory and Industrial Targeting

3.2.1 Economic Base Theory

Economic base theory (EBT) is an easily understood traditional body of thought in the field of regional economic development. As we discussed in Chap. 1, EBT views an economic system as composed of two parts:

(a)   one, called non-basic, is viewed as producing for local consumption; and
(b)   the other, called basic, is viewed as producing goods and services primarily for external consumption, (that is for export from the region).

Examples of basic goods and services include most manufactured products, a variety of services that are provided largely for producers in both upstream and downstream activities, and tourism.

Economic development theorists believe that the critical cornerstone of a regional economic system is its basic economic activities. The reason for this is that income from outside the region comes from these kinds of activities. Thus by expanding (export) base activities, the local regional economy not only expands employment and earnings in the region directly, but also expands employment and earnings indirectly. This occurs because the export base workers spend their income (which comes from outside the region) for goods and services in the local economy. This generates additional employment and a variety of non-basic goods to support their needs. In this way, expanding an economic base activity has a multiplier or growth effect on the local regional economy. As a consequence, the primary focus in sectoral targeting analysis is on basic economic sectors and activities. Non-basic sectors are important when they can be linked to basic sectors in ways that will help transform them in part into new basic sectors. Location quotients (described and illustrated below) are tools that are applied to sectors that are primarily basic in nature.

A complimentary approach is entitled import substitution, whereby goods and services are imported to support basic and, in some cases, non-basic production. With this approach, industry sectors that are insufficiently developed to support local basic activity are targeted for investment and development. By expanding these sectors, the relative importance of basic sectors often can be increased.

3.2.2 Measuring the Economic Base of a Region

A variety of techniques have been developed to separate economic systems into their basic and their non-basic parts. The simplest method is to sort industry sectors into those that are primarily basic and those that are primarily non-basic. While relatively easy to apply, this method of measuring the basic and non-basic parts is very crude.

Subsequent efforts have relied on location quotients, which are measures estimating the importance of industry sectors to the local economy relative to their importance in a larger reference economy; for example, a state or a nation. Two types of location quotients have been developed:

(a) One is measured the minimum requirements approach, where that locality which has the least employment (earnings or some other indicator of scale) in a sector becomes the base against which the same sector in all other regions is compared. Here it is assumed that the locality with the smallest sector under consideration is the minimum required. All other localities will have more than the minimum and that increment above the minimum is viewed as a basic contribution to a locality's economy.

(b) Alternatively, location quotients can be computed in terms of some reference area, (e.g. the nation), whereby the contribution to the basic part of the economy is measured as the part that is greater than the proportional amount found in the reference area.

In Chap. 2 it was shown how various kinds of information about a region's economic structure are needed to inform regional economic development planning and strategy development. Emphasis is often focused on identifying those industry sectors that are fast growing, restructuring and/or underdeveloped:

(a) Fast growing sectors of concern are those that are relatively large sectors that have exhibited rapid recent growth.

(b) Restructuring sectors are those that are relatively large but have experienced little or no growth or decline over the recent past. These are important because often they account for a relatively large proportion of a region's employment base. It is important to determine if there is a way to renew or accelerate these restructuring sectors.

(c) Underdeveloped sectors have relatively low levels of activity but could become large in terms of employment, income generation or production levels in relation to the whole regional economy.

3.2.3 Calculating Location Quotients

For an initial industrial targeting analysis, three types of data are helpful. These are:

(a) measures of size (employment, earnings or regional product);
(b) measures of the change in size (changes in employment, earnings or regional product); and
(c) measures of the relative importance of sectors.

Location quotients are well known measures of the relative importance of sectors compared to their importance in a larger frame of reference as described above. Location quotients (LQ) are computed as follows:

$$LQ_{ir} = (E_{ir}/E_r)/(E_{iN}/E_N) \qquad\qquad (3.1)$$

where:

E_{ir}	=	employment in sector i in region r
E_r	=	total employment in region r
E_{iN}	=	employment in sector i in the reference area (N = national reference area here)
E_N	=	total employment in the national reference area.

Measures of scale other than employment can be used; for example earnings and gross regional product GRP.

3.2.4 An Example: Industrial Targeting in Northern Virginia

A case study application of industrial targeting analysis is a study of the Northern Virginia region in the United States. It is part of the National Capital Metropolitan Region. Northern Virginia, a region of more than 1.5 million inhabitants, grew at a rate of more than 8 percent per year during the 1980s as a result of the development of a high technology services sector. This industry sector was an outgrowth of a number of factors including the enabling presence of a highly educated labour force, the federal defence sector build-up of the 1980s and adoption of a federal policy of increased outsourcing, a hallmark of the Reagan administration of the 1980s (Stough et al, 1997). However, the recession of the early 1990s brought a slowdown to the region's economy and, as a consequence, local officials considered the need for an assessment of their economy and for a new development strategy. The following is a summary (from Stough 1995) of part of the targeting analysis that was performed for this region in an effort to enable it to begin developing its strategy.

The time period for this study was 1988–1993. The beginning year, 1988, was selected because it was the year before the signs of the 1990 recession were recognized. The ending year was selected because it was the year for which the most recent data were available, as well as being a year that was well into the recovery from the recession. It is important to use a time period that spans recessions. A period that begins in a trough leads to over-inflated estimates of change while a period that begins at a peak leads to under estimates of change.

Analyses were conducted for four digit SIC categories and reported in sectoral blocks for:

(a) services;
(b) construction;
(c) manufacturing;
(d) transportation and public utilities;
(e) finance, insurance and real estate;
(f) wholesale trade; and
(g) retail trade.

Table 3.1. Northern Virginia: services sector employment, 1988 and 1993, and location quotient 1993

SIC	Description	Empl. 1988	Empl. 1993	% Change	LQ 1993
	TOTAL	720,898	766,624	6.34	
70	Services total	281,163	334,082	18.82	1.32
8051	Skilled nursing care facilities	2,081	2,565	23.26	0.35
8222	Junior colleges & technical institutes	1,964	2,483	26.43	0.76
8051	Skilled nursing care facilities	2,081	2,565	23.26	0.35
8361	Residential care	1,279	2,347	83.50	0.58
7514	Passenger car rental	1,743	1,972	13.14	2.80
7538	General automotive repair	1,438	1,953	35.81	1.22
7378	Computer maintenance and repair	1,046	1,689	61.47	6.10
8071	Medical laboratories	1,465	1,665	13.65	1.85
8082	Home health care services	745	1,606	115.57	0.80
7991	Physical fitness facilities	1,331	1,587	19.23	1.79
7322	Adjustment and collection services	749	1,562	108.54	n.a.
8732	Commercial non-physical research	2,405	1,522	−36.72	2.14
8331	Job training and related services	909	1,515	66.67	0.90
8231	Libraries	584	1,386	137.33	n.a.
8093	Specialty outpatient facilities, n.e.c.	1,364	1,384	1.47	1.08
8299	Schools and educational services, n.e.c.	554	1,363	146.03	n.a.
7841	Video tape rental	2,181	1,259	−42.27	1.35
7375	Information retrieval services	917	1,258	37.19	3.69
8063	Psychiatric hospitals	968	1,238	27.89	n.a.
8712	Architectural services	1,393	1,198	−14.00	1.32
8399	Social services, n.e.c.	1,167	1,181	1.20	0.81
8744	Facilities support services	860	1163	35.23	2.74
7361	Employment agencies	1,166	1,137	−2.49	n.a.
8059	Nursing & personal care facilities, n.e.c.	595	1,101	85.04	n.a.
7532	Top and body repair and paint shops	1,010	1,014	0.40	0.85
7331	Direct mail advertising services	1,702	1,012	−40.54	1.60
8699	Membership organizations, n.e.c.	1,580	938	−40.63	1.93
7377	Computer rental and leasing	354	912	157.63	12.46
8661	Religious organizations	79	884	1018.99	1.30
7382	Security systems services	164	827	404.27	2.83
7359	Equipment rental and leasing, n.e.c.	599	812	35.56	0.84
7699	Repair services, n.e.c.	916	787	−14.08	0.60
8211	Elementary and secondary schools	33,996	38,928	14.51	0.96
8711	Engineering services	24,714	20,434	−17.32	4.41
8742	Management consulting services	9,725	16,671	71.42	11.75
8062	General medical and surgical facilities	13,774	15,034	9.15	0.51
7371	Computer programming services	10,755	14,696	36.64	n.a.
7011	Hotels and motels	12,510	13598	8.70	1.21
7349	Building maintenance services, n.e.c.	10,906	13,155	20.62	2.48
7363	Help supply services	9,368	10,131	8.14	n.a.
8731	Commercial physical research	8,891	9,864	10.94	5.97
7373	Computer integrated systems design	7,645	7,618	−0.35	n.a.
7374	Data processing and preparation	6,426	6,959	8.29	4.85
7999	Amusement and recreation, n.e.c.	5,908	6,465	9.43	2.47

Table 3.1. (cont.)

8111	Legal services	5,014	5,991	19.49	n.a.
8733	Non-commercial research organizations	5,306	5,581	5.18	4.17
8351	Child day care services	3,709	5,227	40.93	1.92
7379	Computer related services, n.e.c.	2,019	5,173	156.22	8.00
8741	Management services	4,373	5,093	16.46	2.65
8611	Business associations	4,573	5,042	10.26	6.64
7389	Business services, n.e.c.	4,868	4,,883	0.31	1.01
8748	Business consulting, n.e.c.	1,411	4,772	238.20	8.77
7372	Pre-packaged software	3,843	4729	23.05	n.a.
7231	Beauty shops	4,112	4,640	12.84	1.77
7381	Detective and armoured car services	3,756	4,624	23.11	1.43
8221	Colleges, universities & prof. schools	2,788	4,505	61.59	0.31
8021	Offices and clinics of dentists	3,367	3,777	12.18	1.04
8322	Individual and family services	2,379	3,702	55.61	0.84
8811	Private households	2,122	3,662	72.57	1.92
8721	Accounting, auditing and bookkeeping	2,592	3,177	22.57	0.83
7997	Membership sports and recreation clubs	1,817	2,868	57.84	1.42
8211	Civic and social associations	2,228	2,788	25.13	n.a.
8641					
8621	Professional organizations	1,840	2,623	42.55	n.a.

Notes: *Empl.*= Employment, *n.a.*= Data unavailable, *n.e.c.*= not elsewhere classified
Source: Center for Regional Analysis, School of Public Policy, George Mason University; unpublished data and analysis results

Only an illustrative part of the analysis is reported here.

One objective of the study was to identify and evaluate the performance of the primary technology intensive industry sectors. It was generally believed that many of the technology intensive sectors were in the broader services sector. Table 3.1 presents measures of employment in the service sectors for the beginning and ending study years, the percentage change that occurred during that period for each sector, and the location quotient in the ending year (1993) of the study period.

Service sectors that have significant employment tied to producing, using or delivering technology in the field and using technology to solve complex problems are summarized in Table 3.2 and described in detail in Table 3.3. A review of these data reveals that more than 50,000—or about half of the employment—is in computer related services (computer programming, computer integrated systems design, data processing, computer related series, pre-packaged software and computer maintenance and repair). All of these sectors have exceptionally high location quotients, ranging from 4.0 up to 8.0. This means that they are from four to eight times more important in the Northern Virginia economy than they are in the national economy. Further, all of these sectors have grown significantly from about 8 percent for data processing up to 156 percent for computer related services, with the exception of computer integration, which declined by 0.35 percent.

Table 3.2. Technology services in Northern Virginia: employment change 1988–1993 and location quotients

Rank[a]	SIC	Description	No VA Empl. 1993	% Change 1988–93	LQ 1988	LQ 1993
	70	Services total	334,082	18.82	1.24	1.32
2	8711	Engineering services	20,434	–17.32	6.01	4.41
3	8742	Management consulting services	16,671	71.42	9.35	11.75
5	7371	Computer programming services	14,696	36.64	12.90	n.a.
9	8731	Commercial physical research	9,864	10.94	5.85	5.97
10	7373	Computer integrated systems design	7,618	–0.35	n.a.	n.a.
11	7374	Data processing and preparation	6,959	8.29	4.91	4.85
14	8733	Non-commercial research organization	5,581	5.18	4.43	4.17
16	7379	Computer related services, n.e.c.	5,173	156.22	4.05	8.00
17	8741	Management services	5,093	16.46	2.78	2.65
20	8748	Business consulting, n.e.c.	4,772	238.20	n.a.	8.77
21	7372	Prepackaged software	4,729	23.05	n.a.	n.a.
37	7378	Computer maintenance and repair	1,689	61.47	4.92	6.10
38	8071	Medical laboratories	1,665	13.65	1.87	1.85
58	7377	Computer rental and leasing	912	157.63	4.70	12.46
65	7376	Computer facilities management	714	99.44	n.a.	n.a.

[a]In terms of employment in 1993, *No VA Empl.*= Northern Virginia Employment, *n.a.*= Data unavailable. Source: Center for Regional Analysis, School of Public Policy, George Mason University; unpublished data and analysis results

The engineering services sector with 20,434 employees is the largest of the four digit SIC service subsectors, (other than elementary and secondary schools). Engineering services experienced a 17 percent decline between 1988 and 1993. However, its location quotient is 4.41, which means that it is more than four times as important to the local economy as to the national economy. The decline of more than 4,000 employees in this sector is due to the effects of the 1990 recession and perhaps to federal downsizing. Major U.S. Department of Defence weapons development programs use program support contractors classified as engineering services. Thus, as resources for weapons systems are cut so is the support for contractors.

As illustrated above, engineering services is a sector that is quite broad, including such diverse components as civil engineering, marine engineering, design, and petroleum engineering and is closely tied to construction activities. During recessionary periods, construction activity declines, which has a cascading effect upon sectors like engineering and new real estate development that supports it. The engineering services sector began to grow again as the region recovered from the recession in the mid–1990s.

Much of the remainder of the technology services is in the area of management and business services, and consulting. Most of these activities have grown significantly over the study period, with business consulting growing by 238 percent.

The high rate of growth may be due to outplacement and outsourcing in the computer systems, integration and network parts of technology services. This may help to explain in part the flat employment in the computer integration systems sector during the study period. Conditions in the management and business services appear to be tied to the core computer systems and network related sectors.

The purpose of this case study has been to illustrate how industrial targeting analysis is conducted and used to identify the primary sectors in the local economy, and to inform economic development planning. The research was used to inform local decision makers that the core of their economy was in the technology services with particularly strong concentrations in the areas of software engineering, information technology and telecommunications (Stough 1995). As a consequence, industry leaders for the first time understood that these technology service sectors were at the heart of the region's competitiveness, not real estate development and construction. This led to the identification of barriers to the future development of this sector and the adoption of measures to remove these impediments.

Table 3.3. Detailed description of 4 digit SIC technology services in Northern Virginia

SIC	Services
8711	Engineering Services (20,434 employees)
	Designing: ship, boat and machine
	Engineering services: industrial, civil, elect, mechanical, petroleum, marine, design
	Machine tool designers
	Marine engineering services
	Petroleum engineering services.
8742	Management consulting services (16,671 employees)
	Administrative management consultants
	General management consultants
	Human resource consultants
	Management engineering consultants
7381	Computer programming services (14,696 employees)
	Applications software programming: customized
	Computer code authors
	Computer programming services
	Computer programs or systems software development, customized
	Computer software systems analysis and design, customized
	Computer software writers, free lance
	Programming services, computer: customized
	Software programming, customized
8731	Commercial physical research (9,864 employees)
	Agricultural research, commercial
	Chemical laboratories, commercial research: except testing
	Engineering laboratories, commercial research except testing
	Food research, commercial
	Industrial laboratories, commercial research except testing
	Physical research, commercial
	Research and development, physical and biological: commercial
7373	Computer integrated systems design (7,618 employees)
	CAD/CAM systems services
	Computer-aided design (CAD) systems services
	Computer-aided engineering (CA) systems services

Table 3.3. (cont.)

	Computer-aided manufacturing (CAM) systems services
	Local area network (LA) systems integrator
	Network systems integration, computer
	Office automation, computer systems integration
	Systems integration, computer
	Turnkey vendors, computer systems
	Value-added resellers, computer systems
7374	Data processing and preparation (6,959 employees)
	Calculating service, computer
	Computer time-sharing
	Data entry service
	Data processing services
	Data verification service
	Keypunch service
	Leasing computer time
	Optical scanning data service
	Rental of computer time
	Service bureaus, computer
	Tabulating service, computer
7379	Computer related services (not elsewhere included) (5,173 employees)
	Computer consultants
	Database developers
	Data processing consultants
	Disk and diskette conversion services
	Disk and diskette recertification services
	Requirements analysis, computer hardware
	Tape recertification service
8741	Management services (5,093 employees
	Administrative management services
	General management consultants
	Human resource consultants
	Management engineering consultants
	Management information systems consultants
	Manufacturing management consultants
	Marketing consultants
	Operation research consultants
	Personnel management consultants, except employment service
8748	Business consulting (not elsewhere included) (4,772 employees)
	Agricultural consulting
	City planners
	Economic consulting
	Education consulting, except management
	Industrial development planning services, com.
	Radio consultants
	Systems engineering consulting except professional Manager or computer related
	Test development, and evaluation service, education or personnel
	Testing services, educational or personal
	Traffic consultants

Source: Center for Regional Analysis, School of Public Policy, George Mason University; unpublished data and analysis results

3.3 Shift-Share Analysis

A simple descriptive, quick and relatively inexpensive technique for analyzing regional growth and decline over time is shift-share analysis. This technique enables the assessment of a region's overall performance relative to other regions. Focusing on regional employment or output by industry sector, this tool has been used widely since the 1960s to assess a region's performance relative to other regions and to assess the relative importance of an industry sector in a region. It can readily identify a region's industrial sector problems that might require more detailed attention. In particular, shift-share analysis has been especially useful "to demonstrate how industry structure affects regional and local economies, to review regional economic trends, and to advise policy makers on industrial targeting" (Markusen et al. 1991, p.16).

Two relatively recent contributions to the literature on shift-share analysis by Haynes and Dinc (1997) and by Dinc and Haynes (1999) provide comprehensive overviews of the evolution and application of this analytical tool.

3.3.1 The Traditional Shift-Share Model

The traditional shift-share model measures regional growth or decline by decomposing it into three components:

(a) national share (NS): that is, that part of change attributable to overall national trends
(b) industrial mix (IM): that is, that part of change attributable to the industrial composition or mix of the region
(c) regional shift (RS): that is, that part of change attributable to regional advantage or competitiveness.

Typically the analysis uses widely available and easily accessible secondary data, that are typically collected by a national agency (such as the census bureau or a ministry of commerce) on regional employment, income, output, population or other economic factors available at different geographic and responsibility levels—national, state, county, city, locality, etc.

Early shift-share models outlined in Perloff et al. (1960) focused on total regional employment and had only two components:

(a) *Total shift* (TS), expressed as:

$$\text{TS} = \sum_i e_{i,t} - \sum_i e_{i,t-1}(E_t / E_{t-1}) \qquad (3.2)$$

(b) *Differential shift* (DS), expressed as:

$$DS = \sum_i e_{i,t-1}(e_{i,t} / e_{i,t-1} - E_{i,t} / E_{i,t-1}) \qquad (3.3)$$

where:

e_i and E_i respectively are regional and national employment in industry i;

e and E respectively are regional and national total employment in all industries; and

t-1 is the initial period and t the end period (e.g. inter-censal dates) of the analysis.

Dunn (1960) introduced to the model differential rates of growth in individual industries, to give what is known as the 'proportionality effect,' which is equivalent to the industry composition or mix (IM) effect referred to above. Ashby (1967) introduced a three-component model of regional change, incorporating *national share* (NS), *industry mix* (IM), and *regional shift* (RS). Thus:

$$\Delta e_i \equiv e_{i,t} - e_{i,t-1} \equiv NS_i + IM_i + RS_i \qquad (3.4)$$

where:

$$NS_i \equiv e_{i,t-1}(E_t / E_{t-1} - 1) \qquad (3.5)$$

$$IM_i \equiv e_{i,t-1}(E_{i,t} / E_{i,t-1} - E_t / E_{t-1}) \qquad (3.6)$$

$$RS_i \equiv e_{i,t-1}(e_{i,t} / e_{i,t-1} - E_{i,t} / E_{i,t-1}) \qquad (3.7)$$

$$e_{i,t} \equiv e_{i,t-1} + (NS_i + IM_i + RS_i) \qquad (3.8)$$

This classical shift-share model—which has been used extensively by economists, geographers, regional scientists and planners in regional analysis—thus emphasizes not only the role of regional change for a region-specific industry, but also the regional shift or competitive component as a measure of the relative performance of the region for a specific industry. A position shift is interpreted as being associated with the comparative or competitive advantage of the region for that industry, or vice versa. The partition of regional change into the three components—NS, IM and RS—was intended to enable researchers to study the sources of change separately.

3.3.2 An Example: Regional and Axial Shifts in the United States Spatial Economy

In the early 1980s the United States, the Northeast and Midwest regions were loosing out to growth in the Southeast, South, Southwest and West. This was evidenced by population movements as well as shifts in industrial location and employment. The Northeast always was a net out-migration region. What was of significance was the sudden change in pace and destination of population movement beginning in the 1970s. While the West and Southwest showed gains in population, the dramatic growth appeared in the South. At least some observers interpreted this as a direct transfer from the (old) North to the (new) South (Vining et al 1982, pp. 270–278).

Among the many factors hypothesized as contributing to this regional orientation was the controversial notion that an axial shift was occurring in transportation movements. Vining et al (1982) proposed that the natural grain of the geography of the United States is north-south, and that high costs are associated with maintaining the traditional and principal east-west axis of the United States space economy. Given a weakening of conservative values and a liberalizing of institutions in the South after the 1960s, along with such physical improvements as air conditioning, the South evidenced rapid growth from the mid–1960s.

Toft and Stough (1986) desired to learn whether regional and/or axial shifts were having an effect on the geographic pattern of the transport sector. They were particularly curious about north-south shifts in the share of transportation employment among the East North Central states of Illinois, Indiana, Michigan, Ohio and Wisconsin and the East South Central and South Atlantic States (Alabama, Kentucky, Mississippi, Tennessee, Florida, Georgia, North Carolina and South Carolina). Six of the top ten trucking states were included in this group.

To investigate these research questions Toft and Stough used traditional shift-share analysis applied to major transportation industries among the states of the United States for the period 1969–1982. These years were selected because 1969 was characterized by strong economic performance and 1982 by poor economic performance. In business cycle terms, employment changes were thus computed between peak and valley years tending to give very conservative estimates of growth rates. Data were compiled and analyzed on a year-by-year basis to test this belief.

Shift-share analyses were computed for all States and for seven primary transportation industries in order to compare the rates of growth among the different states and in particular to compare the regional competitiveness of the states in these industries. These industry sectors were:

(a) Total Transportation, Communications and Public Utilities (SIC 4)
(b) Local and Interurban Passenger Transit (SIC 41)
(c) Trucking and Warehousing (SIC 42)
(d) Water Transportation (SIC 44)
(e) Air Transportation (SIC 45)

(f) Pipelines (SIC 46)
(g) Transportation Services (SIC 47).

The general pattern among the results was that in all transportation sectors job loss was occurring due to industry mix and further loss in the Northeast and Midwest due to a loss of State competitive share (see Table 3.4). Transportation services were the highest growth part of the transport sector (see Table 3.5). The general trend shows a loss in competitive share from North to South and West even though, in absolute terms, employment remained highest for most transport sectors in the North, because of the heavy industrial concentration there. Two criteria were used to identify high performer States:

(a) the competitive share had to at least equal the national growth component; and
(b) the absolute growth in employment must be at least 2 percent of absolute growth for the industry nationally.

Table 3.4. Employment change in transportation, communications, and public utilities between 1969 and 1982 (SIC 41, 42, 44, 45, 46, 47, 48, 49)

Region, Division State	Empl. 1969	Empl. 1982	Percent change	National growth	Empl. in- dustry mix	Change rela- ted to com- petitive/ State shares	Total[a]
(1)	(2)	(3)	(4)	(5)	(6)	(7)	(8)
NORTHEAST							
New England	196,740	223,509	13.61	62,662	(13,595)	(22,291)	26,769
Maine	12,838	14,899	16.05	4,089	(867)	(1,141)	2,061
New Hampshire	10,009	14,177	41.64	3,186	(692)	1,672	4,168
Vermont	6,397	7,760	21.31	2,037	(442)	(232)	1,363
Massachusetts	106,569	113,047	6.08	33,942	(7,364)	(20,100)	6,478
Rhode Island	14,383	12,188	(15.26)	4,581	(994)	(5,782)	(2,195)
Connecticut	46,544	61,438	32.00	14,824	(3,216)	3,286	14,894
Middle Atlantic	842,916	822,101	(2.47)	268,469	(58,245)	(231,043)	(20,815)
New York	479,919	416,393	(13.24)	152,854	(33,162)	(183,218)	(63,526)
New Jersey	144,221	186,538	29.34	45,934	(9,966)	6,348	42,317
Pennsylvania	218,776	219,170	0.18	69,680	(15,117)	(54,169)	394
NORTH CENTRAL MIDWEST							
East North Central	671,992	712,948	6.09	214,029	(46,435)	(126,639)	40,956
Ohio	173,756	183,251	5.46	55,341	(12,007)	(33,840)	9,495
Indiana	75,787	91,530	20.77	24,138	(5,237)	(3,158)	5,743
Illinois	231,183	227,890	(1.42)	73,632	(15,975)	(60,950)	(3,293)
Michigan	127,829	131,064	2.53	40,714	(8,833)	(28,646)	3,235
Wisconsin	63,437	79,213	24.87	20,205	(4,383)	(45)	15,776
West North Central	280,659	354,972	26.48	89,390	(19,394)	4,317	74,313
Minnesota	63,401	84,260	32.90	20,193	(4,381)	5,047	20,859
Iowa	38,352	45,413	18.41	12,215	(2,650)	(2,504)	7,061
Missouri	104,905	121,367	15.69	33,412	(7,249)	(9,701)	16,462
North Dakota	8,148	13,589	66.78	2,595	(563)	3,409	5,441
South Dakota	8,679	11,858	36.63	2,764	(600)	1,014	3,179
Nebraska	23,639	30,408	28.63	7,529	(1,633)	873	6,769
Kansas	33,535	48,077	43.36	10,681	(2,317)	6,178	14,542

Table 3.4. (cont.)

SOUTH							
South Atlantic	503,877	728,426	44.56	160,485	(34,818)	98,882	224,549
Delaware	9,290	12,320	32.62	2,959	(642)	713	3,030
Maryland	66,700	74,122	11.13	21,244	(4,609)	(9,211)	7,422
Virginia	73,477	100,167	36.32	23,405	(5,077)	8,362	26,690
West Virginia	28,052	29,078	3.66	8,935	(1,938)	(5,969)	1,026
North Carolina	79,231	117,872	48.77	25,235	(5,475)	18,881	38,641
South Carolina	28,984	40,794	40.75	9,231	(2,003)	4,582	11,812
Georgia	84,901	133,486	57.23	27,041	(5,867)	27,415	48,585
Florida	133,242	220,585	65.55	42,438	(9,207)	54,110	87,343
East South Central	160,827	226,833	41.04	51,223	(11,113)	25,896	66,006
Kentucky	38,590	52,,067	34.92	12,291	(2,667)	3,851	13,477
Tennessee	49,477	71,593	44.70	15,754	(3,419)	9,777	22,116
Alabama	49,775	68,783	38.19	15,853	(3,439)	6,595	19,008
Mississippi	22,985	34,390	49.62	7,321	(1,588)	5,673	11,405
West South Central	349,283	592,959	69.76	111,246	(24,135)	156,565	243,676
Arkansas	23,516	33,991	44.54	7,490	(1,625)	4,609	10,475
Louisiana	72,899	122,563	68.13	23,218	(5,037)	31,485	49,664
Oklahoma	48,326	72,427	49.87	15,392	(3,339)	12,048	24,101
Texas	204,542	363,978	77.95	65,147	(14,134)	108,428	159,436
WEST							
Mountain	133,072	249,399	87.42	42,383	(9,195)	83,139	116,327
Montana	10,477	16,231	54.92	3,347	(734)	3,141	5,754
Idaho	10,039	14,713	46.56	3,197	(694)	2,171	4,674
Wyoming	6,473	12,469	92.63	2,062	(447)	4,381	5,996
Colorado	41,272	77,931	88.82	13,145	(2,852)	26,366	36,659
New Mexico	15,047	25,124	66.97	4,792	(1,040)	6,324	10,077
Arizona	23,174	49,465	113.45	7,381	(1,601)	20,511	26,291
Utah	16,447	31,746	93.02	5,238	(1,136)	11,197	15,299
Nevada	10,143	21,720	114.14	3,230	(701)	9,047	11,577
Pacific	534,118	749,412	40.31	170,117	(36,908)	82,085	215,294
Washington	58,492	79,236	35.46	18,629	(4,042)	6,156	20,744
Oregon	39,156	51,815	32.33	12,471	(2,706)	2,893	12,659
California	410,162	573,658	39.86	130,637	(28,342)	61,201	163,496
Alaska	7,449	16,258	118.26	2,373	(515)	6,951	8,809
Hawaii	18,859	28,445	50.83	6,007	(1,303)	4,883	9,586
United States Industry Empl.	3,703,344	4,626,875	24.94				
United States Total Empl.	56,348,479	74,297,252	31.85				

Notes: *Empl.*= Employment, [a]Total is computed by subtracting the 1969 from the 1982 employment figures. Consequently, total figures may vary slightly from the figures obtained by summing National Growth, Industry Mix and Competitive/State Share
Sources: Reproduced from Toft and Stough 1986

The bulk of the high performing states clustered in the South and West, particularly the West South Central and Mountain regions. Most notable states that repeatedly performed well across all transportation industries were Louisiana, Oklahoma, Texas, Colorado and Arizona.

The analyses show that transportation employment grew slower than the national average for the study period. All major transportation industries reflect this pattern except transportation services, which grew at 122 percent compared with a total national employment growth of 32 percent. Even air transportation employment grew slower than the United States total employment.

This was interpreted as pointing to the significance of structural economic changes within the transportation sector, wherein information, coordination, and networking were becoming relatively more important. This observation surfaced economic policy issues such as how a region or a State could capitalize on the services growth component. Superior transportation services were viewed as closely tied to established institutional arrangements and communication/computer systems capability. Telecommunications was viewed as being more of a compliment than a substitute in the transportation growth equation.

The analyses further suggested that regional shifts in the United States space economy were reflected in the changing distribution of employment in the transport sector. While the Northeast and Midwest regions combined still surpassed the South and West for absolute total transportation employment, growth rates in the latter two regions far exceeded the former. This posed a threat to traditional transportation States in the frostbelt, such as Illinois, Indiana, Michigan and Ohio, which were among the top 10 trucking States. The growth rate in the East North Central region for trucking and warehousing was –4.4 percent, whereas directly south and in the East South Central area it was +37.42 percent. The historical competitive position of the New England, Middle Atlantic and East North Central regions in transportation employment was clearly threatened.

While not confirming the axial shift, the results are viewed as congruent with the theory. In each of the transportation industries, the South and the West regions were gaining employment at the expense of the Northeast and Midwest. Only in transportation services, where comparative advantage was viewed by many to hinge on established institutional networks and on computer and communications systems, did the Northeast and Midwest have employment growth comparable with the States to the south. The services component of the transportation industry was, therefore, little affected by the natural north-south grain. However, the high performing growth States were not in the 'old' South as axial shift theory suggested. Rather, they were in the West South Central and Mountain regions. This suggested a diagonal axial shift running from the new south to the near lower west, especially linking states such as Louisiana, Oklahoma, and Texas to Colorado, Arizona and Utah.

The national share (NS) component measures the regional employment change that could have occurred if regional employment had grown at the same rate as the nation. The industry mix (IM) component measures proportional shift due to a difference in industry growth between the region and the nation. The regional shift (RS) component measures the differential shift due to differences in rates of growth of the same industry between the region and the nation as a result of factors such as national resources, other comparative advantages or disadvantages, leadership and entrepreneurial ability, and the effects of regional policy.

Table 3.5. Employment change in transportation services between 1969 and 1982 (SIC 47)

Region, Division State	Empl. 1969	Empl. 1982	Percent change	National growth	Empl. industry mix	Change related to competitive/State Shares	Total[a]
(1)	(2)	(3)	(4)	(5)	(6)	(7)	(8)
NORTHEAST							
New England	4,329	11,428	163.99	1,379	3,887	1,833	7,099
Maine	–	346	–	–	–	–	–
New Hampshire	–	487	–	–	–	–	–
Vermont	109	344	215.60	35	98	102	235
Massachusetts	3,018	6,442	113.45	961	2,710	247	3,424
Rhode Island	276	571	106.88	88	248	(41)	295
Connecticut	926	4,071	339.63	295	831	2,019	3,145
Middle Atlantic	36,869	51,778	40.44	11,743	33,105	(29,938)	14,909
New York	29,043	35,436	22.01	9,250	26,078	(28,936)	6,393
New Jersey	3,291	8,395	155.09	1,048	2,955	1,101	5,104
Pennsylvania	4,535	7,947	75.24	1,444	4,072	(2,104)	3,412
NORTH CENTRAL MIDWEST							
East North Central	13,493	30,483	125.92	4,298	12,115	577	16,990
Ohio	2,289	5,795	153.17	729	2,055	722	3,506
Indiana	644	2,129	230.59	205	578	702	1,485
Illinois	7,000	14,637	109.10	2,230	6,285	(878)	7,637
Michigan	2,809	5,327	89.64	895	2,522	(899)	2,518
Wisconsin	751	2,595	245.54	239	674	930	1,844
West North Central	4,919	10,388	111.18	1,567	4,417	(514)	5,469
Minnesota	1,226	3,390	76.51	390	1,101	673	2,164
Iowa	447	1,097	145.41	142	401	106	650
Missouri	2,616	3,979	52.10	833	2,349	(1,819)	1,363
North Dakota	135	283	109.63	43	121	(16)	148
South Dakota	101	256	153.47	32	91	32	155
Nebraska	900	–	–	287	808	–	–
Kansas	394	1,383	251.02	125	354	510	989
SOUTH							
South Atlantic	11,869	30,621	157.99	3,780	10,657	4,314	18,752
Delaware	–	585	–	–	–	–	–
Maryland	1,556	3,667	135.67	496	1,397	218	2,111
Virginia	873	3,376	286.71	278	784	1,441	2,503
West Virginia	143	–	–	46	128	–	–
North Carolina	977	2,470	152.81	311	877	305	1,493
South Carolina	360	1,878	421.67	115	323	1,080	1,518
Georgia	2,836	3,895	37.34	903	2,546	(2,391)	1,059
Florida	5,267	15,335	191.15	1,678	4,729	3,661	10,068
East South Central	2,123	5,157	142.91	676	1,907	452	3,034
Kentucky	443	1,191	168.85	141	398	209	748
Tennessee	647	2,152	232.61	206	581	718	1,505
Alabama	901	1,368	51.83	287	809	629	467
Mississippi	132	446	237.88	42	119	153	314
West South Central	7,186	25,598	256.08	2,289	6,452	9,661	18,402
Arkansas	146	650	345.21	47	131	326	504
Louisiana	2,415	4,657	92.84	769	2,168	(696)	2,242
Oklahoma	474	1,179	148.73	151	426	128	705
Texas	4,151	19,102	360.18	1,322	3,727	9,902	14,951

Table 3.5. (cont.)

WEST								
Mountain		1,458	8,751	500.21	464	1,309	5,519	7,293
	Montana	92	404	339.13	29	82	201	312
	Idaho	84	504	500.00	27	75	318	420
	Wyoming	44	–	–	14	40	–	–
	Colorado	600	33,263	443.83	191	539	1,933	2,663
	New Mexico	86	513	496.51	27	77	322	427
	Arizona	331	2,062	522.96	105	297	1,328	1,731
	Utah	198	1,078	444.44	63	178	639	880
	Nevada	67	927	1,283.58	21	60	778	860
Pacific		17,687	47,354	167.73	5,633	15,881	8,153	29,667
	Washington	1,644	5,403	228.65	524	1,476	1,759	3,759
	Oregon	770	2,067	168.44	245	691	360	1,297
	California	13,576	35,501	161.50	4,324	12,190	5,411	21,925
	Alaska	–	649	–	–	–	–	–
	Hawaii	1,697	4,383	158.28	540	1,524	622	2,686
United States Industry Empl.		102,117	226,328	121.64				
United States Total Empl.		56,348,479	74,297,252	31.85				

Notes: *Empl.*= Employment, [a]Total is computed by subtracting the 1969 from the 1982 employment figures

Source: Reproduced from Toft and Stough 1986

3.4 Critiques and Extensions of Traditional Shift-Share Analysis

Since the 1960s, shift-share analysis has been used in the United States and in many other nations by regional scientists to examine systematically regional economic data. This includes agencies in the U.S. such as the Federal Reserve Banks and the Department of Commerce.

Applications of the classical shift-share model are diverse in their contexts, and include analysis of:

(a) The impacts of public decision making (Sui 1995).
(b) Migration turnaround (Ishikawa 1992).
(c) Change in occupational composition (Smith 1991);
(d) Analysis of regional productivity (Ledebur and Moomaw 1983; Rigby and Anderson 1993; Haynes and Dinc 1997).
(e) The impacts of transportation on regional growth (Toft and Stough 1986).
(f) Regional changes in employment growth and interregional comparisons (Harrison and Kluver 1989).

Despite its continued widespread use, shift-share analysis has been heavily criticized for having temporal, spatial, industrial aggregation, theoretical content and predictive capability deficiencies (Stevens and Moore 1980; Haynes et al. 1990; Knudsen and Barff 1991; Dinc et al.1998). But there have been many substantial modifications and extensions of the model.

3.4.1 Shortcomings

Dawson (1982) lists six shortcomings of the traditional shift-share model:

(a) Changes in the industry mix in the national economy are not taken into account. This is a weighting problem as changes from the beginning to the end of the period of time over which change is being measured may have quite different weights or opposite signs for the industrial mix and the competitive effects.
(b) Results are sensitive to the degree of industrial and regional disaggregation.
(c) The differential industry component is unstable over time, and the degree of instability varies among industries.
(d) Growth resulting from inter-industry linkage and secondary multi-sector effects are not explicitly isolated but are included in the competitive component (RS), whereas they should be included in the industry mix component (IM).
(e) The differential component (RS) may be influenced by relatively spurious causes, including the incorrect classification of firms, product heterogeneity within firms, and transfers of production between separate sites of individual firms.
(f) The technique provides no information on the capacity of a region to retain growing industries or on how to attract them in the first place (Richardson 1978).

These and other issues are elaborated on below in the sections that follow.

3.4.2 Modifications and Extensions

Despite its shortcomings, the shift-share technique has been defended and the model extended over time. It has been shown that shift-share analysis has both a statistical and a spatio-economic justification. It stands on solid statistical and economic grounds.

Researchers have incorporated the shift-share model into other statistical forecasting methods, including:

(a) Analysis of Variance (ANOVA) models (Berzeg and Koran 1984, 1978).
(b) A multiplicative model of shift-share (Theil and Gosh 1980; Kurre and Weller 1989).
(c) Univariate autoregressive integrated moving average (ARIMA) time series models.
(d) A linear model of shift-share analysis (Knudsen and Barff 1991).

It is worth referring to a number of important ways the traditional shift-share model has been extended to overcome some of its limitations.

A major weakness of the traditional shift-share model is the failure to measure economic change due to the special dynamism of an industry sector in a region. Estaban-Marquillos (1972) separated the differential shift (the competitive effect) into an actual competitive effect and an allocation effect by introducing the concept of homothetic employment, which is the level of employment a region should have if the location quotient for a sector equals one, removing the competitive position of all regional structural influences. Later Arcelus (1984) extended the model further separating both the regional growth effect and the regional industry mix effect into their respective homothetic and differential components. While this extension of the model lost the region-to-region additive property of the traditional shift-share model, Haynes and Machunda (1987) developed a theoretically sound standard procedure for identifying the desired aggregation-disaggregation properties for these extensions of shift-share analyses.

Another trend in shift-share analysis is the use of econometric models developed by Emmerson et al. (1975) and by Berzeg and Koran (1978). These are early forms of the information-theoretic approach developed by Theil and Gosh (1980). The typical econometric model takes the form:

$$Y_{ijt} = a + b_i + (g_j + d_{ij}) + e_{ijt} \qquad (3.9)$$

where:

Y_{ijt} = the observed growth over period t for industry i in regions j, or expressed as $X_{ij,t}/X_{ij,t-1}$ (or its logarithm) with X defined as the activity level,

a = the overall growth effort

b_i = the industrial composition effect

g_j+d_{ij} = the competitiveness effect, which can be further decomposed into the regional effect (g_j) and comparative advantage term (d_{ij})

e_{ijt} = random error.

The sensitivity of the traditional model to the degree of industrial disaggregation has received considerable criticism. But Fothergill and Gudgin (1979) demonstrated that a finer level of disaggregation of industrial classifications made only small differences in outcome measures.

The criticism that the traditional shift-share model takes no account of changes in industrial mix over the time period being studied, and that continuous changes in the size of a region's total employment over time are not taken into account, are addressed by Barff and Knight (1988) who expand the model by calculating the three shift-share effects (NS, IM and RS) for every year of a study period. This dynamic shift-share approach thus adjusts annually for change in industrial mix, continuously updates the region's employment total, and uses annual growth rates. As noted by Kurre and Weller (1989, p.756), this time-series data "can be modeled for forecasting purposes by means of time-series techniques".

A further critique of shift-share analysis is that many economies, such as the United States, are 'open economies', thus the market for many goods and services is international, not just the region or the nation, so that the 'international market' should be the 'relevant reference point' (Sihag and McDonough 1989). Particularly in contemporary terms of globalization where 'international' trade becomes increasingly important to the general economy as well as to certain regional economies—regional growth should not be analyzed relative to the national economy alone. Thus Markusen et al. (1991) have extended the traditional shift-share model by disaggregating both the national growth and the proportionality (industry mix) components into four new components representing trade and productivity effects, while keeping the differential (competitive) shift unchanged because of the lack of reliable data. They disaggregate the national growth component into three sub-markets:

(a) a national export component, capturing the export effect if employment were to expand proportionately to national exports;
(b) a national import component, which is a hypothetical national import effect of national domestic demand shift;
(c) a national domestic market component, which represents a residual.

These are added to:

(d) a productivity component, which is a national labour productivity growth effect.

Markusen et al. (1991, p.35) show that "expanding the components of shift-share analysis to include trade and productivity effects demonstrates that the industrial structures of regions vary widely in their potential to generate employment where trade and productivity gains alter market and production patterns." This is of great importance in an era where trade barriers are dropping and globalization is expanding.

Sihig and McDonough (1991) extended the shift-share model further by including both primary and secondary base economies in the model because the growth of an industry in a region is likely to be fuel to both state and national economies. Such multi-base shift-share models are now common.

3.4.3 Total Factor Productivity Approach

To assess what industry sectors are performing better in a region requires a determination of the productivity of various industry sectors and the sources of productivity change in specific regional sectors.

The traditional shift-share analysis fails to account for the demographic structure of a region and the level of labour force participation rates in the analysis of regional employment change. For example, a region with a lower population growth and labour participation rate relative to the reference area might experience slow growth (or decline) in employment over a period of time, or vice versa. An increase in employment in the region might be seen as an indicator of a good economy, even if the region has a high rate of employment. As Haynes and Dinc (1997) note this issue becomes much more visible in inter-regional comparisons. For example, a region might increase its output while employment is declining in the region because of improvements in production process and the adoption of new technologies. Perhaps data on the number of establishments may be better to use than employment as an alternative measure in a shift-share model, but this then introduces further deficiencies related to size of firms, levels of capitalization, and so on.

Rigby and Anderson (1993) and Rigby (1992) argue that the traditional shift-share model measures the combined effects of output growth and productivity change on employment, and according to the traditional model a region with above average employment growth either has a favourable industry mix or enjoys a competitive advantage over other regions. Thus, a positive (negative) shift may result from above (below) average output growth, below (above) average productivity gains or some other combination of the two, and unless these effects are isolated, regional performance cannot be unambiguously evaluated (Haynes and Dinc 1997, p.10). Rigby and Anderson extend the model by incorporating average labour productivity in an industry or region at a given time, and by estimating change in output when productivity is held constant. Potential changes in employment in the industry that result from variations in productivity are with constant output.

Haynes and Dinc (1997) further improve the Rigby and Anderson extension of the model by separating the contribution of labour and of capital to productivity by employing an approach proposed by Kendrick (1973, 1983, 1984) using total factor productivity (TFP). Productivity is defined as the relationship between outputs of goods and services and inputs of resources and is typically expressed in a ratio form—the ratio of aggregate output to the sum of inputs. Outputs are weighted by their costs per unit in constant prices, and inputs are combined in terms of their share of total costs in constant prices.

The shift-share model thus takes on new forms.

(a)   To investigate employment (L) change:

$$TS_L \equiv NS_L + PS_L + DS_L \tag{3.10}$$

$$NS_L \equiv NS(a_L) + NS(b_L) = \sum E_{ir}(a_{nL} + b_{nL}) \tag{3.11}$$

$$PS_L \equiv PS(a_L) + PS(b_L) = \sum E_{ir}[(a_{inL} - a_{nL}) + (b_{inL} - b_{nL})] \tag{3.12}$$

$$DS_L \equiv DS(a_L) + DS(b_L) = \sum E_{ir}[(a_{irL} - a_{inL}) + (b_{irL} - b_{inL})] \tag{3.13}$$

(b)   To investigate capital (K) change:

$$TS_k \equiv NS_k + PS_k + DS_k \tag{3.14}$$

$$NS_k \equiv NS(a_k) + NS(b_k) = \sum \kappa_{ir}(a_{nk} + b_{nk}) \tag{3.15}$$

$$PS_K \equiv PS(a_k) + PS(b_k) = \sum \kappa_{ir}[(a_{ink} + a_{nk}) + (b_{ink} - b_{nk})] \tag{3.16}$$

$$DS_K \equiv DS(a_k) + DS(b_k) = \sum \kappa_{ir}[(a_{irk} - a_{ink}) + (b_{irk} - b_{ink})] \tag{3.17}$$

In the case of lack of reliable data on capital at the state and industry sector level, to determine the contribution of other factors—such as capital, technology, infrastructure and raw material—to total productivity, the following model is offered as a method for representing its impact on employment change:

$$\Delta EP = \Delta E - TSL \tag{3.18}$$

Where:

(a) ΔE is the actual employment change over time in the region;
(b) TSL is the total shift in labour (employment change) in the region or state resulting from the change in productivity and output; and
(c) The difference between actual change and total shift gives the employment change resulting from the contribution of the other factors to total factor productivity, ΔEP.

Qiangsheng (1997) adds to the Arcelus (1984) extension of the model by introducing a new component, the total industry mix effect (TI), which is equivalent to a differential growth rate between the regional sector and the region's total:

$$TI_{ij} = IM_{ij} + RI_{ij} \tag{3.19}$$

As IM and RI explain two different effects contributing to regional growth, they should be dealt with as separate components:

(a) The regional industry mix (RI) component explains the difference of national growth rate between the region and the nation, or the absolute advantage (disadvantage) of the sector without being influenced by the regional industrial base.
(b) The conventional regional shift component (RS) however explains the relative advantage (disadvantage) of the sector in the region regardless of the region's industrial structure.
(c) The total sectoral mix effect (TI) is thus the net sectoral advantage of the region.

This approach is important as regions typically are differentiated with respect to labour and non capital factors such as infrastructure, and recently studies have shown that there is a strong correlation between productivity and infrastructure. For example, Andrews and Swanson (1995) show how, in the United States, a productivity decline happened after a decline in infrastructure investment; thus if we use standard measures of labour productivity, regions with better infrastructure will enjoy high labour productivity even though labour's contribution to growth in productivity might be very low. Similarly, the industry mix of regions affect their productivity; thus a region with a capital intensive industry mix will enjoy a higher labour productivity than a region with labour intensive industry mix. Therefore, the multi factor productivity (Bronfenbrenner 1985) or total factor productivity (Baumol et al. 1989) approach to increasing regional performance is important, as are approaches using the production function approach—such as the Cobb-Douglas function—that incorporate the technology effect.

As emphasized by Dinc and Haynes in their work cited earlier, this unfolding process to determine the productivity of various regional industry sectors and the sources of the productivity changes is crucial in assessing what industry sectors are performing better in a region. It is also important for two other reasons. First, this approach better assesses the local regional contribution to productivity by sector. Thus it offers an improved indicator for the effect of institutional and other endogenous forces operating in the local economy. This in turn is in keeping with precepts of the new economic growth theory or endogenous growth (Romer 1986, 1990; Karlsson et al. 2001; and Stough 2001). Second, information produced by this as well as other methods improves the quality of the data available to inform local regional economic development policy (Johansson et al. 2001a; and Stough 2001).

3.4.4 An Example of Total Factor Productivity Shift-Share Adaptation: Shifts from the Snow-Belt to the Sun-Belt in the United States, 1960–1990

A comparison of the performance of the Rigby and Anderson (1993) model incorporating productivity adaptations of the shift-share model has been conducted by Haynes and Dinc (1997). The differences between these two models lies in the way productivity gains and losses are computed and allocated. Rigby and Anderson argue correctly that the traditional shift-share model confounds the effects of output growth and productivity change on employment. They extend shift-share to separate these two effects on employment. However, Haynes and Dinc argue that in defining labour productivity as output per unit of labour input, other factors—such as capital, technology, infrastructure and quality of material inputs—are misattributed to labour. Thus Haynes and Dinc, after Kendrick (1973, 1983 and 1984) take total factor productivity (TFP) approach, with TFP defined as the relationship between output of goods and services and inputs of resources. They are able to compute labour productivity and total factor productivity directly, thereby providing the basis for separating labour productivity from productivity due to other factors.

Both teams of researchers have investigated the growth and decline of the same twenty (two-digit SIC) manufacturing sectors in the same twelve states to determine if observed changes in employment were due to changes in output or to labour productivity gains (declines). The analysis by Rigby and Anderson (1993) calculated the national shift components as a residual. Consequently, in order to calculate the national effect on the employment change, the total shift was subtracted from the actual employment change. Haynes and Dinc (1997), as implied above, calculated the national shift separately, and to capture annual change in regional growth or decline they also incorporated in their analysis the dynamic shift-share approach of Barff and Knight (1988).

The period of investigation of both analyses is 1960 and 1990. This period spans the final parts of the post World War II heyday, the decline of the manufacturing belt, and the rise of the sun-belt. The data are from the Annual Survey of Manufactures Geographic Area Series, the Census of Manufactures Geographic Area Series, and Survey of Current Business.

The Rigby and Anderson (1993) model application results appear in Table 3.6, showing how the employment growth in the sun-belt states resulting from the rapid output growth outpaced employment losses due to productivity gains. Among these states, California, Florida and Kentucky experienced employment growth resulting from relatively poor productivity enhancement. For example, if California had improved its productivity at the level of its national competitors it would have lost 1,277 million hours of employment. On the contrary it gained only 162 million hours of employment because of its lagging productivity. Even though the remaining states enjoyed some gains in productivity, none of them outperformed their national counterparts. For example, Arizona—the best performer—would have lost 77 million hours of employment instead of 16 million hours gained if it had improved its productivity at the level of the national economy.

The snow-belt states, in contrast, witnessed an output decline and an accompanying employment loss. In addition, in Massachusetts and New York gains in productivity (although modest relative to the nation) contributed to the employment loss in these states. The remaining four states experienced inferior productivity gains relative to the nation, and had small employment gains resulting from their poor productivity performance.

Examination of the proportional shift shows that none of the states in either region had a higher average proportion of industries experiencing faster output growth than the national manufacturing average. For example, Arizona—the best performer in terms of employment growth—would have gained 58 million hours of employment instead of 33 million hours if it had a more favourable industry mix. Holding output growth constant, the proportional shift indicates that no state contained an industry mix that experienced greater productivity gains than the nation over the study period, although all states did improve their productivity.

Texas—the best performer in productivity—would have faced a 632 million hour decline in manufacturing employment instead of a 599 million hour decline if it had an industry mix that experienced the same productivity gain as national manufacturing had. The differential shift provides additional information on why these states enjoyed different manufacturing employment gains (losses) under the above mentioned conditions. The sun-belt states' industry mix performed better than the snow-belt states' industry mix. Thus, the sun-belt states had a better competitive position.

Regarding individual industry sectors and states, the textile, leather, tobacco and petroleum sectors have been declining in all states. The rubber and plastics industry was one of the driving sectors in all states with the exception of Massachusetts. Massachusetts performed poorly in this sector but gained manufacturing employment in the instruments industry. Michigan and Ohio experienced employment growth in the fabricated metal industry while other snow-belt States lost employment in this sector. The primary metal industry has been loosing employment in all snow-belt states and in California. California also has been loosing manufacturing employment in transportation equipment and miscellaneous industries. The food industry is another sector where all snow-belt states and Kentucky lost employment.

Focusing on the Haynes and Dinc (1997) analysis, a very different picture emerges (Table 3.7). The main difference between the two models lies on the allocation of causes of employment change. Rigby and Anderson attribute all employment change to labour productivity, while Haynes and Dinc attribute it to labour and other factor productivity. Therefore in the comparison of the two models, performance is based on these allocation effects and not on the actual (absolute) numbers, and in this comparison any employment change should be considered within this allocation framework.

Even though both models indicate that the states under investigation have similar performance patterns, employment change in the Haynes and Dinc model, resulting from the labour productivity and output change, is much smaller than estimated by Rigby and Anderson (10–15 percent of the Rigby and Anderson analysis). In other words, in the Haynes and Dinc analysis, the economic performance of these States has been largely driven by improvement of capital stocks, technology, infrastructure and other factors rather than from solely labour productivity and changes in demand.

Additionally, a closer analysis of Table 3.7 reveals how it only Tennessee in the sun-belt improved its labour productivity, even though it is still well below the national average. Tennessee is also the only state that outperformed its national counterparts in output growth. In contrast, all snow-belt states lost employment due to improvement in labour productivity (recall that in the Rigby and Anderson model, only New York and Massachusetts improved their labour productivity), and again this gain is also below the national average. For example, Pennsylvania, the best performer in labour productivity among snow-belt states, would have lost 217.1 million hours of employment instead of 99.9 million hours if it had improved or increased its labour productivity to the national level.

Table 3.6. Rigby-Anderson extension of the shift-share model: snow-belt sun-belt manufacturing employment shifts in the United States, 1960–1990, in million hours

State	NS[a]	NS[b]	Total NS	PS[a]	PS[b]	Total PS	DS[a]	DS[b]	Total DS	Total shift	Empl. change	Empl. change %	Other factor's contribution
Sunbelt													
Arizona	60	-77	-17	33	-34	-1	150	-16	134	117	117	169	105
California	1,162	-1,287	-125	736	-686	50	579	162	741	666	616	36	547
Florida	231	-278	-47	162	-137	25	343	28	371	349	324	117	293
Kentucky	189	-214	-25	79	-93	-14	141	23	164	125	139	54	114
Tennessee	362	-400	-38	141	-177	-36	289	-13	276	202	238	51	204
Texas	541	-630	-89	589	-599	-10	551	-9	541	442	453	64	386
Snowbelt													
Illinois	995	-1,012	-17	518	-365	154	-591	100	-491	-354	-508	-30	-439
Massachusetts	538	-535	3	269	-236	33	-292	-179	-471	-435	-468	-46	-391
Michigan	883	-907	-24	269	-226	43	-476	213	-263	-244	-288	-20	-231
New York	1,316	-1,268	48	540	-520	21	-1,156	-298	-1,454	-1,385	-1,406	-54	-1,271
Ohio	1,053	-1,091	-38	454	-294	160	-593	204	-388	-267	-427	-24	-329
Pennsylvania	1,138	-1,141	-3	351	-415	-64	-822	0	-822	-890	-825	-40	-662

NS = National shift; PS = Proportional shift; DS = Differential shift; Empl. = Employment; [a] denotes that productivity is constant; [b] denotes that output is constant

Source: Rigby and Anderson 1993

Table 3.7. Haynes-Dinc modification of the shift-share model: snow-belt—sun-belt manufacturing employment shifts in the United States 1960–1990, in million hours

State	NS^a	NS^b	Total NS	PS^a	PS^b	Total PS	DS^a	DS^b	Total DS	Total shift	Empl. change	Empl. change %
Sunbelt												
Arizona	5.69	-15.78	-10.10	-4.8	14.2	9.4	4.29	8.20	12.49	11.79	117	169
California	114.6	-254.2	-139.6	-1,01.3	231.2	129.9	9.4	69.6	79.0	69.3	616	36
Florida	22.1	-55.6	-33.6	-21.1	52.1	31.0	2.1	32.1	34.2	31.6	324	117
Kentucky	18.67	-42.20	-23.53	-1,7.12	38.54	21.42	10.72	16.03	26.74	24.64	139	54
Tennessee	35.47	-78.63	-43.16	-32.9	72.7	39.8	39.71	-2.60	37.11	33.76	238	51
Texas	52.4	-125.3	-72.9	-46.9	116.0	69.1	13.2	57.2	70.4	66.5	453	64
Snowbelt												
Illinois	100.6	-194.3	-93.7	-91.5	178.4	86.9	-43.0	-18.5	-61.5	-68.3	-508	-30
Massachusetts	54.3	-101.2	-46.9	-49.3	93.6	44.3	-21.5	-52.7	-74.2	-76.8	-468	-46
Michigan	88.0	-175.2	-87.3	-79.9	157.3	77.5	-25.8	-21.1	-47.0	-56.8	-288	-20
New York	134.2	-238.0	-103.8	-1,23.1	220.6	97.5	-106.4	-22.4	-128.8	-135.1	-1,406	-54
Ohio	105.7	-210.3	-104.7	-97.0	191.1	94.1	-29.2	-57.4	-86.6	-97.2	-427	-24
Pennsylvania	115.2	-217.1	-101.9	-1,05.4	200.1	94.7	-56.4	-99.9	-156.3	-163.5	-825	-40

NS = National shift; PS = Proportional shift; DS = Differential shift; $Empl.$ = Employment; ^{a} denotes that productivity is constant; ^{b} denotes that output
is constant.
Source: Haynes and Dinc 1997, p.215

In terms of the proportional shift both models indicate that all states have a declining industry mix relative to the national manufacturing sectors. However, while the Rigby and Anderson model emphasizes the employment lost in the respective industries from output growth due to productivity improvement and employment gain (both are below the national average), the Haynes and Dinc modification shows that the states have lost employment because of a decline in output, and all states gained employment as a result of below average (national) labor productivity performances in these industries.

Examination of the individual sectors and states reveals that the Rigby and Anderson model overestimates the performance of the sun-belt states. For example, in the Rigby and Anderson model, only Kentucky among the sun-belt states lost employment in the food and kindred product sector. In the Haynes and Dinc model, however, all sun-belt states lost employment in this sector. While textile, apparel and wood product sectors were gaining employment in the Rigby and Anderson model in the sun-belt, almost all states lost employment in these sectors in the Haynes and Dinc model. This result gives support to the conclusion that the United States (and in particular the sun-belt in the decade 1980 to 1990) was either shipping its textile and apparel industries to Third World countries or improved labour productivity in these sectors or both. On the other hand, the Rigby and Anderson model shows that Ohio gained employment in rubber and plastics, while with the Haynes and Dinc model, Ohio lost employment in the rubber and plastics sector.

3.4.5 Shift-Share Analysis of International Trade in the European Union

Dinc and Haynes (1998a) extended earlier work by Markusen et al. (1991) and Noponen et al. (1997) for the application of shift-share analysis to international trade. Here the international trade shift-share model is used to investigate the impact of exports and imports on manufacturing employment change in the countries of the European Union. This case study draws upon an application by Dinc and Haynes (1998b).

The traditional shift-share and its extensions that have been discussed so far ignore international and interregional trade. The international trade models of shift-share focus on assessing international economic forces and thus provide a way for integrating regional analysis into global dynamics.

Markusen et al. (1991) expanded the traditional shift-share model by subdividing the national share and the industry mix components of the analysis into four new components representing trade and productivity effects. The national growth component is decomposed into four parts: a national export; a national import; a national domestic market; and a productivity component. The national export component measures the growth effect if employment were to expand proportionally to national export sales. The national import component estimates the hypothetical national import effect and the national domestic component measures the residual effect of shifts in domestic demand. The fourth component is a measure of national labour productivity growth and shows whether the region's (country's) industrial structure has outpaced or lagged the nation in productivity growth. The industry mix component is decomposed into the same four parts which represent the concentration of a region's (country's) industries. Dinc and Haynes (1998b) extend this framework to capture international trade effects on employment. They assume that exported and domestically consumed goods are produced at the same places by the same workers. Thus they defined actual employment as $E = E_{export} + E_{domestic}$. Any increase in exports and domestic demand will consequently create an increase in employment. At the same time imports can cause decreases in employment because imports will reduce demand for domestically produced goods. They provide thus a model that estimates the effects of export, domestic and import driven employment changes across the various sub parts of the Markusen et al. (1991) and Noponen et al. (1997) extensions to the shift-share model.

In the Dinc and Haynes (1998b) application of the international trade shift-share model to the European Union data from the OECD Statistical Compendium CD (1997) was used. The investigation was divided into two periods: 1980-1986 before the Single European Act (SEA) and 1986-1995 after SEA. The reason for this separation of the analysis is that the SEA defined the point from which liberalized international trade within the EU occurred, although 1993 was the full integration year. Dividing the investigation into two time periods provides a better understanding of international trade's impact on employment change as well as the contribution of integration to trade.

From 1980 to 1995, only four countries increased their manufacturing employment, Denmark, Germany, Greece and Portugal. The largest was Germany with 541 thousand new manufacturing jobs with Portugal following with a 392 thousand gain. The largest looser during this period was the United Kingdom with a decline of 2.5 million manufacturing jobs. France, Italy, Spain, Sweden and Belgium were the other countries with large declines. Declining domestic demand was the force behind these changes although increasing imports played a role. With the exception of the United Kingdom, Sweden, Ireland and Denmark exports made a positive contribution to employment growth. Also, in the United Kingdom, Sweden, and Ireland imports declined thus contributing to employment growth. At the same time this reduced domestic demand and the associated employment loss outpaced the gain from import reduction.

Turning now to the components of the model, the EU share was positive in the export sub-component and negative in imports and domestic demand sub-components. It should be noted that a positive sign for imports, in fact, indicates a decline in employment, on the other hand a negative sign is an indication of declining imports resulting in employment gain. The industry mix component shows a varied pattern, in which only Spain had growing exports and domestic demand and declining imports in industry mix. This component suggests that ten countries out of 13 had an industry mix specialized in domestic demand related sectors.

The country share component is the most important component of shift-share models because it indicates the advantages or disadvantages of the region or country under investigation. Based on this component, six countries had national advantages in exports while seven countries had disadvantages. Among those that had export driven employment advantages Germany, Spain and Portugal had the largest gain because of increasing exports. At the same time the United Kingdom, Denmark and Sweden had disadvantages in export sectors and thus lost employment. On the import side, eight countries reduced their imports and gained employment in manufacturing. The largest reduction in imports occurred in the United Kingdom followed by Italy and Sweden. In this component, overall advantage or disadvantage was determined by the domestic demand sub-component, in which seven countries lost employment due to declining domestic demand. The largest loser was the United Kingdom with a total decline of 1.7 million in employment in manufacturing followed by Spain and Sweden. On the other hand, Germany and Portugal had the largest increases in domestic demand and gained employment, 1.9 million and 449 thousand respectively.

The examination of the findings for the different sub-periods shows different results. Between 1980 and 1986, five countries, Austria, Denmark, Greece, Netherlands and Portugal, increased their manufacturing employment. The largest employment increase in manufacturing was in Portugal, 347 thousand, followed by Greece (125 thousand) and Denmark (101 thousand). Also, the United Kingdom, Italy, France, Germany and Spain lost jobs, 1.65 million, 911, 873, 441 and 369 thousand, respectively. In this period, too, domestic demand was the driving factor in manufacturing employment change. In terms of exports, the United Kingdom, Italy and Germany were big losers. During the same period, four countries, Italy, Belgium, the United Kingdom and Ireland reduced their imports, and thus gained employment.

Following the introduction of the Single European Act of 1986, five countries—Germany, Italy, Denmark, the Netherlands and Portugal—showed a gain in employment in manufacturing. Austria, one of the winners of the earlier period, now lost employment. The biggest losers in the earlier period, the United Kingdom, France, Spain and Belgium, improved their performances in terms of manufacturing employment, although they had employment decline. In this post 1986 period, only in four countries—Germany, Italy, Portugal and Denmark—did domestic demand make a contribution to employment gain, while in nine countries it had a negative impact on employment, the largest being in the United Kingdom with over a one million decline. Again, the domestic demand was the major factor in employment change in this period. In terms of exports, four countries—Austria, Belgium, Sweden and the United Kingdom—had employment losses. The number of countries that gained employment due to import reduction increased to five— Austria, Greece, Ireland, Sweden and the United Kingdom—although their gain was small.

While individual countries and sectors were investigated in this study these parts of the work are not described here individually. However, it is important to note that the largest manufacturing countries are the United Kingdom, Germany and Italy, while the smallest ones are Ireland, Greece and Denmark. The results for the United Kingdom are described to illustrate the analysis results at the level of an individual country (region).

During the whole investigation period, the United Kingdom lost employment in all manufacturing sectors totalling over 2.5 million jobs. Its largest lost was in the fabricated metal products sectors (1.5 million jobs) followed by the textile, apparel and leather products, food, beverage and tobacco products and chemical sectors. In all manufacturing sectors, domestic demand was the driving factor of the decline and thus caused employment reduction. During this period, only three sectors, the wood products and furniture, paper and printing and the basic metal sectors, had employment gains due to increasing exports. At the same time, apparel to declining domestic demand in three sectors, food, beverage and tobacco products, textiles, apparel and leather products and fabricated metal products sectors, imports declined and the United Kingdom increased its employment.

The largest portion of decline in manufacturing employment in the United Kingdom took place between 1980—1986. In this period, it lost employment in all sectors and the order of magnitude of the losses was similar to the earlier period. Domestic demand and exports declined in all sectors during this period and thus caused a large amount of job loss. In only three sectors, the food, beverage and tobacco products, basic metal industries and fabricated metal products sectors, imports declined but the employment gain from it was relatively small.

After 1986, however, the United Kingdom began recovering, although its losses in terms of total manufacturing continued. In three sectors, the wood and furniture, paper and printing and basic metal industries sectors, the United Kingdom gained employment. In this period, with the exception of the chemical products and nonmetallic mineral products sectors, it increased its exports and gained a total of 80.7 thousand manufacturing jobs. However, at the same period domestic demand declined dramatically in all sectors but the paper and printing sector caused a loss of more than one million jobs. Over half of this loss occurred in the fabricated metal products sector, followed by the textile, apparel and leather sector. With the exception of two sectors, the paper and printing and basic metal industries sectors, imports of the United Kingdom declined in all sectors generating 35 thousand jobs. Comparison of these two sub-periods suggests that the United Kingdom benefited from integration by improving its performance in manufacturing, though its manufacturing employment decline continuous.

As noted above, the analyses of the other countries are not presented here with the assessment for the United Kingdom serving as an illustration of the application of the international trade shift-share model. In general, however, the findings of the total analysis showed that domestic demand was the driving force behind trade related employment change in manufacturing in the 13 EU member countries. Exports and imports also had a significant impact. Another pattern emerged from the analysis suggesting that larger manufacturers performed better in the second period, while smaller manufacturers' performances were better in the first study period. This implies that the economic integration has helped large, well-established manufacturers.

At the end of their study, Dinc and Haynes (1998b) indicate that several issues need further attention. For example, international trade was not separated into the within EU vs. outside the EU components. This refinement in the analysis would likely produce deeper understanding about the effect of integration on international trade. In this analysis the productivity impact on employment was not examined, although it was in the model as described above.

3.5 Using the Regional Shift Component as a Dependent Variable to Model Determinants of Spatial Variation in Regional Endogenous Growth: A Case Study of Queensland, Australia

Much focus in regional economic development theory in the last couple of decades has been on endogenous factors in regional growth, as discussed in Sect. 1.4.8 of Chap. 1. However, there is not a universally available variable for endogenous growth which readily enables it to be measured for a region. The regional or differential shift component derived from a shift share analysis does represent a reasonable proxy that might prove useful as a dependent variable in a model of endogenous growth, and it is a measure that is readily calculated using data on the change over time in regional structure that typically is universally available across regions in most nations, such as at the county level in the United States or for Statistical Local Areas (LGAs) in Australia. Stimson et al. (2004) have done that in a case study of Queensland, a large and geographically diverse state in Australia.

3.5.1 The Model

Stimson et al. (2004) use secondary data available from the census to analyze and model endogenous growth across Queensland's regions, as represented by the non-metropolitan LGAs, over the decade 1991 to 2001. Both static point-in-time data derived from the 1991 and 2001 censuses and dynamic change-over-time data for that decade were used to model those factors hypothesized as possible determinants of endogenous growth that may account for spatial variations in endogenous growth in employment across LGAs.

The study used the Haynes and Dinc (1997) method to derive the regional shift component (REG_SHIF) in a shift-share analysis of total employment change in LGAs between the 1991 and 2001 censuses. Because of the large variation in the size of LGAs across rural and regional Queensland – from the most populous LGA, Gold Coast City, with a population of over 418,000 in 2001, to several LGAs in the inland western parts of the state with populations of under 1,000 in 2001 – it was necessary to standardize the REG_SHIF variable to account for the size of the LGA labour force in 1991. This was used as the dependent variable in a model that was developed to identify the determinants of spatial variations, across the space economy of regional Queensland in levels of regional endogenous growth. A set of 27 variables was created to provide the independent variables in the model serving potential factors that might influence the regional endogenous employment performance LGAs. These are discussed below.

A series of variables were compiled to enable testing for the impact on regional endogenous growth of both industrial diversity and size. This was addressed first by compiling an industry specialization index for 1991 (SPEC_91) and for 2001, and then calculating a measure of change in the specialization index over the decade (SPEC_CH). The 2 digit 17 industry sector employment data for LGAs were used for this purpose. Second, a structural change index was computed for 1991 to 2001 (SCI_91-01)), and a measure of change in the structural change index from 1991-1996 to 1996-2001 was compiled (SCI_CH). The log. of the population of a LGA in 1991 (L_POP_91) was used as a measure of size of a region at the beginning of the decade, and the percentage point growth (or decline) in population size over the decade 1991-2001 (POP_CH) was used as a dynamic measure of size. In order to address the effect of proximity to the metro-region, a dummy variable (D_METRO) was used to indicate whether an LGA was adjacent to the Brisbane Statistical Division. To further investigate the nature of the industry structure of LGAs and the effect of industry specialization on regional endogenous growth, for each LGA a location quotient was calculated for three key industry sectors in 1991; namely, for employment in manufacturing (LQ_MAN_91), for property and business services (LQPBS_91), and for personal and other services (LQPER_91). In addition, variables measuring the change in the location quotient of employment in these three industry sectors over the decade 1991-2001 were derived (LQMAN_CH, LQPBS_CH, and LQPER_CH).

To investigate the effect on endogenous growth of labour force participation a simple measure of the unemployment rate in 1991 (UNEMP_91) was used, along with a measure of change in the unemployment rate over the decade 1991-2001 (UNEMP_CH).

It is often proposed that regional growth is enhanced by the existence of skilled workers, the availability of employment opportunities, opportunities for a wide range of skills, and by the existence of higher income jobs. A number of variables were incorporated in the model to assess those effects. First, a variable was derived to measure income levels by taking the log of average annual income for a LGA at 2001 (L_INC_01). Then a series of measures of level of human capital were derived, namely the proportion of an LGA's population with a bachelor or higher degree qualification in 1991 (UNIQUALS_91) and the proportion with a technical qualification (TECHQUALS_91). To measure the effect of shifts in those levels of human capital, two variables were created on the change from 1991 to 2001 in the incidence of those qualifications (UNIQUALS_CH, TECHQUALS_CH).

It has been argued by Reich (1991) and others that the evolution of the knowledge-based economy and of information-intensive activities has led to a *restructuring of occupations* vis a vis skills and functions. Thus, we decided to re-group the census data on the occupational structure of LGAs into three group categories that represent employment in 1991 in Reich's symbolic analyst occupations, in-person service-workers, and routine production workers (SYMBA_91, INPE_91, and ROUTW_91). A measure of change over the decade 1991-2001 in employment in those categories was also calculated (SYMBA_CH, INPERS_CH, and ROUTW_CH).

Some of the research in Australia inquiring into regional performance has investigated the spatial patterns of variations in the context of geographic variables such as remoteness, coastal and inland environments and locations. To incorporate the potential effect of a coastal location on regional endogenous growth a dummy variable was included to indicate whether or not a LGA is located adjacent the coast of Queensland (D_COAST). A further dummy variable was included to test for the effect of spatial proximity to the state capital city Brisbane, which is the third largest metropolitan area in Australia (D_METRO).

Having determined the variables to be included in the general model, a conventional Ordinary Least Squares (OLS) technique was employed to estimate the model of regional endogenous growth across Queensland's rural and regional LGA. Stimson et al. (2004) then employed a stepwise approach to determine a specific model. Each step involves withdrawing one independent variable from the model. That variable deleted had the highest probability that its absolute t-value was greater than 0.05. The model was then regressed and new estimates of the model were obtained. This process occurred until we identified the variable with the highest probability that its absolute t-value was less than 0.05.

One key feature for a reasonable OLS regression is non-constant variance in the residuals. A test was run to evaluate this on the estimated variable. Stimson et al. (2004) also ran tests to determine the existence or not of spatial autocorrelation, but the details of those tests are not included in the summary of the results of the case study provided below.

3.5.2 Patterns of Regional Performance on the Regional Endogenous Growth Dependent Variable

The spatial patterns of variation across Queensland's rural and regional LGAs in their performance on the REG_SHIF dependent variable used in the model as the proxy measure of regional endogenous growth are presented in Fig. 3.1. Rather than using a choropleth map, symbols are used (placed at the centroid of LGAs) to represent the magnitude of an LGA's positive or negative score on the REG_SHIF dependent variable. Those symbols are graduated in scale in order to represent the size category for the population of a LGA. A circle symbol represents a positive score on the REG_SHIF variable, indicating that an LGA has experienced employment growth over the decade on the regional shift component derived from the shift-share analysis, standardized by size of the labour force of the LGA in 1991. A triangle symbol represents a negative score on the REG_SHIF variable indicating loss of jobs over the decade 1991-2001 due to endogenous processes/factors. The map in Fig. 3.1 also uses dark and light shading for the circles (positive) and the triangles (negative) to indicate the magnitude of the endogenous growth effect. Some of the larger regional cities and towns are identified.

Overall, 39 of Queensland's rural and regional LGAs are shown to have experienced endogenous growth in employment over the decade 1991-2001, while a total of 79 LGAs display a decline in employment due to endogenous processes/factors. Figure 3.1 clearly shows that endogenous employment growth is most evident in some of the larger coastal LGAs in the south east corner of the state adjacent to and inland from the Brisbane metro-area, and as well across a few of the inland small population rural LGAs and some of the inland regional centres. Fig. 3.1 also identifies LGAs with negative scores indicating employment decline due to endogenous processes/factors. This is apparent widely across much of western rural and regional Queensland, and as well in many of the LGAs (particularly those with smaller populations) along the coast and in near coastal locations.

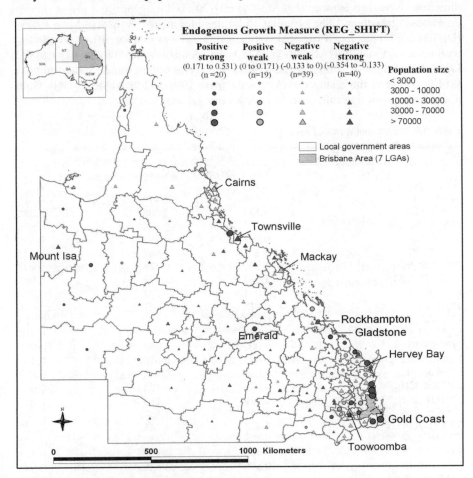

Fig. 3.1. Spatial pattern of LGA performance on the regional endogenous growth variable
Source: Stimson et al. 2004

3.5.3 The General Model

Table 3.8 gives the OLS regression results for the general model. Included at the bottom of the table are diagnostics for the model. As demonstrated by the high adjusted R-squared value (0.88), this model explains much of the variance in the dependent variable (REG_SHIF) measuring the regional shift component of employment change over the decade 1991-2001. Furthermore there are ample degrees of freedom. Those independent variables which exercise a relatively high influence on the dependent variable are identified by the bold type in the table. The variable with the greatest influence on the regional shift variable was found to be population change (POP_CH), which has very significant effect in a positive direction. Average income (L_INC_01) also has a strong positive effect on endogenous growth. The change in unemployment from 1991 to 2001 (UNEMP_CH) has had a strong negative effect. The proportion of the population with university (UNIQUALS_91) and technical qualifications (TECHQUALS_91) in 1991 both have a strong negative influence. The specialization index in 1991 (SPEC_91), and the change in that index from 1991 to 2001 (SPEC_CH), both have a strong positive influence on endogenous growth.

Table 3.8. OLS general model results

	Estimate		t value		Pr(>|t|) value
	Sign	Value	Sign	Value	
(Intercept)	-	2.180	-	1.512	0.134
SPEC_91	+	**0.450**	+	**2.473**	**0.015**
SPEC_CH	+	**0.531**	+	**2.008**	**0.048**
SCI_91-01	-	0.125	-	0.634	0.528
SCI_CH	+	0.098	+	0.593	0.555
L_INC_01	+	**0.535**	+	**3.061**	**0.003**
UNEMP_91	-	0.230	-	0.688	0.493
UNEMP_CH	-	**1.231**	-	**2.775**	**0.007**
L_POP_91	+	0.031	+	1.271	0.207
POP_CH	+	**0.818**	+	**17.187**	**<2e-16**
LQMAN_91	+	0.018	+	0.816	0.416
LQMAN_CH	+	0.041	+	0.963	0.338
LQPBS_91	+	0.000	+	0.002	0.998
LQPBS_CH	+	0.031	+	0.675	0.502
LQPER_91	+	0.006	+	0.148	0.883
LQPER_CH	+	0.016	+	0.433	0.666
UNIQUALS_91	-	**2.270**	-	**2.114**	**0.037**
UNIQUALS_CH	+	1.459	+	1.534	0.129
TECHQUALS_91	-	**1.065**	-	**2.246**	**0.027**
TECHQUALS_CH	+	0.698	+	1.359	0.178
ROUTW_91	-	0.400	-	0.351	0.727
ROUTW_CH	+	0.596	+	0.525	0.601
INPERS_91	-	0.372	-	0.310	0.757
INPERS_CH	+	1.095	+	1.022	0.310
SYMBA_91	-	0.573	-	0.480	0.632
SYMBA_CH	-	0.017	-	0.015	0.988
D_COAST	-	0.031	-	1.389	0.168
D_METRO	-	0.027	-	0.868	0.388

Residual standard error	0.06322	90 degrees of freedom
Multiple R-Squared	0.909	
Adjusted R-squared	0.8817	
F-statistic	33.29	27 and 90 degrees of free- p-value: < 2.2e-16

Source: Stimson et al. 2004

3.5.4 The Specific Model

The results of the stepwise process are given in Table 3.9. This estimates the specific model. The explanatory power of this specific model is very similar to the general model (adjusted R-squared value 0.89). The table shows there are 12 significant independent explanatory variables that affected regional endogenous growth (REG_SHIF). The new additional variables in Table 3.9 - as compared to the general model in Table 3.8 - are: population size in 1991 (L_POP_91), the location quotient for employment in personal and other services in 1991 (LQPERS_91); the change in the proportion of the population with university qualifications from 1991 to 2001 (UNIQUALS_CH); and the change from 1991 to 2001 in the proportion of the labour force who were routine workers and in-personal services workers (ROUTW_CH). Most of these explanatory variables have a significant positive relationship with the dependent variable.

Table 3.9. OLS specific model results

	Estimate		t value		Pr(>|t|) value
	Sign	Value	Sign	Value	
(Intercept)	-	3.432	-	9.398	1.37e-15
SPEC_91	+	0.272	+	2.412	0.018
SPEC_CH	+	0.456	+	2.126	0.036
L_INC_01	+	0.737	+	8.859	2.21e-14
UNEMP_CH	-	1.101	-	3.882	0.000
L_POP_91	+	0.040	+	2.662	0.009
POP_CH	+	0.806	+	22.199	<2e-16
LQPER_CH	+	0.013	+	2.260	0.026
UNIQUALS_91	-	2.587	-	3.190	0.002
UNIQUALS_CH	+	1.886	+	2.390	0.019
TECHQUALS_91	-	1.635	-	4.642	1.00e-05
ROUTW_CH	+	0.797	+	3.661	0.000
INPERS_CH	+	1.178	+	4.077	8.90e-05
Residual standard error	0.06178	105 degrees of freedom			
Multiple R-Squared	0.8986				
Adjusted R-squared	0.887				
F-statistic	77.52	12 and 105 degrees of freedom p-value: < 2.2e-16			

Source: Stimson et al. 2004

However, the change in unemployment from 1991 to 2001 (UNEMP_CH), and the proportion of the population with a university degree in 1991 (UNIQUALS_91) and with technical qualifications (TECHQUALS_91) have a significant negative relationship.

3.5.5 Patterns of Residuals

A feature for a reasonable OLS regression is non-constant variance in the residuals, and the Stimson et al. analysis indicates that this is not a problem with the estimated model in the specific model discussed above and it was demonstrated that there is no visual relationship between the residuals and the fitted regression line.

Figure 3.2 plots the deviation of the residuals for LGAs across regional Queensland from the line of best fit derived from the specific model. There are two outlier LGAs: Whitsunday (negative) located on the central coast barrier reef; and Waggamba located in the inland southern of the state. The positions of the major regional centres are indicated on this plot. It is evident that the large majority of residual scores for the LGAs fall within the +0.05 to -0.05 range.

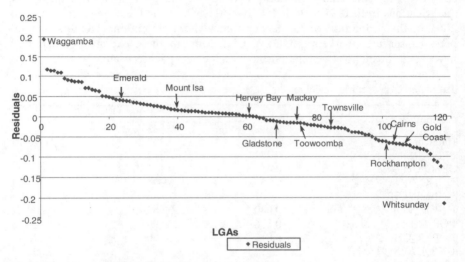

Fig. 3.2. Order of LGA residuals for the specific model
Source: Stimson et al. 2004

From the data underlying the plot in Fig. 3.2, Stimson et al. (2004) then mapped the spatial pattern of the residuals as shown in Fig. 3.3. The map shows that 79 LGAs fall within the +0.05 to -0.05 range of residuals. There are 45 LGAs for which the model estimates are positive or above the regression line of best fit for the specific model. These places are largely found to be located around the Brisbane metro-area in the south east of the state, while some such LGAs are also small places located across the inland areas of the state. The 34 LGAs for which the model estimates are negative or below the regression line within this range are typically larger coastal centres or small rural places on the coast, and as well some of them are found in the inland south-east parts of the State.

Figure 3.3 also identifies those LGAs for which the model estimates are well above or well below the regression line. There are 17 LGAs with residuals greater than +0.05, and these are located mainly around Cairns in far north Queensland and across the rural areas of the south east part of the state inland from the coast. There are 20 LGAs with negative residuals greater than -0.05 and they include the coastal regional centres of Cairns and Rockhampton, the Gold Coast, some coastal rural places, and some isolated largely indigenous places around the Gulf of Carpentaria and the Torres Strait islands.

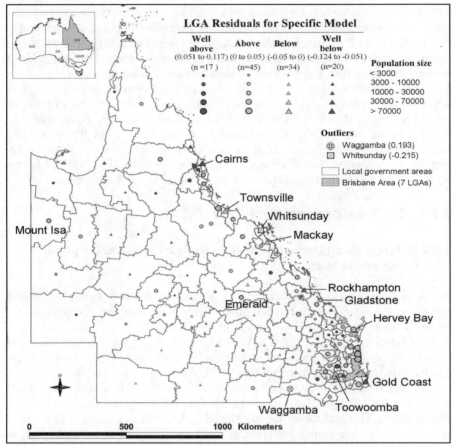

Fig. 3.3. Spatial pattern of LGA residuals for the specific model
Source: Stimson et al. 2004

3.5.6 Implications of the Model Results

The modelling conducted by Stimson et al. (2004) showed that there are clear spatial patterns of variation in LGA performance on the dependent variable (REG_SHIF) measuring regional endogenous employment growth across Queensland's regional LGAs over the decade 1991-2002. The modelling conducted to identify the key factors that might explain the endogenous growth performance of LGAs using the OLS regression and stepwise regression techniques (the general model and the specific model) indicated that a reasonably small number of independent variables are important, with the model resulting in high R squared values that explain over 88 percent of the variance in REG_SHIF. The results tend to lend support to importance of variables relating to industrial structure, population size, and human capital as explanatory factors in explaining differential endogenous growth performance. Population growth (POP_CH) emerges as a particularly strong positive factor, as is the level of income at the beginning of the decade period analysed (L_INC_01). Population size at the beginning of the period is also important. But in addition, change in levels of unemployment (UNEMP_CH) as well as change in the concentration of occupational employment in Reich's routine production workers (ROUTW_CH) and in-person service workers (INPERS_CH), along with change in the concentration of employment in personal service industries (LQPER_CH), are also found to be significant determinants of regional endogenous growth across Queensland's regional LGAs. The modelling also identifies the significance of industrial specialization (SPEC_91) at the beginning of the period and of change in industrial specialization (SPEC-CH) towards greater diversification, as processes that impact endogenous growth.

3.6 Data Envelopment Analysis (DEA)

3.6.1 Differentiating Between Relative Productivity and Regional Competitiveness

For policy purposes it is important to be able to differentiate between industry sectors on the basis of their relative productivity in comparison to other regional sectors so as to ascertain regional competitiveness.

Data Envelopment Analysis (DEA) is a mathematical programming technique known also as the non-parametric frontier efficiency approach. It measures the relative efficiency of a decision making unit (DMU)—which, in regional analysis, are the regional industrial sectors—relative to other DMUs with the simple restriction that all DMUs lie on or below the efficiency frontier (Seiford and Thrall, 1990). This technique was introduced by Charnes et al. (1978) and now is applied widely in regional performance analysis (Charnes et al.1989; Ali and Lerm 1990; Ali et al. 1993; Chang et al. 1995; Sarafoglou and Haynes 1990; Haynes et al.1990a; Haynes et al. 1990b; Dinc and Stough 1997).

The DEA technique generalizes the Farrel (1957) single output/input technical efficiency measure to multiple output/multiple input cases by constructing a single virtual output to a single virtual input relative efficiency measure. DEA optimizes on each individual observation, with an objective of calculating a discrete piecewise frontier determined by a set of efficient DMUs. DEA does not require assumptions about functional form, unlike parametric approaches that assume an explicit functional form for the technology as well as often for the distribution of efficiency terms. In reality DEA estimates a best practice frontier for a DMU.

Below two commonly used DEA models are discussed:

(a) the CCR (Charnes et al.) ratio model (Charnes et al. 1978);
(b) the BCC (Banker et al.) model (Banker et al. 1984).

Assume that there are n DMUs to be evaluated. Each of them consumes different amounts of i inputs and produces r different outputs, that is, DMUj consumes χ_{ij} amounts of input to produce γ_{rj} amounts of output. It is assumed that these inputs, χ_{ij} and outputs, γ_{rj}, are non-negative, and each DMU has at least one positive input and one positive output value (Seiford and Thrall 1990).

Common notation used in what follows can be summarized as:

j	:	DMUs, j = 1 ....n
r	:	outputs, r = 1.........t
i	:	inputs, i = 1 ....m
γ_{rj}	:	the value of the rth output of the jth DMU
χ_{ij}	:	the value of the ith input for the jth DMU
ε	:	a small positive number
S_i, σ_r	:	slacks corresponding to input and output respectively [30]
λ_j	:	weight of DMU in the facet for the evaluated DMU [30]
μ_r, v_i	:	virtual multipliers for output and input respectively [3 ε]

The characteristics of the CCR ratio model are the reduction of the multiple-output/multiple-input situation, for each DMU, to a single virtual output and a single virtual input ratio. For a given DMU this ratio provides a measure of efficiency which is a function of multipliers (Charnes et al. 1978). The mathematical formulation of the model is:

$$\text{Max } h_k = \sum_r \mu_r \gamma_{rk}$$
$$\text{s.t} \sum_i v_i \chi_{ik} = 1$$
$$\sum_r \mu_r \gamma_{rj} - \sum_i v_i \chi_{ij} \le 0 \qquad (3.20)$$
$$\mu_r, v_i \le 0$$

This is the output oriented formulation of the CCR model. The objective is to find the largest sum of weighted outputs of DMU_k while keeping the sum of its weighted inputs at the unit value, thereby forcing the ratio of the weighted output to the weighted input for any DMU to be less than one. The dual formulation of the model is:

$$\min h_k = \Theta_k - \varepsilon \left(\sum_r \sigma_r + \sum_i s_i \right)$$
$$\text{s. t.} \sum_j \gamma_{rj} \lambda_j - \sigma_r = \gamma_{rk}$$
$$\sum_j \chi_{ij} \lambda_j - \Theta_k \chi_{ik} + \varsigma_i = 0 \qquad (3.21)$$
$$\lambda_j, \sigma_r, \varsigma_i \ge 0$$

A similar formulation of the input oriented CCR model can be written when the dual is:

$$\max h_k = \beta_k - \varepsilon \left(\sum_r \sigma_r + \sum_i s_i \right)$$
$$\text{s. t.} \sum_j \chi_{ij} \lambda_j - \varsigma_i = \chi_{ik} \tag{3.22}$$
$$\sum_j \gamma_{ij} \lambda_j - \beta_k \gamma_{rk} - \sigma_r = 0 \text{ and } \lambda_j, \sigma_r, \varsigma_i \geq 0$$

Banker et al. (1984) take into account the effect of returns to scale within the group of DMUs to be analyzed. The purpose here is to point out the most efficient scale or size for each DMU and at the same time to identify its technical efficiency. To do so, the BCC model introduces another restriction, convexity, to the envelopment requirements. This model requires that the reference point on the production function for DMU_k will be a convex combination of the observed efficient DMUs. The dual formulation of the model is written as:

$$\min h_k = \Theta_k - \varepsilon \left(\sum_r \sigma_r + \sum_i s_i \right)$$
$$\text{s. t.} \sum_j \gamma_{rj} \lambda_j - \sigma_r = \gamma_{rk}$$
$$\sum_j \chi_{ij} \lambda_j - \Theta_k \chi_{ik} + \varsigma_i = 0 \tag{3.23}$$
$$\sum_j \lambda_j = 1$$
$$\sigma_r, \varsigma_i \geq 0$$

In all the above DEA models, a DMU is efficient if and only if $\theta^* = 1$, and all slacks are zero.

Applying the DEA model involves three-stages (Golany and Roll 1989):

(a) Definition and selection of DMUs to be analyzed, with all units performing similar tasks with similar objectives under the same set of market conditions, and should use the same kinds of inputs to produce the same kind of outputs.
(b) Determination of input and output variables to be used in assessing the relative efficiency of selected DMUs.
(c) Application of one of the DEA models and analysis of results.

3.6.2 Intertemporal Examination of the Relative Efficiencies of DMUs

The DEA models discussed and referenced above are static in nature. Neither the inputs nor the outputs involve time in any essential manner, though window analysis developed by Charnes et al. (1985) offers an approach for the intertemporal examination of the relative efficiencies of DMUs.

Assume we have data about a given DMU industry and all the inputs it uses and outputs it produces over a long period of time. Since this DMU has been using similar inputs and producing similar outputs from the beginning under the same market conditions, let each specified time period be a DMU, DMU_t. Each year is treated as a separate DMU, and the DEA optimizes on each observation (year) to obtain a best fit to the pertinent year for evaluating its performance. The result will be the efficiency of a given DMU in that year, relative to itself over time. As emphasized by Dinc and Haynes (1999), "it is assumed that technological change is embodied in the input side of the DMU, that is, new capital investments or expenditure on equipment and training". This application of DEA has important policy implications when applied, for example, to evaluate the management of any DMU or performance of an industry sector over time. Also, the DEA model provides information about the objective values of inputs and outputs, and it is possible to use this information for limited projection purposes because those objective (efficient) values of inputs and outputs could produce future efficient forecasts.

A more comprehensive approach for investigating the intertemporal aspects of DEA has been explored but on a limited basis (Sengupta 1995; Fare and Grosskopf 1996). In a case study on Fairfax County, Virginia, in the United States, below, an extended approach to intertemporal analysis of local government efficiency is examined in an effort to illustrate the use of DEA in an economic development related context.

The utilization of window analysis offers one way to examine the change in efficiency of a DMU over time. This method makes it possible to learn about the efficiency of a DMU over time relative to similar DMUs. However, knowing efficiency of a DMU relative to similar DMUs may not be sufficient to develop appropriate policies or adopt best practices. Every DMU has its own context (endowments, resources—both physical and human), and in most if not all cases, it is impossible to obtain information from efficient DMUs or to study them in any depth, that is, competing DMUs may not want to share the underlying factors responsible for their achieved efficiency. Additionally, some variables used in the analysis may need to be omitted because of measurement problems or excessive data collection costs. In cases such as these it may provide better policy relevant insight to evaluate a given DMU against itself.

The underlying conditions for the application of a DEA model are that all DMUs under consideration should perform similar tasks with similar objectives under the same set of market conditions and should use the same kind of inputs to produce the same kind of output. Evaluation of any DMU relative to its past performance is therefore possible as it satisfies these conditions (applications of the analysis of the efficiency of a single DMU over time appear in Cooper et. al.

1995; Button and Costa 1997). With such applications each year is treated as a separate DMU, with DEA optimizing on each observation (year) to obtain a best fit to the pertinent years for evaluating its performance (Cooper et. al,. 1995). The result of such analyses will be the efficiency of a DMU each year relative to itself over time. In short, the efficiency of the DMU is benchmarked against itself over time. In such applications it is assumed that technological change is embedded in the DMU's inputs; that is, new capital investments or expenditure on equipment and training.

As demonstrated in the following application for Fairfax County in the Washington metropolitan region of the United States, intertemporal DEA provides important policy implications. For example, it is possible to evaluate the management of a DMU or performance of an input or output sector over time. Further, such applications provide information about the objective values of inputs and outputs making it possible to utilize this information for limited projection purposes to provide potentially efficient forecasts over limited future time periods. This gives administrative units and policy makers the opportunity to estimate future inputs and outputs needed to achieve efficiency. Finally, the intertemporal approach can be combined with any of the various models outlined above–for example, CCR or BBC. The BBC approach is utilized below.

3.6.3 An Example: Using DEA to Analyze Government Local Efficiency in Fairfax County, Virginia

To illustrate the proposed model we will apply it to Fairfax County, Virginia in the United States. There are two reasons for this selection:

(a) First, in the decade to the early 1990s, Fairfax County was one of the top local jurisdictions in the United States in terms of per household and per capita income. This is important because one would like to know how efficiently one of the wealthiest counties of the U.S. is performing.

(b) The second relates to data availability. The data used in this analysis come from the following sources: employment data are from the United States Bureau of Economic Analysis; government expenditure and revenue data are from the Comparative Report of Local Government Revenues and Expenditures, Commonwealth of Virginia; and GRP and investment data are from the Centre for Regional Analysis, School of Public Policy at George Mason University.

The analysis covers a 23 year time span, from 1970 to 1992. This time period was selected because of data availability and because this time span includes several business cycles. In evaluating overall efficiency of Fairfax County, each year is assumed to be a separate DMU, and the following measures have been used as the inputs and the outputs of this economy.

Two output and three input measures are specified in the model (see Table 3.10). The output measures are:

(a) Gross Regional Product (GRP): this is a measure of the revenues of almost all economic activities in the region and therefore is the best single measure of the aggregate welfare associated with a regional economy.

(b) Local government revenues: this measure represents local government's revenues generated by its own resources (taxes, fees and charges), and can be seen as a proxy for the outcome of government expenditure and general economic transactions that take place in the region.

Table 3.10. The input and output measures in the Fairfax County economy, 1970–1992

Year	GRP change	Labour cost change	Government Expenditures	Annual Investment	Government Rev. change
1970	943.936	82.313	431.798	872.291	40.316
1971	560.316	268.805	459.501	984.938	23.631
1972	572.793	455.911	516.758	1,239.203	48.841
1973	697.380	590.604	535.192	1,467.233	15.724
1974	55.230	20.496	552.102	1,287.036	14.425
1975	322.190	133.873	612.640	1,116.684	51.638
1976	641.190	380.487	658.106	1,253.310	38.783
1977	568.680	256.749	656.085	1,459.274	−1.724
1978	1,051.350	429.578	701.660	1,637.481	38.875
1979	728.240	419.180	706.315	1,783.291	3.970
1980	876.930	290.287	727.126	1,687.256	17.752
1981	205.030	284.719	768.639	1,685.529	354.410
1982	401.750	319.693	800.115	1,560.148	34.956
1983	681.670	700.786	831.087	1,863.321	49.361
1984	1,947.340	915.086	870.933	2,468.252	40.873
1985	1,324.210	964.816	908.319	2,959.826	59.726
1986	1,602.720	953.884	975.077	3,415.203	55.183
1987	1,816.960	1,059.585	1,033.302	3,720.349	55.345
1988	2,024.820	1,211.586	1,138.462	4,143.699	55.577
1989	843.280	766.243	1,251.086	4,277.670	157.175
1990	144.130	301.246	1,318.916	3,815.891	73.703
1991	523.620	−162.457	1,442.786	3,157.676	78.776
1992	679.470	451.591	1,376.213	3,234.637	−32.332

Source: Center for Regional Analysis, The School of Public Policy, George Mason University; unpublished data and analysis results

The three input measures are specified as:

(a) Labour: either cost of labour or number of workers (or hours worked) can be used. In this analysis the former is used and represents the payments to employees (self-employed persons are excluded).

(b) New investments: represents total annual private investments made during the year and includes residential and non-residential structure and equipment investments in the county.

(c) Government expenditures: represents total expenditures made by the local government annually including infrastructure, education, operation (police, fire protection, health and welfare) expenditures for the county; that is, quality of public school system, safety or utilization of public money. All above variables are in millions of constant 1987 dollars.

Some may argue that any involvement of investments in the process may cause lagged output because the results of some investments, especially lumpy ones, can often only be seen after a period of time. To investigate this situation, a one-year lag to investment and government expenditures was applied. The results of the lagged analysis are slightly different form those of the original analysis.

Because of the lack of information about capital stock and infrastructure, the value of annual absolute change of the above variables (note that investment and government expenditures, themselves are specified as change variables) are used to analyze the efficiency of Fairfax County. The change in the output is attributed to the change in inputs that is, additional employment, investment, etc. The underlying assumption here is that each additional unit of the inputs will create an increase in the outputs or vice versa. This can be expressed as:

$$\Delta Y = f\ (\Delta L,\ \Delta K,\ \Delta M,\ \Delta G) \tag{3.24}$$

However, this process–in most cases–involves negative values because of decreases in inputs or outputs as in the case of 1977, 1991 and 1992. This outcome violates the non-negativity constraint of DEA models. In such cases, this problem may be solved by a simple arithmetic transformation, that is, by adding the same amount to both sides of the equation for a given DMU.

In the projection application of the new DEA model we employed the United States economy data produced by Jorgenson (1995), which covers a 39-year period between 1948 and 1986. This provides a larger time series, therefore improving the ability to forecast with this model. In this model the inputs are capital and labour and the output is the GNP. All values are in 1982 constant dollars. This process involves a three-stage application. In the first stage, the new DEA model is applied to the data set. In the second stage, by using the objective values of the inputs and outputs obtained in the first stage the inputs and outputs for the next several years are forecasted. The third stage employs the forecasted values of the inputs and outputs for analyzing the relative efficiency during the forecast period.

An input oriented BCC (variable returns to scale) model is employed in the analysis because returns to scale may be increasing or decreasing over time. The input oriented model provides better information about the input side over which decision-makers have much better control. The software package used for the analysis is the Integrated Data Envelopment Analysis System (IDEAS) software.

Table 3.11 shows the efficiency and inefficiency scores of Fairfax County. In this table IOTA represents the input efficiency, and THETA represents the output efficiency. Inefficiency scores measure the difference between the objective values and the observed values of a given variable (input or output). A close look at this table reveals that Fairfax County was 100 percent DEA efficient in seven of

the twenty-three years. In the remaining years it performed poorly in terms of relative efficiency. The comparison of IOTA and THETA indicates that output efficiency scores were relatively higher than input scores. That means in most years
Fairfax County has obtained the objective output but it has used its resources in an
inefficient way–that is, it employed more workers or invested more than needed.

Table 3.11. Efficiency and inefficiency scores of Fairfax County, 1970–1992

Year	Efficiency		Inefficiency				
	IOTA	THETA	GRP change	Gov. rev. change	Labour change	Gov. expenditures	New investment
1970	1.000	1.000	0.000	0.000	0.000	0.000	0.000
1971	0.247	0.939	383.620	16.690	–186.500	–27.700	–112.650
1972	0.503	0.951	363.810	0.000	–323.740	–25.230	–118.630
1973	0.184	0.807	246.560	24.600	–508.290	–103.390	–594.940
1974	1.000	1.000	0.000	0.000	0.000	0.000	0.000
1975	1.000	1.000	0.000	0.000	0.000	0.000	0.000
1976	0.352	0.696	302.750	1.540	–298.180	–226.310	–381.020
1977	0.207	0.655	375.260	38.600	–174.440	–227.310	–586.980
1978	0.560	0.682	0.000	1.500	–258.120	–222.850	–594.350
1979	0.198	0.611	215.700	36.350	–336.870	–274.510	–911.000
1980	0.144	0.594	67.010	22.570	–207.980	–295.330	–814.970
1981	–0.502	0.562	738.910	4.910	–202.410	–336.840	–813.240
1982	0.049	0.559	542.190	5.360	–237.380	–368.320	–687.860
1983	0.576	0.606	204.580	0.000	–566.770	–327.640	–734.570
1984	1.000	1.000	0.000	0.000	0.000	0.000	0.000
1985	0.685	0.815	0.000	0.000	–440.860	–168.310	–897.120
1986	0.766	0.848	0.000	0.000	–229.630	–147.910	–1,052.920
1987	0.829	0.954	0.000	0.000	–152.890	–108.020	–980.520
1988	1.000	1.000	0.000	0.000	0.000	0.000	0.000
1989	1.000	1.000	0.000	0.000	0.000	0.000	0.000
1990	–0.398	0.729	746.170	0.000	–81.500	–356.830	–1,505.460
1991	1.000	1.000	0.000	0.000	0.000	0.000	0.000
1992	0.081	0.299	264.470	7.990	–369.280	–1009.070	–2,362.350

IOTA= input efficiency; *THETA*= output efficiency; *GRP*= Gross Regional Product;
Gov.rev.= government revenues. Source: Center for Regional Analysis, School of Public
Policy, George Mason University; unpublished data and analysis results

Inefficiency scores indicate that in sixteen years during the investigation period
Fairfax County had excess labour, government expenditure and investment. In
three years –1981, 1982 and 1990– input efficiency were negative which reflects
anomalous conditions or inconsistencies (Ali and Lerme 1990). This may be
explained by the recession in these years. Further, Fairfax County is highly
dependent on federal government expenditures and contracts. Any reduction in
these expenditures could also affect efficiency. This situation is not reflected in
the model.

The sensitivity of each measure is ascertained by performing additional DEAs.
In each of these analyses one variable from the original analysis is removed. If the
efficiency values of the new analysis are larger than the value of the original

analysis this variable signifies a potential weakness for the region or the DMU. The converse implies that this variable is a competitive strength (Ali and Lerme 1990).

In the Fairfax model, the sensitivity analysis using lagged inputs (see Table 3.12) does not reflect any substantial weakness or strength of the above measures in the economy during the study period.

Table 3.12. Fairfax County efficiency and inefficiency scores (lagged inputs)

Year	Efficiency				Inefficiency		
	IOTA	THETA	GRP change	Gov. rev. change	Labour change	Govern expenditures	New investment
1970	1.000	1.000	0.000	0.000	0.000	0.000	0.000
1971	0.252	0.951	383.620	16.690	−186.500	−47.270	−42.670
1972	1.000	1.000	0.000	0.000	0.000	0.000	0.000
1973	0.149	0.744	246.560	24.600	−508.290	−132.230	−409.580
1974	1.000	1.000	0.000	0.000	0.000	0.000	0.000
1975	1.000	1.000	0.000	0.000	0.000	0.000	0.000
1976	0.358	0.742	302.750	1.540	−298.180	−228.110	−287.060
1977	−0.236	0.662	375.260	38.600	−174.440	−277.020	−423.690
1978	0.554	0.659	0.000	1.500	−258.120	−223.760	−519.000
1979	0.167	0.548	215.700	36.350	−336.870	−317.130	−807.860
1980	0.134	0.544	67.010	22.570	−207.980	−321.780	−953.670
1981	−0.421	0.529	738.910	4.910	−202.410	−342.600	−857.640
1982	−0.0843	0.500	542.190	5.360	−237.380	−384.110	−855.910
1983	0.646	0.657	204.580	0.000	−566.770	−343.280	−534.530
1984	1.000	1.000	0.000	0.000	0.000	0.000	0.000
1985	0.684	0.786	0.000	0.000	−440.860	−185.980	−685.560
1986	0.756	0.855	0.000	0.000	−229.630	−131.550	−1,027.690
1987	0.799	0.897	0.000	0.000	−152.890	−100.500	−1,217.790
1988	1.000	1.000	0.000	0.000	0.000	0.000	0.000
1989	1.000	1.000	0.000	0.000	0.000	0.000	0.000
1990	−0.348	0.715	746.170	0.000	−81.500	−356.460	−1,687.280
1991	1.000	1.000	0.000	0.000	0.000	0.000	0.000
1992	0.0485	0.262	264.470	7.990	−369.280	−1,122.920	−2,328.060

IOTA= input efficiency; THETA= output efficiency; GRP= Gross Regional Product; Gov.rev.= government revenues. Source: Center for Regional Analysis, School of Public Policy, George Mason University; unpublished data and analysis results

However, it does reflect, how government revenues and expenditures play an important role on the efficiency of the county. The investment measure is relatively more important than the labour measure. These results can be checked by looking at the relative prices (weights or multipliers) of these variables.

The results of the projection application show that the United States economy is efficient in all years during the forecast period, as expected (see Table 3.13).

Table 3.13. The United States economy input, output, efficiency and inefficiency scores and the objective values of variables

Year	Output	Capital input	Labor input	Efficiency			Inefficiency			Objective values	
				IOTA	THETA	Out put	Capital	Labor	Output	Capital	Labor
1948	2,148.51	4,161.49	2,104.48	1.000	1.000	0	0.00	0.00	2,148.51	4,161.49	2,104.48
1949	2,173.32	4,401.65	2,090.76	1.000	1.000	0	0.00	0.00	2,173.32	4,401.65	2,090.76
1950	2,285.40	4,574.55	2,149.20	1.000	1.000	0	0.00	0.00	2,285.40	4,574.55	2,149.20
1951	2,414.13	4,855.10	2,244.43	0.996	0.996	0	-18.79	-8.68	2,414.13	4,836.31	2,235.75
1952	2,474.37	5,075.35	2,281.89	0.993	0.993	0	-32.28	-14.51	2,474.37	5,043.07	2,267.38
1953	2,541.33	5,220.00	2,319.31	0.996	0.996	0	-20.38	-9.05	2,541.33	5,199.62	2,310.26
1954	2,536.32	5,400.24	2,309.39	0.991	0.997	0	-171.80	-6.60	2,536.32	5,228.44	2,302.79
1955	2,661.89	5,555.65	2,379.65	1.000	1.000	0	0.00	0.00	2,661.89	5,555.65	2,379.65
1956	2,723.92	5,815.04	2,445.54	0.987	0.989	0	-111.03	-26.25	2,723.92	5,704.01	2,419.29
1957	2,789.82	6,019.06	2,497.01	0.983	0.985	0	-157.46	-35.62	2,789.82	5,861.60	2,461.39
1958	2,810.48	6,197.88	2,519.13	0.977	0.982	0	-286.87	-44.54	2,810.48	5,911.01	2,474.59
1959	2,946.24	6,253.14	2,600.70	0.987	0.987	0	-80.30	-33.40	2,946.24	6,172.84	2,567.30
1960	3,013.44	6,438.40	2,659.21	0.981	0.981	0	-116.05	-47.93	3,013.44	6,322.35	2,611.28
1961	3,135.03	6,601.45	2,762.48	0.978	0.978	0	-139.42	-58.34	3,135.03	6,462.03	2,704.14
1962	3,306.82	6,731.12	2,896.27	0.98	0.980	0	-128.32	-55.21	3,306.82	6,602.80	2,841.06
1963	3,456.33	6,922.65	2,999.18	0.984	0.984	0	-110.19	-47.74	3,456.33	6,812.46	2,951.44
1964	3,627.99	7,146.15	3,102.53	0.991	0.991	0	-63.20	-27.44	3,627.99	7,082.95	3,075.09
1965	3,820.33	7,398.89	3,224.46	0.996	0.996	0	-22.48	-9.80	3,820.33	7,376.41	3,214.66
1966	4,023.03	7,730.67	3,357.02	1.000	1.000	0	0.00	0.00	4,023.03	7,730.67	3,357.02
1967	4,183.63	8,090.66	3,488.04	0.997	0.997	0	-23.33	-10.06	4,183.63	8,067.33	3,477.98
1968	4,352.14	8,397.35	3,612.70	1.000	1.000	0	0.00	0.00	4,352.14	8,397.35	3,612.70
1969	4,477.80	8,744.48	3,732.87	0.992	0.996	0	-33.53	-42.41	4,477.80	8,710.95	3,690.46
1970	4,560.45	9,113.46	3,785.66	0.983	0.983	0	-146.42	-60.82	4,560.45	8,967.04	3,724.84
1971	4,663.42	9,376.72	3,828.35	0.987	0.987	0	-118.64	-48.44	4,663.42	9,258.08	3,779.91
1972	4,841.11	9,693.54	3,908.55	0.997	0.997	0	-22.03	-8.88	4,841.11	9,671.51	3,899.67
1973	5,037.51	10,107.72	4,036.82	1.000	1.000	0	0.00	0.00	5,037.51	10,107.72	4,036.82
1974	5,079.50	10,621.69	4,110.01	0.979	0.979	0	-221.4	-85.67	5,079.50	10,400.29	4,024.34
1975	5,144.95	10,999.38	4,140.81	0.975	0.975	0	-264.42	-99.54	5,144.95	10,734.96	4,041.27
1976	5,343.64	11,195.39	4,236.35	0.989	0.989	0	-121.64	-46.03	5,343.64	11,073.75	4,190.32
1977	5,543.38	11,520.03	4,342.91	0.997	0.997	0	-32.29	-12.17	5,543.38	11,487.74	4,330.74
1978	5,723.89	11,948.68	4,442.90	1.000	1.000	0	0.00	0.00	5,723.89	11,948.68	4,442.90
1979	5,835.88	12,442.08	4,542.71	0.990	0.990	0	-116.66	-42.59	5,835.88	12,325.42	4,500.12
1980	5,861.59	12,882.25	4,567.77	0.980	0.980	0	-245.30	-86.98	5,861.59	12,636.95	4,480.79
1981	5,907.89	13,137.18	4,561.17	0.984	0.984	0	-208.88	-72.52	5,907.89	12,928.30	4,488.65
1982	5,955.55	13,412.05	4,646.80	0.972	0.972	0	-368.99	-127.8	5,955.55	13,043.06	4,518.96
1983	6,119.06	13,565.86	4,711.21	0.983	0.983	0	-219.59	-76.26	6,119.06	13,346.27	4,634.95
1984	6,357.46	13,835.90	4,811.65	1.000	1.000	0	0.00	0.00	6,357.46	13,835.90	4,811.65
1985	6,459.55	14,335.60	4,831.42	1.000	1.000	0	0.00	0.00	6,459.55	14,335.60	4,831.42
1986	6,545.77	14,844.47	4,861.42	1.000	1.000	0	0.00	0.00	6,545.77	14,844.47	4,861.42

IOTA= input efficiency; *THETA*= output efficiency;
Source: Jorgenson (1995), Table 8.6

The practical implication of this application of the forecast objective inputs and outputs projection (Table 3.14) is that it establishes the input targets for efficient output at some future time. This empirical analysis showed that the proposed model can be a useful tool for examining any given entity's performance ranging from corporations to government and regional economies. It may also be applied to investigate the performances of a political unit over time for comparison purposes; that is, performances of different political parties (leadership) in a given jurisdiction over time.

The projection application of the model shows it may provide better guidance for future planning and decisions and may establish input targets for efficient outputs of DMUs under consideration.

In addition, the analysis reveals that one of the wealthiest counties of the United States in fact has not used its resources efficiently most of the time during the investigation period. The findings suggest that being one of the wealthiest counties does not necessarily mean performing efficiently over time.

Table 3.14. Forecasted values of inputs and output in the U.S. economy

Year	Output	Capital	Labour
1987	15,277.21	4,920.64	6,652.35
1988	15,660.96	4,990.33	6,766.73
1989	16,013.19	5,063.78	6,884.09
1990	16,345.14	5,138.57	7,002.59
1991	16,664.04	5,213.84	7,121.52
1992	16,974.54	5,289.29	7,240.62
1993	17,279.64	5,364.79	7,359.78
1994	17,581.26	5,440.33	7,478.97
1995	17,880.64	5,515.86	7,598.16
1996	18,178.58	5,591.41	7,717.36
1997	18,475.60	5,666.95	7,836.56
1998	18,772.02	5,742.49	7,955.77
1999	19,068.06	5,818.04	8,074.97
2000	19,363.85	5,893.58	8,194.17

Source: Center for Regional Analysis, School of Public Policy, George Mason University; unpublished data and analysis results

3.7 Conclusions

Despite the criticisms and problems of economic base theory, and of location quotient, shift-share analyses, and DEA modelling, nonetheless they remain important analytical tools for regional and industry sectoral analysis.

Dinc and Haynes (personal communication) emphasize how:

... the key technique used for performing shift-share analysis is the decomposition of the regional growth effects into contributing components which are dealt with separately. Such characteristics seem most attractive when we come to study the regional or local structure that is nested in a higher level system. Therefore, the shift-share technique has been used to analyze the employment and industrial development of a region, especially as a means to evaluate the effectiveness of certain regional policies oriented towards employment and economic growth. Social and regional scientists apply this technique to the study of social and demographic change, occupational and industrial transformation, ethnic and sex compositions, income and employment growth, import and export growth, and so forth. It is also an important tool for statisticians to report on estimated regional and local data given limited national and/or base regional data, or given limited regional data to estimate national statistics.

It is important that all variables be properly defined and specified in applying shift-share models, and that their relationships be correctly explained or explicitly hypothesized. Data must be homogeneous, which means the same digit Standard Industrial Classification (NAIC Classification since 2000) codes for both national and regional data must be used. Shift-share assumes proportionality of a regional industry sector to the regional economy, but the regional economy is related to the national economy by three location quotients for:

(a) the industry sector to the national whole;
(b) the national corresponding industry sector; and
(c) the regional whole.

This makes shift-share analysis more meaningful than the single location quotient used in economic base theory, which uses just the first quotient. Shift share analysis can also be linked to the input-output model for a region (discussed in Chap. 4) by building a sector to sector relationship, but this is difficult given the complexity of inter-industry purchases. Thus, the sector to total relationship is used as a proxy, but this proportion may not be a precise measure of national change as many factors can change this balance including industrial restructuring, productivity change, and changes in demand and technology. However, reforms to national sectoral relationships and the use of time series dimensions correct for some of these distortions in shift-share models. In this context, the time series addition brings shift-share analysis into an econometric framework, providing greater flexibility in modelling variable relationships and an ability to test the significance of those relationships. Thus, demographic and a whole range of other economics and social data variables can be added as additional indicators of regional change, particularly for conducting policy impact assessment. And the incorporation of the DEA approach enabling investigation of the differentiation between relative productivity (efficiency) and regional competitiveness of industry sectors between and within regions in an inter temporal context provides a powerful tool for evaluating the efficiency of a local region across a range of output and input measures benchmarked against national performance and that of other regions.

4 Traditional Tools for Measuring and Evaluating Regional Economic Performance II: Input-Output Analysis

4.1 Measuring Regional Inter-Industry Linkages

In Chap. 3 the discussion focused on tools for analyzing regional performance over time relative to other regions and on assessing which industry sectors are performing well or poorly in a region, both relative to other regions and relative to other industries within the region. But a further important consideration in regional economic analysis is the inter-industry linkages existing within a region and the evaluation of the impacts of changes (growth or decline) in one industry sector on other sectors. Typically this has been done through a technique known as input-output analysis.

Input-output analysis (I/O) is a methodology named for a modelling framework developed by Nobel laureate Wassily Leontief for work he undertook in the 1930s (Leontief 1941). Thus input-output models are also referred to as Leontief models. Given that the primary purpose of input-output models is to model the interdependence among industrial sectors in an economic system (national or regional), it is not surprising that such models are also referred to as interindustry models. Regional input-output models describe both transactions between the region and the rest of the world and among activities within the region. These models produce a multiplier index or ratio that measures the total effect or impact of an increase in demand on employment or income. They can also be used for predicting and forecasting the impacts of the potential future performance of a regional economy and changes in interindustry transactions.

4.2 Compiling a Regional Input-Output Table

It is useful to know how to compile a *regional input-output (I/O) table* and describe different sources of exogenous and endogenous effects.

4.2.1 The I/O Transactions Table

The foundation of input-output analysis is a set of accounts describing transactions among major industrial or economic sectors. The sectors are typically composed of the following (after Hoover 1971, p.225):

(a) *Intermediate*: private business activities within the region outside the region.
(b) *Households*: individuals and families residing or employed in the region who may be considered as both buyers of consumer goods and services and sellers of labour.
(c) *Government*: state, local and national public authorities, both within and outside the region.
(d) *Outside world*: activities (other than government) and individuals located outside the region.
(e) *Capital*: the stock of private capital, including both fixed capital and inventories.

Table 4.1 illustrates the basic I/O table of these relationships. This is called a transactions table, which is simply a matrix of the inputs and outputs for the various activities and, in particular, for the intermediate sector. Typically the inputs and outputs for the other activities are suppressed because obtaining the data necessary to complete the transactions table is difficult, and usually are not available. Consequently, the lower right hand part of the transactions table is left blank. All transactions are typically converted to a common metric, such as the dollar value of flows.

4.2.2 Input Coefficients and Multiplier Effects

In order to undertake meaningful analyses it is necessary next to convert the transactions table to a table representing each sector's inputs (or outputs under an alternative formulation) to its relative proportion of total regional inputs. These input coefficients from the transactions table appear in Table 4.2. They are useful for evaluating the summed effects of vertical (cross activity) linkages in the region being examined. For example, assume a US$1,000 increase in export sales on the part of industry sector A, which we will call paper products. For the paper products industry to provide the additional products it will have to spend additional amounts for inputs both from other intermediate sectors and from primary supply sectors. For example, the paper products industry will spend an additional US$442 for labour, US$12 for inputs from industry B, and US$233 for industry C's inputs. Further, each of the input sectors will now have to purchase more inputs in order to provide the additional input to paper products, for example industry C will have to spend US$233 x .032 for additional inputs from paper products (industry A) and US$233 x .323 for additional inputs from industry B, and so on. As these individual sectors feel and respond to the demand from the increased output from the paper products industry sector, its purchases in the region will increase.

Table 4.1. Simple form of regional input-output table

		Intermediate sector, by industry				Final demand sectors				Output Totals
		A	B	C	D	Households (consumer goods sales in region)	Government (sales to governments)	Outside (exports)	Capital (gross private investment, including additions to inventories)	
Intermediate sector, by industry:	A	300	400	100	500	1,600	500	200	700	4,300
	B	50	200	1000	300	100	200	100	900	2,850
	C	1,000	200	100	700	100	300	200	500	3,100
	D	0	800	200	500	700	0	0	400	2,600
Primary supply Sectors										
Households (labour services)		1,900	300	1000	400					
Government (public services)		200	100	200	100					
Outside (imports)		200	300	300	0					
Capital (capital consumption and withdrawals from inventories)		650	550	200	100					
INPUT TOTALS		4,300	2,850	3,100	2,600					

This chain of effects, known as indirect effects, will continue until the echoing of effects gradually leak totally out of the region. The total effect will be at most no more than two or three times the size of the initial demand increase (that is, US$1,000).

The ratio of the final demand increase to the initial demand increase is called the export multiplier. The export multipliers calculated for the case of an externally driven increase in demand for US$1,000 for paper products appear in Table 4.3.

Table 4.2. Input coefficients for activities in the intermediate sector[a]

Purchases (in Dollars) from:	Per Dollar's worth of gross output in			
	A	B	C	D
Intermediate sector				
A	0.070	0.140	0.032	0.192
B	0.012	0.070	0.323	0.115
C	0.233	0.070	0.032	0.269
D	0.000	0.281	0.065	0.192
Primary supply sectors				
Households	0.442	0.105	0.323	0.154
Government	0.047	0.035	0.065	0.038
Outside	0.047	0.105	0.097	0.000
Capital	0.151	0.193	0.064	0.038
TOTAL[b]	1.000	1.000	1.000	1.000

Notes: [a]The input coefficients have been calculated from the illustrative data provided in the preceding table. [b]Colums do not always add exactly to totals, because of rounding off

Table 4.3. Total direct and indirect effects of an increase in final demand[a]

Total added sales (in US$) by intermediate activities	Per dollar of increased sales to final demand by:			
	A	B	C	D
A	1.118	0.289	0.157	0.661
B	0.126	1.234	0.439	0.439
C	0.297	0.284	0.501	0.477
D	0.068	0.452	1.400	0.400
Total[b]	1.609	2.259	2.612	1.977
Added purchases (in US $) by all intermediate activities from primary supply sectors:				
Households	0.614	0.419	0.532	0.427
Government	0.079	0.092	0.108	0.086
Outside	0.095	0.171	0.167	0.059
Capital	0.215	0.317	0.193	0.179

Notes: [a]The total direct and indirect effects have been calculated from the illustrative data provided in the preceding table. [b]Colums do not always add exactly to totals, because of rounding off

The values in this table show "the amount by which each processing activity's sales are increased as the ultimate result of a dollar's increase in the final demand sales of any intermediate activity" (Hoover 197, p. 229). The effects of a change in demand are highest for the activity that experiences the increase in external demand, because that increase in demand is part of the total effect (direct and indirect). Consequently, it should be no surprise that figures on the diagonal in the table are by far the largest.

This introduction to input-output analysis necessarily has been short to provide the space needed to lay out a full application for this methodology to a real planning and analysis case developed below. The uninitiated reader will find a more detailed description of input/output models in any of a number of texts on the subject (for example, Hoover 1971; Miller and Blair 1985).

4.3 Exogenous and Endogenous Effects

In the above introductory discussion of input-output analysis, a regional economy is viewed as experiencing an *exogenously* driven increase in demand; that is, from outside the region. Such increases can be motivated by any number of causes including, for example, favourable currency changes, a disproportionate increase in a general sectors demand at the national or global level (for example, plastics in the 1960s, semi-conductors later in the 20th Century), and positive business cycle changes. Historically, exogenously generated increased demand has been seen as the primary driver of regional change in general and for input-output analysis in particular.

However, demand can also arise from *endogenous* forces, that is, from within the region. This is a major point made by the new growth theorists (for example, Romer, 1986; 1990) as well as other regional scientists (Stough 1998; and Karlsson et al. 2001). How can or does demand increase endogenously? As with an exogenously derived demand increase, there are a variety of ways demand can occur from within a region. For example, a change in the regional institutional structure, such as a regulatory change that reduces production costs (such as reduced taxes), would make the firms in a region more competitive with producers outside of the region. Alternatively, learning by doing on the job may lead to new process and/or product innovations that make local regional firm(s) more productive and, in turn, create a relatively larger external market. A very different type of endogenous effect may be generated by a process known as import substitution, whereby a local region invests in the development of a local producer of an input to a local production process of a basic good. By eliminating the need to import this input more of the return on output is retained locally.

In summary, increased demand can, in effect, be created from within a region as well as external to the region. This is an important consideration when regional development policy is formulated. When the objective is to increase regional product, both endogenous and exogenous policies must be considered. Local regions usually have more ability to affect such increases through policies targeted to local factors. Consequently, at a local level an endogenous perspective is often crucial to the formulation of appropriate development policy. This will become obvious in the context of a case study discussed later in this chapter.

4.4 Using Input-Output Analysis to Measure Impact of an Economic Activity

4.4.1 Direct, Indirect and Induced Effects

An important area of application of input-output modelling is to measure and evaluate the impact of expansion or contraction of an existing industry (or even a firm) and to predict (or estimate) its expansion or the impact of a proposed new development on a region's economy. Typically this is achieved by generating I/O tables to measure three impacts or effects:

(a) *Direct*, which are those economic categories directly affected by the business activities of the industry, or from a development.
(b) *Indirect*, which are those impacts caused by inter-industry purchases of goods and services as they respond to changes in the business activities of the industry, or from a development.
(c) *Induced*, which are those impacts created through household spending of those employed directly and indirectly through the business activities of the industry, from a development.

4.4.2 An Example: Measuring the Contribution of the Newport News Shipbuilding to the Virginia Economy, 1996 to 2020

In the United States, Newport News Shipbuilding (NNS) is Virginia's largest manufacturing employer, and performs the design, construction, overhaul and re-fuelling of nuclear powered aircraft carriers and submarines for the United States Navy. NNS also performs design and construction of commercial ships for domestic and international customers.

The Centre for Regional Analysis (CRA) at George Mason University investigated the economic impact of NNS's business activities on Virginia's State economy (Arena et al. 1996). Using input-output analysis, economic impacts were measured in terms of employment, personal income, and state and local taxes. The analyses and findings of the study relied on data provided by NNS, state and local economic data, and IMPLAN input-output software that generated a transactions table and an input coefficient table. The analysis was for 1996, with forecasts to the year 2020 assuming 1.5 per cent growth in employment annually, 1997 to 2000. The direct, indirect and induced effects were measured. The analysis measured the employment, personal income, employee composition, business taxes, and state and local income tax impacts of this industry. This study showed that:

(a) 45,516 Virginia jobs were attributable to NNS in 1996;
(b) Individuals holding these jobs earned over US$1.5 billion in wages, interest and dividends 1996;
(c) In 1996, Virginia employers tied to NNS disbursed over US$1.5 billion in wages and benefits for these jobs;

(d) In 1996, NNS business activities resulted in US$107 million of Virginia business taxes; and
(e) In addition to the US$107 million in business taxes, Virginia also received US$63 million in state and local income tax in 1996.

Table 4.4 shows the forecasted effects of this assumed growth of the NNS on the Virginia economy, and Fig. 4.1 illustrates the type of graphics that can be produced to demonstrate an effect (in this case employment) on a selected group of industry and service categories. The study forecasts these effects on 64 industry sectors, plus the household sector.

Table 4.4. A summary of all impacts of the NNS on the Virginia economy, 1990 to 2000

Category	Impact	
Employment		
	Direct	17,924
	Indirect	7,470
	Induced	16,122
	Total	41,516
Personal income (in US$)		
	Direct	936,130,039
	Indirect	240,216,422
	Induced	416,018,534
	Total	
Employee compensation, includes fringes (in US$)		
	Direct	912,108,791
	Indirect	222,792,364
	Induced	372,298,834
	Total	1,507,199,989
Business taxes (in US$)		
	Direct	9,200,000
	Indirect	26,291,559
	Induced	71,013,847
	Total	106,505,406
State and local income taxes (in US$)		
	Direct	38,446,601
	Indirect	9,126,475
	Induced	14,985,565
	Total	62,558,641

Source: Arena et al.1996

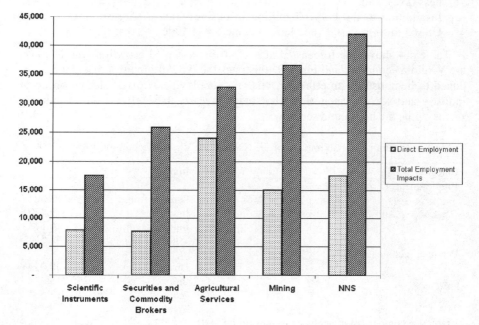

Fig. 4.1. Employment impact comparisons of the NNS on selected industry sectors in Virginia, 1996–2000
Source: Arena et al. 1996

4.5 Using Input-Output Analysis as a Predictive Tool

Input-output analysis, as described above, works through three phases:

(a) from a description of relations among different intermediate sectors and other input providers (transactions table) to
(b) measures of relative inputs (input coefficient table) to
(c) the computation of inter-industry and provider ratios called multipliers (export base multiplier table).

By examining past patterns of regional growth, it is possible to estimate changes in demand in the regional economy by industry sector. Such changes may be used to produce a base line forecast of regional economic change. These forecasts are built up sector-by-sector and year-by-year and as such provide much more defensible estimates than various simple linear or even non-linear trend line forecasts. In advanced input-output modelling, multipliers are adjusted on the basis of previous experience; that is, dynamic multipliers are computed to further enhance the quality of the forecasts.

Once a base line forecast has been derived it is possible to undertake scenario analysis. Such analysis may be aimed at estimating how an external change; for example, national growth rates will impact the regional economy and selected sectors and or input providers. Alternatively, analysis may serve to simulate the effect of an endogenous policy change, for example, tax modification, regulatory change or investment strategy.

4.6 A Forecasting Application of Input-Output Analysis: An Economic Development Strategy for the Brisbane South East Queensland Region, Australia

A case study of the Brisbane-South East Queensland (SEQ) regional economy in Australia demonstrates the use of an input-output model to examine the future potential effects of possible regional and national policy interactions.

Input-output modelling was used as a predictive tool to inform regional economic development strategy planning to evaluate potential impacts and outcomes of alternative future development scenarios for this region into the future (Blakely et al. 1991). As part of a broader strategic planning study (Stimson 1991b) looking at the future development potential of the Brisbane SEQ region, input-output analysis was employed as one tool in a broader analytical framework to estimate the economic implications of a number of strategic development options and opportunities for the Brisbane-SEQ region over a planning period of 20 years from 1991 to 2011. As such this case study shows how input-output analysis complements a variety of tools often used in development strategy studies.

The Brisbane-SEQ region is Australia's most rapidly growing metro region, characterized by a poli-centric structure, incorporating the capital city of Brisbane at its centre, the Gold Coast tourist centre to the south, the Sunshine Coast to the north, and the old mining and industrial centre of Ipswich to the west. Internal migration, largely from other states in Australia, is the main driver of growth in the region. Demographic forecasting indicates the population of the region will likely increase from 1.5 million in 1991 to between 2.8 to 3.1 million by 2011.

4.6.1 The Methodology

The methodology consisted of:

(a) A *projection component*, which provides projections of the Brisbane-SEQ regional economy in a 'hands-off' development scenario, which is the path the regional economy, was expected to follow in the absence of major strategic planning initiatives.
(b) An *impact component*, which provides estimates of the expected impact of a number of alternative strategic economic indicators which might accompany the implementation of those scenarios.

The procedures used in the Brisbane-SEQ region study embraced a combination of conventional shift-share analysis and input-output tables to provide estimates of structural change in the regional economy over time. The shift-share component enabled growth in industry gross output to be attributed to national growth factors, to industry-mix factors, and to region-specific or differential factors. Extensions to shift-share analysis and the use of input-output tables enables the estimation of these effects within each cell of the input-output table, thus allowing a more detailed study of the effects of regional economic structural change.

Three phases were involved in the construction of the I/O tables showing the potential structure of the Brisbane-SEQ regional economy in the years 2001 and 2011 (see Fig. 4.2):

(a) *Phase 1:* The identification and measurement of the components of structural change in the Brisbane-SEQ regional economy between the years 1985/86 and 1989/90 as a possible source of data for the projection of the regional economy to the two reference years 2001 and 2011.
(b) *Phase 2*: The construction of a 1989/90 I/O table of the Brisbane-SEQ region, consistent with the existing 1985/86 table. The 1989/90 base year I/O table is used as the base table for the projection of the regional economy over 11 years to 2001 and over 21 years to 2011 (the reference years).
(c) *Phase 3*: The projection of each component of structural change to obtain projections of the expected size and structure of the Brisbane-SEQ regional economy in the two reference years, 2001 and 2011.

4.6.2 Phase I: Components of Economic Structural Change

The first phase of the study involved the identification and estimate of the components of economic structural change in the region over the period 1985/86 to 1989/90. These estimates are reflected in subsequent phases 2 and 3.

The general approach to the identification of past structural change is represented in Fig. 4.2. The base year table was the 1986/86 interindustry I/O table for the region. The components of structural change refer to those estimated to have occurred between 1986 and 1990. For the projective phases, the base year interindustry table for year 2001 was the 1990 table, and for year 2011 it was the projected 2001 interindustry table. The components of structural change reflect the growth assumptions of the model.

In all phases each component has a similar meaning; phase 1 attempts to estimate the contribution of that component to changes in economic structural change in the Brisbane-SEQ region in the period 1985/86 to 1989/90; phases 2 and 3 show the expected contribution of that component to future regional economic change to the two reference years 2001 and 2011.

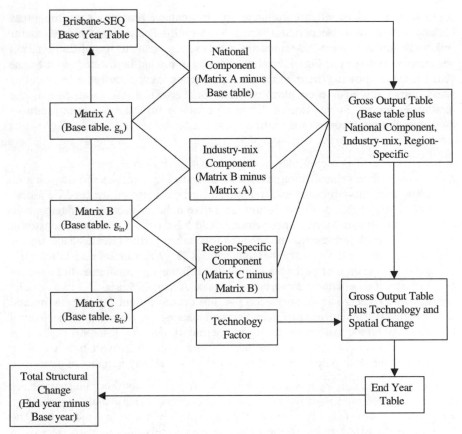

Fig. 4.2. Analytical structure for estimation of structural change 1985/86 to 1989/90 and projection to years 2001 and 2011 for the Brisbane-SEQ regional economy
Source: Blakely et al. 1991

What follows is a description of the model as applied to the Brisbane-SEQ case study (see City of Brisbane 1991; Strategy Paper No 2 for a detailed account of the methodology and the resultant I/O tables).

The national component represents that part of the transactions change, either in total or in each cell of the I/O table, attributable to the growth of the Australian economy as a whole. This recognizes the fact that some of the growth which had occurred in the Brisbane-SEQ regional economy occurred because of the economic environment of the national economy; that is, changes in national standard of living, balance of payments, monetary and fiscal policy, and so on. Economic factors affecting the national and state economies influence substantially the performance of the Brisbane-SEQ economy. While each region of a nation reacts differently to the national economy as a whole, there is a sense in which each region in a national environment exhibits characteristics of the national economy simply because it is part of that economy. The national component is calculated according to the estimated growth rate of the gross national output over the period 1985/86 to 1989/90. The national growth factor (g_n) was estimated at 1.099, representing an average growth rate of 2.4 per cent per annum. Matrix A in Fig. 4.2 is calculated by applying the factor g_n to the base year I/O table, showing the transactions which would have occurred in each cell if gross output levels of each industry sector had grown at the same rate as the national economy. Subtraction of the base year table from the A matrix produces the national growth component matrix.

The industry-mix component measures the effect of the composition of industry in the Brisbane-SEQ region, due to the fact that, nation-wide, some sectors grow more rapidly than others. A region specializing in slow growth industry sectors will show a negative industry-mix component, while a region favoured by a high share of rapid growth sectors will show a positive industry-mix effect. Matrix B in Fig. 4.2 estimates the transactions which would have occurred if each industry at the regional level had reacted in the same way to economic circumstances in the period 1985/86 to 1989/90 as that industry at the national level. Matrix B is derived by the scaling of each industry by the estimated national growth factor for that industry ($g_{i,n}$). The subtraction of Matrix A from Matrix B produces the industry-mix effect. The Brisbane-SEQ region contains eight faster-growing and eleven slower growing industries in terms of national growth rates. The overall industry-mix effect is positive during the period 1985/86 to 1989/90, suggesting that those industries in the region with a faster national growth rate are more important in terms of output than those growing more slowly in national terms.

The region-specific component effects arise from the fact that industry sectors grow at different rates in different regions. This occurs because some regions have a comparative advantage over others for the growth of individual industries. The region-specific effect refers to the extent to which individual industry sectors in the Brisbane-SEQ region has grown faster or slower than the same industries at the national level because of the economic environment in the region. Matrix C is estimated by the scaling-up or down of each sector by the regional growth factor for that sector ($g_{i,r}$). The region-specific component is the difference between Matrix B and Matrix C. This reflects the differences between the growth rates of industries at the national and regional levels. In the period 1985/86 to 1989/90, eight industry sectors in the Brisbane-SEQ region, are estimated to have grown faster than their national counterparts, and eleven more slowly. The region is shown to have a comparative advantage as a service centre and for some manufacturing activities.

The technology component reflects a number of influences, such as changes in technology of production, input substitution and importing patterns. It reflects the changes which occur within each column of the interindustry I/O table, or the changes in the process of production and local input purchases. The technology component is estimated by the identification of changes in the I/O table column structure since 1985/86 or the relative change in each cell of the I/O table.

4.6.3 Phase II: The Base Year I/O Table

The second phase in the study involved the preparation of a 1989/90 interindustry I/O table for the Brisbane-SEQ regional economy, to be used as the base year table for projections of future regional economic structure. The base year I/O table is constructed to represent the current state of industry in the region rather than as a 'snapshot' of the regional economy in 1990. As current data were not readily available it was necessary to construct the I/O table using expected growth rates of industry based on the identified components of structural change estimated in the first phase and to validate these estimates from industry sources where possible. The resulting base year I/O table is given as Table 4.5, which is included here to illustrate the detail typically generated in a regional I/O table.

4.6.4 Phase III: The Reference Year

The third phase of the study involved the derivation of interindustry I/O tables showing the expected size and structure of the Brisbane-SEQ regional economy in the two reference years 2001 and 2011. This involves the projection of each component estimated in phase 1, and the layering of each of these components on the (now) base-year 1989/90 table (Table 4.5) to obtain the respective projected I/O tables of the Brisbane-SEQ region for the two future reference years 2001 and 2011.

The derivation of the national and regional industry growth rates for incorporation into the projection model represents an important but complex issue. The various growth factors, (g_n, $g_{i,n}$, and $g_{,i,r}$ in Fig. 4.2, developed in phase 1 could not be applied uncritically to project the Brisbane-SEQ economy for the 11 year and 21 year projection periods (to 2001 and 2011 respectively). The growth rates derived in phase 1 are accepted as a basis for future growth rates, but were modified in terms of current and expected future economic circumstances and expected developments in technology and consumption patterns. These include population projections, forecast trends and industry and professional opinions. The projected rates of change incorporated into the projection model are conservative, but they do incorporate available opinions on likely market developments. In this sense, the projected growth rates and the resulting projected interindustry I/O for the two reference years were not forecasts in any sense, rather they represent a consensus view on the most probable nature of the future Brisbane-SEQ regional economy.

Table 4.5. 19-sector interindustry I/O table for the Brisbane-SEQ region, 1989/90 (1989/90 prices) (A$'000)

Sector	Animal ind.	Other Agric.	Fo- restry/ Fishing	Coal/ Petrol. Mining	Other Mining	Food Manuf.	Wood, Paper Manuf.	Machinery/ Appliances	Metal Prod. Ma- nuf.	Non- Met.Ma nuf.
Animal Industries						147,865				
Other Agriculture	24,271	314,888	59			152,861				
Forestry/ Fishing			37			9,925	19,730			
Coal/Petrol/ Mining				2,371	296	1,727	4,019		6,923	6,775
Other Mining				1,278	2,183				989	46,316
Food Manufacturing	129	2,409	1,585	348	1,267	901,831	426	1,108	1,192	849
Wood Manufacturing	13	82,208	93	964	1,863	5,085	448,287	8,768	3,047	3,854
Machinery/ Appliances	3,124	11,280	5,214	5,646	5,771	14,690	5,502	20,856	9,284	1,336
Metal Manufacturing	918	2,362	530	2,942	5,292	5,784	560	210,747	91,992	62,498
NonMetal Manufacturing	789	4,932	66	1,419	1,377	13,848	723	425	14,293	32,790
Other Manuf.	21,686	18,909	9,255	9,569	10,093	572,616	20,119	39,575	10917	29,321
Electricity/Gas	7,069	5,873	135	2,058	3,052	81,081	33,255	57,375	42,644	27,569
Building/ Construction	1,336	1,852		3,233	2,175	13,934	2,151	27,651	14,873	25,263
Trade	16,351	25,304	17,516	2,899	3,752	126,401	35,741	39,685	39,477	14,335
Tpt/Com	8,956	22,412	2,975	18,378	4,771	214,298	5,958	7,553	43,604	103,221
Finance	13,801	44,899	17,233	14,272	23,516	692,735	122,018	173,051	14,3726	153,749
Pbl.Adm	4,278	14,523	6,192	1,006	3,180	19,215	1924	3,487	5,389	3,863
Com Serv	1,260	7,033	282	465	2,044	14,175	3,126	2,646	2,029	2,290
Rec/Pers.Services	202	4,698	322	159	1,616	59,167	1,610	8,291	2,870	2,673
TOTAL	104,182	200,582	61,493	67,006	72,247	3,047,238	719,699	601,219	433,249	516,703
H-hold	32,464	95,167	24,694	90,757	122,101	475,045	463,661	368,207	384,127	149,708
Exports	5,202	136,425	9,399	1,862	17,411	287,682	135,438	261,690	408,209	155,471
Imports	15,027	123,533	11,106	7,504	32,751	1,252,907	490,023	515,315	372,693	145,436
TOTAL	52,693	355,125	45,199	100,123	172,263	2,015,634	1,089,122	1,145,212	1,165,029	450,615
TOTAL	156,875	555,707	106,692	167,129	24,4510	5,062,872	108,821	1,746,431	1,598,278	967,318
Employment	1,994	4,592	714	1,659	2,503	19,166	20,166	14,030	15,578	5,573

Source: Blakely et al. 1991

Table 4.5 (cont.)

Other Ma-nuf.	Electr./Gas	Building/Constr.	Trade (Retail)	Transp./Comm	Finance	Public Admin.	Commun. Services	Recr/Pers. Serv	TOTAL	H-Hold
									147,865	327
125	293	99	11,406	83	462	385	465	1,065	223,462	130,440
		2,,837	24,957	2,399	344		132	48	60,408	610
18,257	86,418	701	1,798	3,009	589	1,366	273	1,255	138,378	
1,256	127	150692	209	1,156	2,168	161	249	712	207,497	
1,634	434	8,270	423,345	7,460	36,494	3,606	19,326	11,917	1,423,629	1,683,985
20,245	1431	171,211	24,234	45,366	146,406	49,003	71,803	3,831	1,007,714	23,327
29,891	10,914	22,644	150,753	60,643	23,322	44,028	26,728	16,391	468,017	183,463
2,224	5,344	346,616	220,552	12,406	13,914	12,803	15,003	9,563	1,022,050	27,018
1,413	1,878	171,647	201,123	19,151	3,666	2,181	5,466	5,107	482,295	14,530
53,513	10,881	25,024	256,252	115,464	51,307	21,169	47,994	23,623	1,347,285	683,061
66,274	1,867	7,593	7,502	11,152	89,928	9,648	20,198	19,932	494,202	129,052
36,842	3,529	3,064	9,926	95,833	53,622	28,845	27,605	4,991	356,725	167,416
73,815	3,258	219,231	132,334	186,485	125,366	27,540	94,481	37,761	1,221,732	3,419,188
94,509	7,120	221,959	120,357	155,451	292,731	72,432	95,068	40,698	1,532,452	1,062,868
301,970	10,245	298,859	498,558	205,503	1,356,862	148,357	289,495	146,969	4,655,819	3,962,407
11,865	1,121	16,307	11,731	15,444	29,429	2,576	17,689	4,005	173,224	55,592
3,767	795	12,744	11,765	14,479	76,370	14,699	38,436	11,516	219,920	1,601,205
29,887	823	6,384	24,095	13,292	135,119	11,622	21,574	48,203	387,159	1,756,963
747,485	146,479	1,685,882	2,130,897	964,776	2,438,098	450,421	794,585	387,588	1,5569,829	14,901,452
295,466	361,542	1,825,690	2,460,227	1,507,601	1,602,226	1,130,252	2,760,515	773,136	14,922,586	
759,084	587,728	1,693,678	1,510,628	987,283	4,265,177	483	914,011	907,201	13,044,062	55,1380
1,933,573	461,275	487,588	417,079	471,527	3,324,868	61,687	226,896	972,790	11,075,579	935,688
2,988,123	1,410,545	4,006,956	4,387,934	2,966,411	9,192,271	1,192,422	3,901,422	2,405,127	39,042,227	1,487,068
3,735,608	1,557,024	5,692,838	6,518,831	3,931,187	11630369	1,642,843	496,007	2,792,715	54,612,056	16,388,520
12,627	11,067	55,482	149,098	56111	78365	43,496	120,250	48,254	660,725	

The derivation of each of the components built into the projection may be summarized as follows.

(a) The *national growth factor* assumed for the period 1989/90 to 2000/2001 is 1.256, representing an average annual growth rate of the national economy of 2.1 per cent over eleven years. The growth factor assumed for the period 2001 to 2011 is 1.195, or an average annual growth rate of 1.8 per cent. As in the Phase I calculation, the national component can be expected to be a substantial, and possibly the dominant, factor for the Brisbane-SEQ regional economy, and for all sectors within the economy.

(b) The *industry-mix factor* assumed national growth rates of industry sectors are given in column (2) of Table 4.6 for the period 1990 to 2001 and in column (7) for the period 2001 to 2011. Columns (3) and (8) of Table 4.6 give the expected relative values of the industry-mix component in the respective time periods. The sign of each entry in the column indicates whether each sector is expected to contribute positively or negatively to the industry-mix effect. The industry-mix trends identified in phase 1 could be expected to continue, leading to a positive industry-mix contribution to the growth of the Brisbane-SEQ region in the two reference periods. Generally the service sectors of the economy are expected to provide the greatest positive industry-mix effects in the future, with some manufacturing and primary industries growing significantly less than the national growth factor.

(c) The *region-specific factor* uses the expected growth rates of industries in the Brisbane-SEQ region in the reference period are given in column (4) of Table 4.6 for the period 1990 to 2001, and in column (9) of that table for years 2001 to 2011. The relative size of the region-specific effect, in terms of the difference between the assumed growth rate of each industry at the national and regional levels is given for the two-time reference years (2001 and 2011) respectively in columns (5) and (10) of Table 4.6. A positive entry in these columns reflects the assumption that the industry identified will grow more rapidly at the regional level than at the national level. If those trends were to continue, the Brisbane-SEQ region would have a locational advantage for most manufacturing industries and for most of the service sectors, and show lower than national growth rates for the primary sectors.

(d) The *technology factor* is taken into account within the structural changes represented in the columns of the inter-industry I/O table and reflects changes in the purchasing patterns of the industries of the Brisbane-SEQ region. These changes reflect the assumption that the industries within the region will purchase, for example, more of their inputs from within the region as the diversity of the regional economy is increased. In particular, it could be expected that relatively increased purchases from the finance sector will occur, as well as from the transport, communication and personal services sectors.

Table 4.6. Assumed annual rates of growth for projection model for the Brisbane-SEQ region, 1990–2001, 2001–2011

Sector	1990–2001 (annual percentage growth rates)					2001–2011 (annual percentage growth rates)				
	National economy growth rate	National sector growth rate	Industry mix component	Regional sector growth rate	Region-specific component	National economy growth rate	National sector growth rate	Industry mix component	Regional sector growth rate	Region-specific component
	(1)	(2)	(3)	(4)	(5)	(6)	(7)	(8)	(9)	(10)
Animal Industries	2.1	1.20	-0.90	-0.03	-1.23	1.8	0.95	-0.85	-0.02	-0.97
Other Agriculture	2.1	2.00	-0.10	1.30	-0.70	1.8	1.65	-0.15	1.10	-0.55
Forestry, Fishing	2.1	0.85	-1.25	-0.02	-0.87	1.8	0.65	-1.15	-0.01	-0.66
Coal, Petroleum Mining	2.1	2.25	0.15	1.10	-1.15	1.8	2.00	0.20	0.65	-1.35
Other Mining	2.1	1.80	-0.30	1.40	-0.40	1.8	1.50	-0.30	1.10	-0.40
Food Manufacturing	2.1	0.80	-1.30	1.30	0.50	1.8	0.60	-1.20	1.20	0.60
Wood, Paper Manufact.	2.1	1.60	-0.50	1.80	0.20	1.8	1.40	-0.40	1.60	0.20
Machinery/Appliances	2.1	-0.03	-2.13	-0.02	-0.05	1.8	-0.02	-1.82	-0.01	-0.03
Metal Products Manufacturing	2.1	1.45	-0.65	1.50	0.05	1.8	1.25	-0.55	1.30	0.05
Non-Metal Product Manufacturing	2.1	1.00	-1.10	1.60	0.60	1.8	0.80	-1.00	1.35	0.55
Other Manufacturing	2.1	2.40	0.30	1.95	-0.45	1.8	2.15	0.35	1.80	-0.35
Electricity/Gas	2.1	2.20	0.10	2.15	-0.05	1.8	1.90	0.10	1.95	0.05
Building, Construction	2.1	2.15	0.05	2.35	0.20	1.8	1.95	0.15	2.00	0.05
Trade (Retail etc.)	2.1	0.95	-1.15	1.40	0.45	1.8	0.75	-1.05	1.30	0.55
Transport, Communication	2.1	2.00	-0.10	2.25	0.25	1.8	1.70	-0.10	1.95	0.25
Finance	2.1	3.10	1.00	2.80	-0.30	1.8	2.70	0.90	2.70	–
Public Administration	2.1	2.05	-0.05	1.75	-0.30	1.8	1.80	–	1.60	-0.20
Community Services	2.1	2.20	0.10	2.30	0.10	1.8	2.00	0.20	2.10	0.10
Recreation, Personal Services	2.1	2.90	0.80	2.70	-0.20	1.8	2.70	0.90	2.60	-0.10

Source: Blakely et al. 1991

4.6.5 The Projected Inter-industry I/O Tables for the Reference Years 2001 and 2011

From these projected estimates, inter-industry I/O tables for the Brisbane-SEQ region may be derived for the two reference years 2001 and 2011. Given that change in the level and the mix of economic activity is inevitable in any economic system, it is possible to estimate the expected changes in the nature of the Brisbane-SEQ regional economy in the future. The projected interindustry I/O tables for the years 2001 and 2011 provide several types of summary information as to the nature of the regional economy of the future. For example, it is possible to calculate changes in the expected size of regional economic indicators and the contribution in the future of each industry in the region to these indicators.

These contributions reflect the assumed national and regional growth rates listed in Table 4.6 and the effect of those assumptions on the estimated industry-mix in the region in the future. These projected I/O tables are not reproduced here for space reasons. However, from them it is possible to estimate the potential changes in the structure and performance of the Brisbane-SEQ regional economy under this 'hands off' scenario through to the two reference years 2001 and 2011.

4.6.6 Economic Development Choices and Building Future Scenarios

In this case study, identifying industry sector choices for a regional economic development strategy for the Brisbane-SEQ region involved more than the traditional economic studies, such as I/O analysis, to determine the current state of the regional economy and then to project that state to determine the shape of the future regional economy based on the past. While the options and opportunities facing the Brisbane-SEQ regional economy (or any regional economy) over a period of twenty or so years are bound by national economic performance and by regional population dynamics, the growth and shape of a regional economy may also be determined by the choices the community seeks to make and the policies government seeks to pursue through regional planning strategies and regulations.

Brisbane-SEQ region was seen by the researchers undertaking this study (Blakely et al. 1991; Stimson 1991b) to be likely to face three economic development choices. These are summarized below:

A *'Hands-Off' Choice*. This choice assumes the pursuit of a non-interventionist role in which current policies would continue, and where market forces would influence how individual industrial sectors developed. Such a 'hands-off' policy may be disadvantageous. There would still be economic growth resulting from the normal market forces in an expanding national and regional economy. On the basis of current trends (in 1990 when the study was conducted), however, it was judged unlikely that this expansion would be enough to provide sufficient employment opportunities for the greatly increased work force in the region. It was judged to be more likely that, for a number of reasons, the increase in population would be accompanied by local structural unemployment unless an active strategy was undertaken to increase the economic activity in the region faster than the rate of population growth. Without such positive steps, social dislocation and severe welfare problems could result from a shortfall in employment. Even if sufficient regional opportunities were created, the 'hands-off' strategy would also need to ensure that sufficient additional output would be generated to increase per capita gross regional product. If economic activity in the region were to fail to achieve greater output per capita, then it can be expected that the standard of living measured by real per capita income would deteriorate in the Brisbane-SEQ region.

A Concentrated Development Choice. The second choice assumes a concentration on development of a few key industry sectors by attracting new enterprises, and encouraging development in selected existing firms in that sector, with the aim of seeding economic growth through the flow-on effects. Because of the narrow focus on a limited range of the industrial sectors and the emphasis on attracting new industries, the success of this option is judged to be vulnerable to economic downturn in the national economy, international competition and the reluctance of new business to relocate to the Brisbane-SEQ region. This option also would require the provision of large sums of capital funding, with a high risk of failure if new entrants were not attracted to Brisbane-SEQ region.

A Broadening and Deepening of Economic Structure Choice. This third choice assumes a strategy that will seek to broaden and deepen the existing economic structure of the Brisbane-SEQ region by removing barriers to entry and promoting a climate conducive to restructuring, emphasizing the current strengths and natural advantages of the regional economy. A broader-based economy is judged to be better able to withstand fluctuations in the national and international economies, and as the economy develops new industries will find it easier to locate in the Brisbane area. While this option would require capital outlays to provide the climate for expansion, the amount will not be as great as under the previous option and the risk of failure is lower.

4.6.7 Assessing Community Industry Sector Preferences

In order to determine community reactions to various future industrial choices, targeted samples of the business and professional communities in the Brisbane-SEQ region were surveyed regarding their reactions to various industrial sectors and to assess their community's potential for development. These took two forms:

(a) The holding of a series of panel discussions with a number of industry sector leaders—drawn from the business and commerce, manufacturing, research and development, tourism and government sectors of the city and the region—to identify the strengths, weaknesses, options and opportunities for the regional economy and its industry sectors.

(b) The collection of data through a questionnaire survey of leaders from these fields in which they were asked to make judgments based on their assessments of local resources and the potential of 17 industrial sectors using five-point Likert scales to assess the current capacity and future capability of the region according to five factors—namely, finance, political and headquarter potential, environment and image. These data were used to develop a matrix which averaged out the weightings for the 17 industry sectors across the respondents so as to provide a weighting for each sector on each factor to show the relative perceived strengths or weaknesses of each sector. The results are summarized in Fig. 4.3.

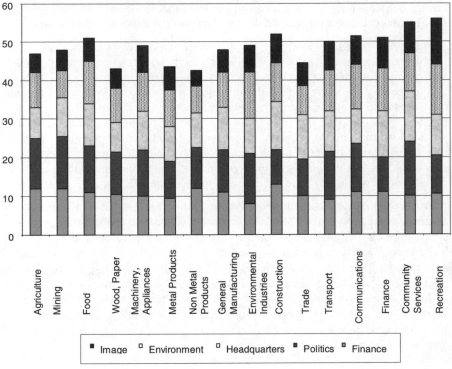

Fig. 4.3. Business sector preferences for industry sector development for the Brisbane-SEQ region
Source: Blakely et al. 1991

The data thus collected indicated the following for the Brisbane-SEQ region:

(a) Old manufacturing raw materials based industry is seen as a dominant economic resource.
(b) There are no real barriers for financing most of the raw materials based industries as well as transport and communications.
(c) There is less financial and political capacity in the metals and metals related industry.
(d) There is more capacity to support its past than resources to develop its future.
(e) Environment and image are the most significant factors with respect to community reactions, with the food, communications, community services, transportation, recreation (tourism) and finance sectors having the highest ratings.

Based on the results of this survey, business sector preferences indicated that the tourism/recreation, financial management, construction, environmental industries and food manufacturing have strongest support. These industrial groupings potentially could lead to economic development scenarios emphasizing regional services, tourism/education and/or transportation/distribution industries.

4.6.8 Future Development Scenario Building for the Brisbane-SEQ Region

Using the results of the base year and the two reference year I/O regional tables, plus the results of the survey data appraising the structural analysis of the Brisbane-SEQ region's economy which identified strengths and weaknesses, and the assessments of sectoral capacity and potential, it was possible to generate scenarios that outlined potential economic development options for the Brisbane-SEQ regional economy. These take advantage of Brisbane-SEQ region's strategic location, the role of Brisbane City as the capital of the State of Queensland, its historical development as the major service centre for the rapidly growing Brisbane-SEQ metro region, and the fact that the Brisbane-SEQ region is well located to provide links between Australia and the Asia-Pacific region.

The scenarios developed were based on three important assumptions:

(a) demographic forecasts (prepared for the region by the Applied Population Research Unit at the University of Queensland) showed population growth in the region from 1.81 million in 1990 to 2.21 million in 2001 and 2.6 million in 2011;
(b) a scenario would be regional in focus; and
(c) the quality of lifestyle (weather, housing, recreation, culture, physical environment, etc.) would be a significant factor in the growth of the Brisbane-SEQ region.

The scenarios developed were not mutually exclusive; rather they tended to complement each other. While they are not additive in their effects, they do have a synergy effect.

The five scenarios developed are summarized below:

Scenario A: Development of an international intermodal transportation centre (Gateway Ports). This scenario capitalizes on the strategic position of Brisbane in relation to the Asia-Pacific region, the rest of Australia and the State of Queensland. Brisbane City has an unusual strategic concentration of transportation infrastructure the area at the mouth of the Brisbane River—a seaport, an airport (an international and domestic terminal), a reasonably well-developed road network and rail system, all of which have spare capacity, together with large areas of relatively underutilized land. The scenario combines those advantages to develop a modern multimodal transportation facility with the capacity to manage air and surface transport within Australia for the South Pacific region.

Scenario B: Development of a technology and education based sector. This scenario proposes the expansion of the emerging education, technology and knowledge-based industry in Brisbane. It would use the resources of the Technology Quadrangle of the four universities located in the region and the existing research and development base in the region by developing innovation centres and related technology infrastructure. It builds on existing technology rather than relying on attracting new firms to the region. There is an emphasis on promoting and developing technologies related to the existing areas of strength, and applying technology to existing natural resources.

Scenario C: Continued development of Brisbane as a regional service centre. This scenario assumes that continued development of Brisbane City's service and administrative sectors would be required to support the growth of the region's population and the expansion of its manufacturing and service base. In this scenario, Brisbane City's role as a regional government centre is the basis for economic growth. Public service, arts, culture, and community services (sports, conventions, and media) would be based in Brisbane City. Brisbane City's share of the regional population is seen to decrease substantially, but its headquarters role will be enhanced.

Scenario D: Brisbane-SEQ as an events and destination centre. The central idea of this scenario is to build on the region's natural features and its image as a city attractive to residents and visitors. The scenario builds Brisbane as the hub of travel in the region. Along with the major international tourism area of the Gold Coast and the domestic tourism area of the Sunshine coast, the region would become a major destination for conventions and for tourism in general. The Central Business District (CBD) would also become the headquarters for the regional tourism industry, providing the management and supporting infrastructure. A concerted program to attract major events to the region, in addition to continued growth of tourism, is an important element of this scenario.

Scenario E: Greater emphasis on adding value to the State's natural resource output. In this scenario, Brisbane City and the wider SEQ region strengthens its capacity to process or manufacture the raw materials that are mined or grown in the State of Queensland. This scenario depends on the region's growth as the primary driving force for the expansion of the manufacturing base. As the regional manufacturing sector expands its capacity to add value to raw materials to meet regional markets, the potential to export to the rest of the nation and overseas also expands.

4.6.9 Projected Scenario Impacts and Regional Economic Development Strategy Formulation

The expected additional contribution to regional gross output by adoption of each of the above scenarios is shown in Table 4.7. Similar expected additional contributions by each scenario to regional employment by industry are shown in

Table 4.8. The results of the impact analysis are summarized in Table 4.9 which shows that the expected employment level of 827,470 persons employed in the region in the 'hands-off' situation in the year 2011 could be augmented by 135,900 persons if Scenario A were successfully implemented, and by 128,169 persons employed if Scenario B were implemented.

These five scenarios can be subjected to the same process of input-output analysis of the Brisbane-SEQ regional economy, in order to assess their potential economic impacts by evaluating likely changes to gross regional output, employment levels and income levels for the two reference years 2001 and 2011. In carrying out this analysis, the industry sector growth rates are adjusted to take into account the likely variations in industry sector activity levels that potentially could result from the implementation of each scenario as part of the process of formulating a regional economic development strategy.

It is appropriate either to treat each of the five scenarios as stand alone options, both to assess the impact and to consider the strengths and weaknesses of each scenario, or to look at the overall impact of a strategy that might encompasses an integrated use of the various scenarios. As Table 4.9 shows, by the first reference year 2011, Scenario A (development of an international intermodal transportation centre) potentially offers the greatest opportunity both for employment and additional gross regional product. Scenario B (expansion of the region's technology and education based sector) ranks second, followed by Scenario D (Brisbane as an events and destination centre). Scenarios C and E are similar in creation of employment opportunities, although their ranking is reversed for the creation of additional gross product.

By the second reference year 2011, Scenario B will likely overtake Scenario A in additional regional output, although it still ranks second in employment opportunities. Scenario D remains ranked third.

By comparing the five scenarios and the inter-industry data derived from the projected I/O tables, and taking into account the additional household income generated both in total and per additional employment opportunity, it is possible to identify the following priorities for the future economic development of the Brisbane-SEQ region.

(a) Early expansion of the transportation related industries (Scenario A) could provide rapid shorter term growth, overtaken eventually by the technology and education scenario. This suggests that technology sector capital investments and an emphasis on human resource development, provides a longer lead time for income generation than would investment in the transportation sector, but it also offers greater longer-term potential for the economic growth of the region.

(b) The services Scenario C provides lower growth potential than Brisbane's historical development as a service centre would suggest, probably because the sector is already well developed in the region and could cope with the rapid population growth and demand for services without a proportional growth in employment, that is, higher productivity and output levels could be achieved.

Table 4.7. Expected additional contributions of development scenarios to industry and regional gross output in the Brisbane-SEQ region, 2001 and 2011 (AS$ milion)

	Year 2001									
Scenario	A		B		C		D		E	
Sector	Amount	%	Amount	%	Amount	%	Amount	%	Amount	%
1.Animal Industries	6.9	0.1	4.7	0.1	17.8	0.6	7.1	0.2	8.9	0.3
2. Other Agriculture	29.2	0.5	20.9	0.5	48.6	1.6	21.7	0.6	61.6	2.3
3. Forestry, Fishing	5.7	-	2.7	-	2.7	-	4.0	0.1	3.0	0.1
4. Coal, Petro-leum Mining	7.3	0.1	5.0	0.1	6.3	0.2	4.3	0.1	3.9	0.1
5. Other Mining	60.3	-	3.6	-	17.6	0.6	3.3	-	11.1	0.4
6. Food Manufacturing	308.7	4.8	211.3	5.2	797.8	26.1	317.0	8.8	399.9	14.8
7. Wood, Paper Manufacturing	95.9	1.5	64.6	1.6	36.7	1.2	102.9	2.9	31.4	1.2
8. Machinery/ Appliances	71.3	1.1	43.4	1.1	56.7	1.9	37.3	1.0	75.9	2.8
9. Metal Prod. Manufact.	70.4	1.1	39.6	1.0	44.1	1.4	41.0	1.1	37.6	1.4
10. Non-Metal Prod. Manufact.	46.6	0.7	23.0	0.6	92.0	3.0	27.6	0.8	22.1	0.8
11. Other Manufacturing	232.6	3.7	150.0	3.7	528.2	17.3	141.6	3.9	259.2	9.6
12. Electricity/ Gas	67.3	1.1	45.9	1.1	45.4	1.5	40.7	1.1	34.1	1.3
13.Building, Construction	106.8	1.7	68.7	1.7	257.1	8.4	50.4	1.4	35.2	1.3
14. Trade (Retail etc.)	893.9	14.0	366.3	9.0	191.1	6.3	579.6	16.1	485.7	17.9
15. Transport, Communications	267.3	19.9	457.5	11.3	149.9	4.9	366.8	10.2	223.8	8.3
16. Finance	2,035.1	31.9	852.3	21.0	492.9	16.1	920.4	25.6	765.3	28.2
17. Public Administration	190.0	3.0	180.3	4.4	94.3	3.1	36.9	1.0	72.8	2.7
18. Community Services	673.0	10.6	1,093.9	27.0	83.5	2.7	453.6	12.6	86.0	3.2
19. Other Services	258.4	4.1	420.9	10.4	94.5	3.1	445.2	12.4	93.6	3.5
TOTAL	6,372.8	100	4,054.8	100	3,057.2	100	3,601.1	100	2,711.3	100

Note: Rounding errors occur.
Source: Blakely et al. 1991

Table 4.7. (cont.)

Scenario	A		B		C		D		E	
Sector	Amount	%	Amount	%	Amount	%	Amount	%	Amount	%
1.Animal Industries	21.5	0.2	57.4	0.4	37.1	0.6	189.7	0.2	20.3	0.4
2. Other Agriculture	90.5	0.7	161.3	1.2	106.2	1.7	65.2	0.9	101.5	1.8
3. Forestry, Fishing	16.1	0.1	13.9	0.1	7.5	0.1	10.6	0.1	7.9	0.1
4. Coal, Petro-leum Mining	25.3	0.2	23.7	0.2	18.3	0.3	15.4	0.2	12.6	0.2
5. Other Mining	24.0	0.2	17.6	0.1	34.8	0.6	13.2	0.2	18.2	0.3
6. Food Manufacturing	854.1	6.6	3,090.9	22.9	1,138.3	18.1	672.1	8.9	671.3	12.2
7. Wood, Paper Manufacturing	313.1	2.4	249.5	1.8	127.3	2.0	237.4	3.1	109.8	2.0
8. Machinery/ Appliances	183.8	1.4	143.9	1.1	103.6	1.6	100.9	1.3	122.4	2.2
9. Metal Prod. Manufact.	227.6	1.8	165.9	1.2	127.3	2.0	133.4	1.8	110.3	2.0
10. Non-Metal Prod. Manufact.	142.1	1.1	102.1	0.8	137.1	2.2	84.4	1.1	63.8	1.2
11. Other Manufacturing	694.2	5.3	836.0	6.2	810.1	12.9	429.2	5.7	481.6	8.8
12. Electricity/ Gas	221.2	1.7	229.8	1.7	135.3	2.1	135.4	1.8	106.0	1.9
13.Building, Construction	324.4	2.5	266.7	2.0	345.7	5.5	168.9	2.2	115.5	2.1
14. Trade (Retail etc.)	1,701.1	13.1	1,168.6	8.6	534.7	8.5	1,049.1	13.8	794.9	14.4
15. Transport, Communications	1,862.3	14.3	1,092.8	8.1	442.5	7.0	708.7	9.3	467.8	8.5
16. Finance	4,220.2	32.5	3,191.2	23.6	1,533.7	24.3	2,210.9	29.1	1,684.9	30.6
17. Public Administration	254.1	2.0	249.6	1.8	125.9	2.0	74.7	1.0	100.2	1.8
18. Community Services	1,108.7	8.5	1,518.3	11.2	256.5	4.1	754.5	9.9	248.8	4.5
19. Other Services	702.4	5.4	941.9	7.0	281.1	4.5	711.1	9.4	265.8	4.8
TOTAL	12,986.9	100	13,520.9	100	6,303.0	100	7,593.8	100	5,503.7	100

Year 2011

Table 4.8. Expected additional contributions of development scenarios to industry and regional employment in the Brisbane-SEQ region, 2001 and 2011 (number of jobs)

	Year 2001									
Scenario	A		B		C		D		E	
Sector	Amount	%	Amount	%	Amount	%	Amount	%	Amount	%
1.Animal Industries	76	-	52	-	195	0.8	77	0.2	98	0.3
2. Other Agriculture	176	0.2	126	0.2	293	1.2	131	0.3	371	1.3
3. Forestry, Fishing	38	-	18	-	18	-	26	-	20	-
4. Coal, Petroleum Mining	69	-	47	-	59	0.2	41	-	37	0.1
5. Other Mining	60	-	34	-	168	0.7	31	-	105	0.4
6. Food Manufacturing	1,124	1.4	769	1.3	2,905	11.9	1,154	2.5	1,456	5.1
7. Wood, Paper Manufacturing	980	1.2	660	1.1	375	1.5	1,052	2.3	321	1.1
8. Machinery/ Appliances	582	0.7	354	0.6	462	1.9	304	0.7	619	2.2
9. Metal Prod. Manufact.	629	0.8	354	0.6	394	1.6	366	0.8	336	1.2
10. Non-Metal Prod. Manufact.	250	0.3	123	0.2	493	2.0	148	0.3	119	0.4
11. Other Manufacturing	742	0.9	478	0.8	1,684	6.9	451	1.0	826	2.9
12. Electricity/ Gas	414	0.5	283	0.5	279	1.1	250	0.5	210	0.7
13.Building, Construction	925	1.2	596	1.0	2,228	9.1	437	1.0	305	1.1
14. Trade (Retail etc.)	19,848	25.2	8,134	13.8	4,243	17.4	12,869	28.1	10,785	38.0
15. Transport, Communications	16,699	21.2	6,028	10.2	1,974	8.1	4,833	10.6	2,949	10.4
16. Finance	12,315	15.6	5,158	8.7	2,983	12.2	5,570	12.2	4,631	16.3
17. Public Administration	4,639	5.9	4,404	7.5	2,303	9.4	900	2.0	1,779	6.3
18. Community Services	15,301	19.4	24,870	42.2	1,899	7.8	10,313	22.5	1,955	6.9
19. Other Services	3,964	5.0	6,459	11.0	1,451	5.9	6,831	14.9	1,437	5.1
TOTAL	78,831	100	58,947	100	24,406	100	45,785	100	28,358	100

Note: Rounding errors occur
Source: Blakely et al. 1991

Table 4.8. (cont.)

Year 2011

Scenario	A		B		C		D		E	
Sector	Amount	%	Amount	%	Amount	%	Amount	%	Amount	%
1.Animal Industries	232	0.2	621	0.5	402	0.8	202	0.3	220	0.4
2. Other Agricul-ture	443	0.3	788	0.6	519	1.0	319	0.4	496	1.0
3. Forestry, Fishing	110	-	95	-	51	-	72	-	54	0.1
4. Coal, Petro-leum Mining	232	0.2	217	0.2	168	0.3	141	0.2	115	0.2
5. Other Mining	219	0.2	160	0.1	317	0.6	120	0.1	166	0.3
6. Food Manufacturing	3,008	2.2	10,884	8.5	4,008	7.7	2,367	2.9	2,364	4.5
7. Wood, Paper Manufacturing	2,942	2.2	2,343	1.8	1,196	2.3	2,230	2.8	1,031	2.0
8. Machinery/ Appliances	1,523	1.1	1,192	0.9	858	1.7	836	1.0	1,014	2.0
9. Metal Prod. Manufact.	1,843	1.4	1,343	1.0	1,030	2.0	1,080	1.3	893	1.7
10. Non-Metal Prod. Manufact.	722	0.5	519	0.4	697	1.3	429	0.5	324	0.6
11. Other Manu-facturing	2,041	1.5	2,458	1.9	2,382	4.6	1,262	1.6	1,416	2.7
12. Electricity/ Gas	1,202	0.9	1,248	1.0	735	1.4	736	0.9	576	1.1
13.Building, Construction	2562	1.9	2,106	1.6	2,730	5.3	1,333	1.7	912	1.8
14. Trade (Retail etc.)	36,476	26.8	25,059	19.6	11,465	22.1	22,497	28.0	17,044	32.8
15. Transport, Communications	22,575	16.6	13,247	10.3	5,363	10.4	8,591	10.7	5,671	10.9
16. Finance	22,034	16.2	16,662	13.0	8,008	15.5	11,543	14.4	8,797	16.9
17. Public Administration	5,712	4.2	5,611	4.4	2,830	5.5	1,679	2.1	2,253	4.3
18. Community Services	22,563	16.6	30,898	24.1	5,220	10.1	15,355	19.1	5,064	9.7
19. Other Services	9,485	7.0	12,718	9.9	3,796	7.3	9,602	11.9	3,589	6.9
TOTAL	135,921	100	128,169	100	51,774	100	80,393	100	51,999	100

Table 4.9. The potential impact of the five scenarios and the 'Gateway Strategy' on the Brisbane-SEQ regional economy in 2001 and 2011 compared to the likely outcome of the 'hands-off' scenario

	'Hands-off' Regional economy	Additional contributions of development scenarios					
		Scenario A	Scenario B	Scenario C	Scenario D	Scenario E	Gateway strategy
Projected year 2001							
Output Level ($m)	67,998	6,373	4,055	3,057	3,601	2,711	9,681
Household Income ($m)	18,554	1,876	1,427	618	1,026	636	2,794
Employment (persons)	753,246	78,831	58,947	24,406	45,785	28,358	116,301
Projected year 2011							
Output Level ($m)	82,111	12,986	13,520	6,303	7,593	5,503	20,123
Household Income ($m)	22,300	3,494	3,334	1,372	2,006	1,292	5,315
Employment (persons)	827,420	135,921	128,169	51,774	80,393	51,999	205,640

Source: Stimson 1991b

(c) Lower per capita household income generation in Scenarios A, D and E reflects the structure of those industry sectors, with high capital to labour ratios, lower wage structures with less dependence on professional or skilled workers, and probably an existing industry structure that can expand to meet market demand by addition of lower paid workers without requiring similar growth at management or technical levels.

(d) While adding value by transformation of the State's mining and rural outputs under Scenario E offers good potential for growth, there appears to be real limits on expansion of the manufacturing sector imposed by the small local markets (although these could grow) by the limitations of the international market, extreme wage price competition in these sectors (from China, for example, and by the region's historic lack of an extensive manufacturing base.

4.6.10 An Integrated Strategy

The process of industry analysis and consultation undertaken during the development of these potential scenarios for economic development of the Brisbane-SEQ region and their evaluation permits the identification of the relative strengths and weaknesses of the regional economy. Potential options for economic development

can then be proposed, based on taking advantage of those strengths, taking correc-
tive action to overcome weaknesses (such as infrastructure deficiencies), within
the opportunities and trends revealed by sectoral and market analysis.

The development of a proposed strategy for the future economic development
of the Brisbane-SEQ region followed an eclectic selection of the actions and
policies required to achieve the benefits of each scenario. This occurred through a
process of evaluating comparative advantages, markets, infrastructure needs,
structural and technological change and likely investment. Although the five
scenarios are evaluated separately, the resulting strategy options are presented as a
cohesive economic plan covering a wide range of actions and policies which
overlap the various scenarios (Stimson 1991a).

While an industry or sector is often isolated for the purposes of evaluation,
separation of particular sectors or industries does not imply that a sector or
industry would be subjected to actions or policies that will not affect the wider
economy.

The economy of any region is an integrated network of industries where
increased activity in any one industry or sector is reflected through flow-on effects
(the economic multiplier) throughout the economy. These flow-ons result from
increased demand for inputs into the expanded industry (for example, labour,
capital, raw materials or components, transport, business services) and from the
increased outputs which generate wealth through the distribution of profits,
dividends and wages.

The economic development strategy that was proposed for the Brisbane-SEQ
region provided for a wide range of actions and initiatives across all five
scenarios, and promoted the expansion in a number of industries that would create
even greater flow-on effects throughout the whole economy, including those
service sectors that were not directly the subject of any of the proposed actions.

The projected future development of the regional economy may be evaluated
for the potential impacts of implementing such a strategy, using input-output
analysis based on the total range of measures suggested by a strategy. But it
should be pointed out that these results—as illustrated in the summary data in
Table 4.9 showing the potential impacts on output level, household income and
employment—assumes the full and successful implementation of a proposed
strategy.

If these scenarios were to be actively pursued, then it is estimated that the
potential enhancing impact of the strategy on the level of output, on wages and
salaries, and on employment in the Brisbane-SEQ Region for the two reference
years 2001 and 2011, over and above the growth that would likely come from the
'hands-off' scenario, could be of the magnitude illustrated in Table 4.9. This also
shows how a composite set of strategies embracing the economic development
and employment enhancing impacts of the above five scenarios might be
combined through the formulation and implementation of what was called the
Gateway Strategy for Economic Development. The potential enhancing effects are
quite dramatic, generating an additional $20.1 billion to the GRP, an additional
$5.3 billion household income, and an extra 205,600 jobs, compared to letting
things progress without strategic planning as in the 'hands off' scenario.

It is important to note that up to the year 2001 it is Scenario A (the Brisbane Ports Area development) that potentially has the most dramatic enhancing effect on improving regional economic development and employment; while in the next decade to 2011 it is Scenario B (the technology and education based sector development) that potentially has the major enhancing effect. This indicates the long lead time to investment in technology development and enhancement (including human resources development) to produce major impacts in terms of regional performance, but the potential outcomes are dramatic. The necessary strategic actions, many of which relate to the timely development and facilitation of 'smart' as well as basic infrastructure services, (including transportation infrastructure) are outlined in detail as part of The Brisbane Plan project (see Stimson 1991b).

As a result of these analyses, it was proposed that the Brisbane-SEQ region adopt a Regional Economic Development Strategy called the Gateway Strategy, which made use of Brisbane's key position as a link between the emerging economic strengths of the Asia-Pacific region and Australia, as a major point of entry to the fastest growing region in Australia and its advantageous position as a transportation hub for passenger and cargo services.

Much of the growth in economic activity and employment would be likely to occur through the transportation, manufacturing, technology and knowledge-based industries, as well as tourism, and other service sector activities. The secondary (multiplier) effects represent opportunities and flow-ons that would be distributed through the regional economy to many sectors and locations. What these expected results indicate is the importance of exploring the 'industry cluster and linkages' that potentially could emerge in the Brisbane-SEQ region to enhance its development. This notion is developed in Chap. 6.

The largest actual employment growth industry sectors are seen to be most likely to be in the community services, trade (retail and wholesale), finance and business services, manufacturing, transport and communications, and personal services sectors. The primary sector host likely would experience very weak growth, although growth rates in public administration and utilities (gas, water, electricity) would likely be lower than regional growth rates.

But the 'Gateway Strategy' identified potentially would maximize the most appropriate and likely components of each of the five individual development scenarios by building on their synergy through the flow on effects that activity in one sector creates. Under the 'Gateway Strategy', economic growth could create more demand for industrial, commercial and retail space that would be widely dispersed across the Brisbane-SEQ Region, including:

(a) the CBD and inner city fringe;
(b) the inner suburbs;
(c) regional and sub-regional and district centres;
(d) industrial areas.

This would create both greater actual geographic dispersal of activities and jobs, as well as enhancing agglomeration in and around some existing nodes and the creation of new ones at strategic locations which would likely be places that are highly accessible to the regional transport infrastructure network. Fig. 4.4 conceptualizes the relationship between the economic activities that could be enhanced under the 'Gateway Strategy', and the location of activity and employment in agglomerations at specific nodes as well as across dispersed locations.

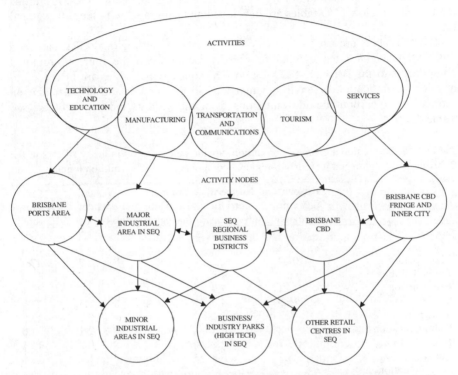

Fig. 4.4. Employment opportunities effects of the 'Gateway Strategy' for the economic development of the Brisbane-SEQ region
Source: Stimson 1991a

4.7 Using Sub-Regional Input-Output Analysis to Measure Impact Assessment

The above example demonstrating the use of I/O analysis as an analytical tool in regional development scenario evaluation is restricted to the aggregate regional scale. It is possible, however, to disaggregate such analysis to investigate inter-industry flows between sub-regional components of a region, and to measure the potential impact of a development on an event on those inter-industry structures and flows. Further work conducted in the Brisbane-SEQ region in Australia has applied I/O analysis at a disaggregated level for sub-regional impact analysis provides a good example of such an approach (McGovern 1997). In this case four sub-regional I/O tables are developed for the northern, southern, western growth corridor sub-regions and for Brisbane City at the centre of the Brisbane-SEQ region. The base tables for the I/O analysis are derived from an existing I/O table, similar to that identified in the previous section. Location quotient analysis was used to derive scale factors in each of the four sub-regional economies. From this the four sub-regional I/O tables are developed and calibrated with the full regional table.

The I/O tables are used to provide an estimate of sub-regional GRP based on 1998/90 data and inter-regional flows of trade between the four sub-regional components of the Brisbane-SEQ region. The sub-regional tables are then used to develop an indication of environmental load in the region, and in particular an estimation of carbon dioxide emissions from the various industry sectors (see Table 4.10).

Table 4.10. Sub-regional estimates of carbon dioxide emissions by industry sectors in the Brisbane-SEQ region and its four sub-regional components 1989/90 (million tonnes)

	Brisbane City core sub-region	Northern corridor sub-region	Southern corridor sub-region	Western corridor sub-region	SEQ Region total
Agriculture	0.4	1.9	1.0	1.7	5.0
Mining	0.1	0.1	0.5	0.7	1.4
Manufacturing	102.4	12.2	20.2	12.7	147.5
Public Utilities	6.1	4.8	3.2	3.9	17.9
Construction	20.1	7.9	19.1	2.0	49.1
Trade	15.9	4.2	6.4	1.5	27.9
Transport	9.3	1.1	2.0	0.7	13.1
Communications	0.2	0.2	0.1	2.1	2.0
Fire	13.7	2.1	3.9	0.7	13.1
Public Administration	8.3	0.8	1.0	1.2	11.3
Community Services	11.3	2.9	2.7	1.0	17.0
Personal Services	7.5	2.3	4.7	0.7	15.2
Sum	197.1	39.6	64.9	26.9	328.4

Source: McGovern 1997

Following this, and using the sub-regional I/O tables, estimates of the change of CO_2 emissions in the South East Queensland region at 2001 for three economic development scenarios are derived (using some of the development scenarios derived from the analyses in the case study for the Brisbane-SEQ region discussed in Section 4.5 above). The three scenarios included the development of Brisbane-SEQ region as an international inter-modal transport centre, a technology and education centre, and the extension of the status-quo as a major service centre. The second option demonstrated the lowest CO_2 emissions. The sub-regional tables have proved to be particularly useful in estimating the spatial distribution of final demand, intra-regional trade flow and carbon dioxide emissions.

This case study demonstrates the potential value and detail that I/O analysis can achieve to develop estimates and test scenario outcomes for economic and environmental impacts on a regional and sub-regional level.

4.8 New and Emerging Extensions to Input-Output Models and Conclusions

Updating input-output models has been a tedious process as reflected in the fact the U.S. input-output table updates usually lag five or more years behind the present. However, the value of a time-series of input-output tables would have significant value in the study of structural change and in advising the policy process. Yet the time and cost involved has resulted only rarely in the development of such a series. Recent work by Jensen et al. (1988), Sonis et al. (1995), Israilevich et al (1997), Sonis and Hewings (1998), van der Linden et al. (2000), and Sonis et al. (2000) has developed a method whereby updating of the input-output table is achieved by estimating directly changes in highly linked sectors and then using partial updating to estimate analytically the other less linked sector cell values. This has made it possible to reduce the time and cost in updating input-output tables with little loss in accuracy.

Hewings and others at the Regional Economic Applications Laboratory at the Chicago Federal Reserve Bank, in collaboration with the University of Illinois, have developed a methodology that enables them to represent the major forward and backward linkages evidenced by an input-output table in a graphic form that they call economic landscapes. When a series of input-output tables for the same region is computed using the new decomposition and updating procedures discussed above it is possible to graphically compare changes in the structure of the an economy over time. In the initial application they reveal in pictorial form for the Chicago region over a 20 year period a hollowing-out process whereby intrametropolitan dependence is found to be replaced by dependence on sources of supply and demand outside the region. Further, they show the unfolding of a complex internal transformation of the economy as local sources of manufacturing inputs are replaced by a much greater dependence on local service activities. In short, they show the impact of the rise of the service revolution and the effect of globalization on the Chicago economy. Additional examples of this input-output based landscape analysis appears in papers by Hitomi et al. (2000), Haddad and Hewings (1999, 2000), Sonis and Hewings (1999)

A shortcoming of the assessment of input-output analysis in this chapter is the failure to examine some other significant new developments over the past 10 years in the use of input-output models in an interregional context. Such approaches have been developed and go under the heading of computable general equilibrium modelling. Early discussions of this approach appear in the later work of Leontief and at the regional level by Hewings (1985), Horridge et al. (1993), and Harrison and Pearson (1994, 1996). The interested reader is referred to these sources for an introduction and illustration of the methodology.

Finally, it is important to note that attempts have been made to construct integrated economic-environmental and even energy input-output models with some of the early work focusing on the development of the CEASE Model by Lakshmanan et al. (1985). More recently Fritz et al. (1998) examine the interactions between polluting and non-polluting sectors.

Thus, input-output analysis represents a powerful tool not only for analyzing inter-industry interaction effects within a region, but also as a predictive tool to estimate impacts of industry changes, including policy interventions.

5 Path Setting: An Approach to Regional Economic Development Futures

5.1 New Approaches in the Contemporary Era

So far in this book the focus has been on discussing the evolution of what might be regarded as traditional approaches to regional economic analysis and to regional and local economic development planning that occurred up to the 1980s and into the 1990s. Previous chapters have overviewed the development of a number of standard quantitative tools for measuring and evaluating change overtime in the performance of a region. Attention now turns to consider the emergence of a new set of approaches to regional economic development which focus on how regions might position (or reposition) themselves to be competitive in the current era of globalization with its attendant attributes of rapid change and uncertainty. This has been leading researchers to consider the way regions might set *paths* and develop *strategies for the future*.

This chapter considers the concept of *futures for regions* and how *futures* might be analyzed. It describes recent trends in regional economic development strategy facilitation. It then proceeds to discuss emerging synthetic strategic frameworks for regional economic development strategy formulation, which includes techniques of analysis through which regions might contemplate their futures. These include *industry cluster analysis* (ICA) and *multi sector analysis* (MSA). Contemporary thinking on future strategies for regional economic development has been evolving since the early 1990s and have increasingly brought approaches from management sciences to the field of regional science and in particular the public policy field of regional economic development.

5.2 Futures

5.2.1 Futures as Paths

In thinking about setting paths and formulating strategies for the future economic development of a region it is useful to consider how different paradigms and analytical techniques have been used. For many regional economic development planners there has been a tendency to think of the future as an extension of the past or the uniformitarian view. History can provide important clues about the future; but the future is not just an extension of the past (Minkin 1995; Gibson 1996), nor is it necessarily something that can be viewed as being just an extrapolation of current trends (Hamel and Prahalad 1994). The future remains largely unexplored territory that is not easily predicted. Marshall McLuhan, the noted Canadian critic and commentator, provides this insightful comment:

If we look at the present through a rear view mirror, we march backward to the future. (Marshall McLuhan cited in Crainer 1997, p. 131)

The inference here is that the future should involve looking through the windscreen and anticipating the road ahead, not looking at the rear view mirror expecting the road a head to be much the same as the route that has passed.

Forecasting futures is something that fascinates many social scientists. It is of increasing importance to firms, corporations and organizations, including governments, as well as regions as they seek to strategically position themselves to achieve competitive advantage for their productive efforts. Considerable reward may be reaped by those who can accurately predict the future. Popular management experts and writers such as Naisbitt (1994) and Toffler (1980) continue to provide a recipe of best sellers on the future. But history can and has proven many of these 'prophets of the future' wrong in their predictions. However, the desire to know about and anticipate the future tends to be strong in all of us.

Since the dawn of civilization, humankind has been concerned with predicting and planning for the future. Forecasting techniques continue to improve in their sophistication and accuracy, and knowledge and experience of predicting futures has improved dramatically. However, efforts at predicting the future continue to have a high and possibly unacceptable degree of error. Much of this is due to the way we think about the future, and an unwillingness to move beyond entrenched paradigms that do not provide us with the tools needed to adequately prepare and manage the future. A danger in extrapolating the future is that we might miss it completely and end up in the position of playing catch-up (see Fig. 5.1). When placed in the position of catch-up, it can take considerable reserves and effort to recover and join the leading pack in the race to the future. Planning for the future requires new tools and new ways of thinking about the future and of analyzing it (Godet 1991). It also requires extensions and modifications of older techniques and methods. In sum, there is a need to learn to anticipate the future with greater reliability (Godet 1994), and in so doing become better at shaping and capturing opportunities as they emerge.

As regional economies become more globally integrated, the need to improve planning the future and to prepare better strategies to manage the future assumes increased importance. In the 1940s, researchers such as Maslow (1954) showed how planning for the future is used to provide for basic needs, to impart knowledge, to accumulate wealth, to push for order, to strive for equity, and to improve quality of life. A primary reason for why humans and organizations plan is a belief that it is possible to act consciously to influence the future in directions that are desirable. However the future cannot be prescribed unless agents are in full control of the events that will dictate it.

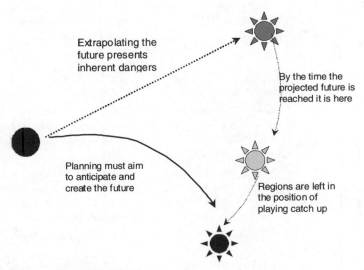

Fig. 5.1. The danger of extrapolating the future is playing catch-up

Seldom is this possible, as planning for the future takes place in largely uncontrolled environments. This has become particularly acute in the contemporary era of globalization with rapid and uncertain change. Thus we require frameworks for planning in which decisions are made and events are played out which are neither precisely determined by the past and the present, nor known in advance. If planning is to be a relevant tool for managing the future of regions under rising conditions of uncertainty, frameworks need to be developed that are both flexible and capable of accommodating the changing needs and values that will shape the future. The role of planning for futures thus needs to become less concerned with defining or predicting events and outcomes and more concerned with the processes of anticipating and managing the future. That requires a significant shift in the paradigms and practices used by those responsible for setting paths and planning for the future—especially in the field of regional economic development.

5.2.2 Futures as States of Uncertainty

The extent that any agent or organization can control and shape the future is limited because much of the future lies outside human control; and what can be controlled is predominantly short-term or pre-determined by natural laws. The future may be viewed as a series of 'future states' over which there are different levels of control and ability to predict. A typology of these states is proposed and plotted along a curve by degrees of predictability and controllability as shown in Fig. 5.2.

Each of these states has an influence upon the way plans are created and formulated to build a strategy for the future. The precise position of each future state on the curve in Fig. 5.2 is never fully known, and some states might overlap along the curve.

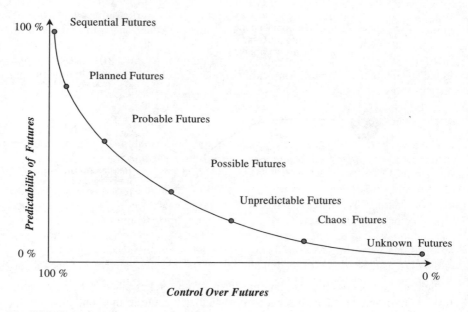

Fig. 5.2. Futures states

Sequential or Routine Futures. These are the most predictable futures states. They are controlled by natural laws, events, edicts, and patterns that can be predicted in advance with reasonable accuracy. Routine and habitual procedures, cosmic events, and fixed term events are predictable outcomes that seldom change. Much planning is based upon the study of routine futures. However, sequential or routine futures are not absolute. They often change imperceptibly over time, as in the aging process. For regional economic development, routine futures are important as they set patterns of investment and consumption, which provide the basis for short-term planning. Routine futures can be manipulated for competitive advantage. One of the successes of Japanese business strategy in the 1980s was the ability to understand routines in economies in different countries and to successfully orchestrate changes in consumer patterns to grow future markets.

Planned Futures. These futures are more predictable, involving events, activities and developments that are planned, with outcomes that are reasonably predictable, and over which a relatively high level of control can be maintained. Planned futures involve the applications of programming or scheduling. The main indeterminate elements of planned futures are time, impact, success and scale. Construction projects, budgets, fixed-term capital investments are examples of planned futures where expected outcomes or returns are reliable. Much of our strategic and detailed planning for regional economic development has been based on planned futures.

Probable Futures. These futures relate to outcomes that may eventuate from a situation that is either planned or unplanned. Forecasting may be used to determine the form, geographic location, timing and duration of future events or outcomes. The risk of failure typically is less than 50 percent. Analyzing probable futures involves the use of scenario and sensitivity testing to determine likely outcomes (Godet 1990; Robbins 1990). The random nature of the activities predicted means that the realization or magnitude of probable future outcomes is not certain. Probable futures analysis is used in regional economic development to determine 'what if' type scenarios where regions are under threat–for example, when a defense base or major industry closes.

Possible Futures: These futures involve simulating outcomes that are neither easily tested nor measured. Many assumptions are built into the modeling of possible futures. The use of heuristic programming (discussed later in this chapter) and less empirical techniques may be applied in modeling possible future scenarios. For example, extensive use has been made of game theory (Van Neuman and Morgenson 1943) and Monte Carlo linear programming techniques to simulate future outcomes, and optimization modeling techniques can be used to predict possible futures according to a predetermined goal of efficiency or equity. Possible futures are usually highly unpredictable, given the combinations of indeterminants used in simulating future outcomes. Analyzing possible futures becomes an important tool in strategic planning for regional economic development in the future. Therefore, organizations need to develop contingency strategies for managing futures that are likely to become more chaotic and unpredictable (Brown and Eisenhardt 1998). Possible futures provide an important framework in which contingency planning can operate in the future. Regional economic development futures strategy planning seems to be moving in this direction.

Unpredictable Futures. These futures are events over which there may be little or no control. Some events may be known but lie too far into the future to anticipate, and their form, time, magnitude and duration are unpredictable. The reality is that much of the real future of regions lies in this state. Many of the frontiers of science are dealing with unpredictable futures, but unpredictable futures are not necessarily uncontrollable. What may be controllable are some activities leading up to an event. As our knowledge improves in the future, unpredictable states might move towards a higher level of predictability and control.

Chaotic Futures. Some future events will be highly or totally unpredictable and uncontrollable or chaotic*s*. They can be highly dangerous–for example, the Chernobyl nuclear reactor disaster in Russia in the early 1990s and the collapse of the eastern seaboard electricity supply network in the United States in 1965 were unpredictable and uncontrollable events at the time. Major natural disasters, the collapse of financial systems, and revolutions are past and future chaos events in the realm of the unknown. Chaos theory suggests that uncontrollable states eventually stabilize around attractor states, but they can become unstable again rapidly.

Chaos theory is being used in the business world for strategic planning (Crook 1996) to understand possible implications caused by future failure of networks and business systems. It is likely that chaos futures will command close attention by regional economic development strategists as global economic systems become more dynamic, and as the potential for major disruption to trade commerce and information systems becomes greater as the result of a growing dependence on technology.

Unknowable Futures. These futures fall into the void of the complete unknown. They are futures that are beyond knowledge and experience. As a system moves toward the future, many elements of unknown futures will be realized in more highly predictable and controllable states. Much of the thinking in the realm of unknown futures lies in the imagination. Einstein (cited in Crainer 1997, p. 78) is credited with saying, "imagination is more important than knowledge". Imagination leads to creative thinking and the realization of dreams and possibilities for the future. Imagination is an important element of futures building for economic development. It is embedded in inventiveness, creativity and innovation and enables regions and communities to create and invent new futures. Without imagination, Walt Disney would never have created an entertainment empire that brings pleasure to millions around the world. Imagination and striving to innovate have driven much of the creative thinking in the re-engineering of Silicon Valley in the southern San Francisco Bay Area of California, enabling the region to turn unknown technology futures into realized futures. Imagination recognizes the realities of present resources, values and organization structures and their value in shaping the future.

5.2.3 Futures as Frontiers of Opportunity

Increased knowledge and improved forecasting makes it possible to gain a better understanding of, and sometimes control over events and outcomes that will shape the lives of people and the communities of the future. While predicting economic futures is difficult, it will become easier to anticipate and manage economic futures. To do this however, requires that individuals and organizations learn new techniques to anticipate futures, and to adopt new ways of thinking about strategy and risk management (Porter 1991).

The French researcher Michel Godet has written extensively on anticipating futures (Godet 1991, 1994; Godet and Roubelat 1996) arguing that reason and intuition need to be combined to enable us to anticipate and plan strategically for the future. Intellectual and emotional appropriations are seen as being essential if anticipation is to crystallize into effective action. Godet and others (Mintzberg 1994; Perrott 1996; Schoemaker 1995) have called for a new strategic thinking about the future.

What appears to be emerging is that forecasting will become secondary to anticipating and managing possible futures. If futures can be better anticipated, then it will be possible to build better and more reliable *paths* to the future. To do this

requires understanding that the future may have many emerging *frontiers* and potential economic and social regional landscapes, and that the role of strategy for
regional economic development will be to continually create paths through these
landscapes and extend their frontiers.

This concept of looking at the future as a series of evolving or transforming
landscapes with their leading edge frontiers presents a different perspective on the
future and the way to plan and analyze it. The elements that form the landscape of
futures frontiers are both beneficial and harmful. Some landscape elements pose a
threat and or may represent disaster to individuals, businesses, corporations, government, cities or regions. Other elements might create a new stage of opportunity
or possibility.

The major *frontiers* and the forces that create future regional landscapes might
be seen as including:

(a) advanced technologies;
(b) building design;
(c) transport and communications;
(d) knowledge-based networks;
(e) social/cultural values including livability and equity;
(f) sustainable development; and
(g) governance and institutional arrangements.

Past events have shaped many elements of the regional landscape in these frontiers, and new elements are continually being added. The elements defining these
futures form what might be called the *architecture* of regional futures. The term
'architecture' is used here to describe the props that give form to or shape the future. The architecture of the regional future is both hard and soft. Built infrastructure, including buildings and telecommunications are hard elements of future architecture. Services such as education, health and cultural amenities are soft
elements of future architecture. And knowledge, information, values and beliefs,
and social capital are elements of future architecture which, like hard and soft
elements of architecture, will affect significantly the way society is shaped and
functions in the future. Many current elements of architecture will be lost in time,
while others will be transformed or re-engineered many times over.

There will never be perfectly clear images of the edge of future landscape of
regions. Like space, the frontiers that are created are the outcomes of processes
that are evolving continuously. This is what makes planning for the future so difficult, as targets are continually moving and the changing regional structure and
form make it difficult for organizations and institutions in regions and thus for the
regions themselves to develop clear and permanent *paths in the economic and social landscape of the future*. Both the landscape of the regions and the paths to the
future along which regions are progressing are continually evolving and changing.
Subsequently, paths to the future need to undergo continuous transformation to fit
the changing terrain of economic and social landscapes and their physical infrastructures. Certain variables of regional landscapes change fast but might have little long term impact on structure, while others, like changes in social capital, values, etc, are difficult to achieve but might have huge impacts on directionality.

The role of strategy formulation and planning is to assist in mapping the changing digital nature of regional landscapes, and to constantly reposition paths for business, organizations, institutions and communities for them to harvest opportunities of the future and shape the regional landscapes while avoiding those pitfalls that may lead to impending disaster or sub-optimal performance and outcomes.

What is seen from an individual, organizational or regional development perspective of the future is not a gallery of future landscapes, but rather a collage of selectively combined landscape features of the region. That collage is a unique perspective of how an organization, community or region sees and maps its future operating landscape. The collage will be composed of landscape elements from different frontiers. Many factors—such as space and place, the level of technology development, the size of population—will fix the elements of future economic development paths. Others elements of the economic landscape will be shared with other regions—for example, common markets and infrastructure. These shared elements of landscape provide unique opportunities for leveraging infrastructure, resources and information for mutual benefit. The proposition is that in order to be economically competitive in the future, a region or community needs to learn to anticipate the future by building an *architectural collage* that will define the landscape in which it operates.

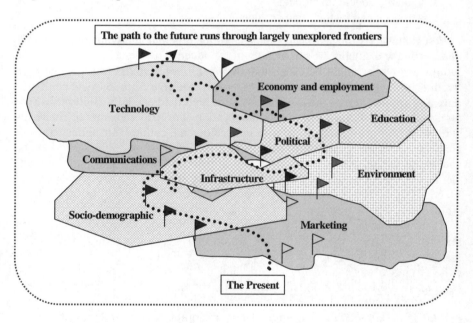

Fig. 5.3. The path to the future through a collage of frontiers

Figure 5.3 represents a conceptual collage of the economic landscape through which a *path to the future* is shown. The path traverses many overlapping landscapes and will fit somewhere between the marker posts that define the key elements of architecture in each landscape that shapes the future. Organizations will need to understand clearly the changing conditions in these landscapes before setting a path for the future, and they will need to constantly redefine the path as conditions in the landscape change.

This approach to thinking about the future—in terms of frontiers and landscapes, anticipating, defining architecture, and managing the future—represents a different paradigm to those currently used by most regional economic development planners. Much regional economic development planning continues to rely on an old paradigm of projecting and setting visions for the future, and of developing goals, strategies and actions to get to a future which, all too often, is never realized. It has been estimated that about one-half of the new job types that will be created in the next two to three decades are still to be invented, and that the technology required to support these has not been invented.

Thus, how can one realistically project and plan for economic futures? It is important to understand that the future environment in which businesses, communities and regions operate cannot be predicted accurately. Thus, it will be necessary to learn to anticipate the future in terms of probable outcomes and to develop strategies suitable for managing them. To anticipate the future new ideas to improve our understanding and to create pictures of the frontiers and landscapes that will shape the future environments in which we live, work, recreate and travel will need to be imagined and created.

Some recent approaches have been developed that may to help to shape the future are discussed in the following parts of this chapter.

5.3 Analyzing Futures

In Chap. 1 we discussed how theories and techniques of regional economic development have evolved through a series of shifting paradigms and structural revolutions (Kuhn 1996). Most techniques used to analyze and develop understanding about the future of regions involve the type of quantitative methods of analysis as discussed in Chaps. 3 and 4, and that is so primarily because little empirical data exists on 'the future' in the field of regional economic development. This places regional economic development analysts and planners in the difficult position of having to either 'guess' the future or to examine 'possibilities' about the future. One important role of economic development strategy is to influence or shape future possibilities for the potential advantage of a region or local community. Some of the techniques used by regional economic planners and managers to analyze futures are discussed below.

5.3.1 Futures by Extrapolation

The idea of the future as an extension of the past dates back many centuries. Ancient civilizations knew that the methodical keeping of records and the systematic analysis of events enables trends or patterns to be identified that might be used to make predictions about the future. Predicting futures was initially concerned with the events of the seasons and cosmos (Brownowski 1973). The German mathematician and astronomer Friedrich Gauss was one of the first persons to develop mathematical theories on probability, which still provide much of the basis of trend forecasting. Two centuries ago, Malthus was one of the first people to use mathematical extrapolation to predict that the world would run out of food as population increased; but it did not. This illustrates the dangers of using simple extrapolation to project the future.

The post-industrial school of thought has had a major impact upon regional economic development in the post war era. Neo-Malthusian theories have made extensive use of extrapolation, regression and other statistical trend analysis techniques to project the future. Authors such as Bell (1976) and Simon and Khan (1984) have strongly expressed the view that future change could be controlled by planning based on forecasting and extrapolating from the past. In the post World War II era up to the 1970s planners and policy makers tended to view the economic future with optimism, a view entrenched firmly in the belief of technological determinism. The key to the future was to extrapolate technological change and to allow it to shape technology research and development to fit defined future values and expectations. It was the age of the second industrial divide driven by the emergence of information technologies, biotechnology and new materials (Bluestone, et al. 1981). It was a vision and partial reality of seemingly limitless resources. While authors such as Carson (1962), Whyte (1968) and Commoner (1972) were warning of pending ecological disasters and that resources were not unlimited, the paradigm of the post- industrial school continued to remain optimistic until the OPEC oil shock of 1974 when the idea of continuous exploitation of resources extrapolated into the future was checked. A more somber and even pessimistic forecasting of the future began to emerge, and by the late 1970s the idea that the future could be extrapolated with confidence was under increasing challenge. Extrapolating the future lost further credibility after the second oil shock in 1981, by which time the neo-Malthusian school of thought, which presented a more pessimistic future, began to emerge.

During the 1980s, models to extrapolate the future became more complex, incorporating constraint factors such as resource limitations, technology change and market change. A wide range of techniques involving linear programming and extrapolation using multiple regression analysis have been used to develop projections for testing futures. Linear programming is a valuable tool when applied to scenario analysis, and it can be used to solve a wide range of complex planning and resource allocation problems. Ferguson and Sargent (1958) define linear programming as:

... a technique for specifying how to use limited resources or capabilities of the business (or region) to obtain a particular objective, such as least cost, highest margin, or least time, when those resources have alternative uses. It is a technique that synthesizes for certain conditions the process of selecting the most desirable course of action or number of available courses of action, thereby giving management information for making a more defective decision about resources under its control (p. 3).

Andrew (1997) has applied linear programming techniques to examine the effect of constraints to the development of accommodation-centered tourism for Cornwall in Britain. The results suggest that this accommodation-centered approach to tourism for the expansion of industry may not be an optimal strategy for the economic development of that region. Rather, an accommodation-centered strategy might have a negative impact on indigenous industries. While tourism has an important role in generating positive external balances, there may be a trade-off between the generation of these balances and other forms of economic development in the region.

Sensitivity analysis has been used to test a range of outcomes for predicted futures. It enables risks and elements of the unknown or unpredictable to be included in the analysis of futures. However a problem with sensitivity analysis is that the range of sensitivities is often too broad to provide meaningful information for detailed planning. Nevertheless, it is useful for setting frameworks for economic strategy in which economic development is managed and encouraged to operate within agreed parameters.

In Chap. 4 it was shown how input-output analysis was used for extrapolating regional economic futures. However, input-output analysis relies on historical data to generate the coefficients used in developing the economic matrices. In some cases, data used in developing national and local tables may be more than ten years out of date. Developing a coefficient structure for the future of regional economies is difficult when changes in trade, technology, employment and transactions need to be factored into the coefficients. Techniques are available to address some of these problems. The process known as 'hollowing out', where a significant number of local transactions are out-sourced, has been observed in recent research on the economy of the State of Queensland in Australia (West 1998). But the extent of future hollowing of regional economies is highly speculative. The changing nature of work—especially the reclassification of work as the result of outsourcing by firms and government—is compounding problems in input-output modeling and creates difficulties in defining possible future structures for regional economies. While such problems should not undermine our confidence in input-output modeling as a tool for futures analysis, they do suggest the real value of input-output analysis might be more as an analytical and management tool for establishing working hypotheses rather than as a predictive tool.

5.3.2 Futures by Heuristics

Heuristic problem solving is often used for arriving at the best combination of activities to determine a probable series of outcomes. Heuristic programming is an advanced field of computational science, which has been applied to futures evaluation (Wilde 1996). Heuristic problem solving also involves less sophisticated methods of evaluation than that used for mathematical modeling (Hodgetts 1982). It relies on 'rules of thumb' and the use of 'trial and error' methods, and involves a degree of intuition, relying on experienced judgment for solving problems when predictable outcomes are not easily anticipated or capable of calculation. It is probably true to say that a significant amount of strategic planning for regional economic development relies on heuristic problem solving—particularly when time and data are constrained.

Heuristic programming involves a range of qualitative assessment techniques, including the widely used Delphi and Nominal Group Techniques, and round tables/panel discussion groups.

Delphi. This is a heuristic programming technique that uses a combination of qualitative and quantitative methods to arrive at a consensus on a problem or a set of predicted outcomes. It is used extensively for technology forecasting, and is receiving increasing use in economic development planning.

The Rand Corporation in the United States originally developed the Delphi technique. The approach involves using pools of experts that are presented with a series of questions by an investigator or planning analyst. Delphi processes often will involve several rounds of questioning until a consensus is achieved on the problem or issues to be addressed. The application of Delphi techniques can significantly improve the confidence of projected outcomes for planning processes (Bordecki 1984).

Research by Lui (1996) on tourism forecasting in Hawaii demonstrates a higher level of accuracy for forecasts using Delphi than for conventional modeling techniques. She provides an interesting case study because it used both Delphi and computer model forecasts for tourism in Hawaii between 1984 and 1994. The analysis included demand for accommodations, visitor numbers and revenues. The actual figures for 1994 were then compared against the two forecasts. Delphi techniques were shown to be between 15 and 25 percent more reliable than the computer modeling techniques used.

Roundtables and *Nominal Group Technique (NGT).* Roundtables are a particularly useful means of analyzing futures. Initially developed for use by business corporations, roundtables have been used extensively in local economic development to address issues and problems facing regions. In Australia, in recent years the Centre for Regional Enterprise Development (CARED) has convened roundtables bringing together business, governments, and community development organizations in regions from around the nation to discuss key issues affecting economic development—including futures and strategy. The main purpose of these roundtables is to generate projects and ideas of economic benefit to communities and regions and to mobilize resources to implement projects from individuals or organizations participating in the roundtables.

For many parts of the developing world, access to advance forecasting tools and reliable information needed to operationalize models is much more limited than it is in the developed nations. Heuristic programming thus represents a valuable tool for making judgments about future outcomes and strategy in these situations. It may not always achieve the best and most reliable answers, but it does allow planners and decision-makers to obtain answers that will be more reliable than those derived from pure 'guesswork'. In some cases, heuristic programming techniques have proved more reliable than the forecasts derived from using advanced modeling techniques. A modified version of the NGT is used in the application of the MSA technique in Chap. 7 is and not elaborated on in detail here.

5.3.3 Futures by Historical Analysis: A Time Basis for Forecasting

Historical analysis was developed to assist diplomats to address crises and can be used to seek solutions to economic and other problems (Davidson 1998). The method uses time as a basis of forecasting. Courses of events from the past are identified and can be used to recreate the issues for a current or future concern. The technique has parallels with the concepts of economic and technology cycles discussed below.

The historical analysis technique is based on the recognition that the problems or pending events do not have to be treated as if they had never happened in the past, and on a process where attempts are made to identify issues that may be applicable to a current problem or concern using analogies from the past. If patterns can be identified, it may be possible to predict outcomes for the future. Historical analysis involves seven stages:

(a) defining the context of the event of action for which a decision has to be made;
(h) looking for historic precedents;
(i) analyzing factors leading up to the event;
(j) analyzing the factors surrounding the current event;
(k) identifying if there are congruent factors between past and present events;
(l) looking at congruence factors to see if they are relevant to the current event; and
(m) predicting the course of outcome to be taken.

5.3.4 Futures as Cycles

Economic cycles are of major research interest to economists. Trade, business, product and economic cycles have been studied extensively for many centuries. In the 17th Century, William Perry observed economic cycles and the need to plan for them through levying taxes. The theory of trade cycles was extensively developed by Ricardo and other economists in the 19th century, and was later explored in detail by Mitchell (1927).

There are also other cycles that have a significant influence on economic development. For example, the Russian economist Kondratiev (1935) was one of the first to identify technology cycles. He developed the theory that global economic growth and development occurs in waves (see Fig. 5.4.)

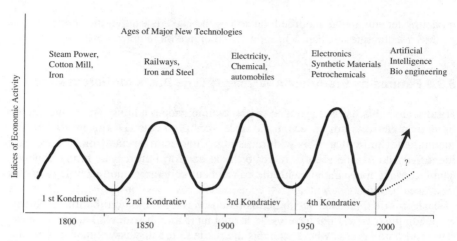

Fig. 5.4. Kondratiev's Technology Cycles
Source: Derived from Dicken 1992, p. 99

There have been four K–waves, each lasting approximately 50 years and being associated with significant technology changes around which other innovations in production, distribution and organization cluster and ultimately spread through an economy. Such diffusion stimulates economic growth. Each wave has a specific geographical impact: K–1 in the United Kingdom and France; K–2 in Germany and the United States; K–3 in the United States and the Netherlands; K–4 in the United States, Japan and Sweden. In the late 20th century we entered a new K–5 cycle which involves artificial intelligence and bioengineering.

Authors such as Dicken (1992), Speigel (1991) and Pearce (1997) have drawn attention to longwave cycle theory and its contribution to ideas on economic development and futures scenarios. Degreene (1993a) argues that management plans, policies and decisions need to fit the given stage of societal evolution. He used Kondratiev cycles/structures to examine the life cycle of collective intelligence and knowledge for decision-making processes. Degreene further conceptualizes the development of the contemporary K-5 Kondratiev cycle as an evolving informational field. That work is highly theoretical, but it provides a useful basis for thinking about long-term strategy and futures. The theory and analysis of long wave cycles also has value in long term strategic planning for cities and regions (Hall 1988).

Economic and technology cycles can be traced in the historic development of many regions. Technology cycles have been recognized in the long range planning of Hong Kong. In developing the Hong Kong 2047 Strategy, the process involved examining the long-term cycles of technology and development leading to a peak of the 5th Kondratiev in the decade 2030–2040 (Enright 1995). That long-term assessment was necessary in order to develop strategies and identify the hard and soft infrastructure ('architecture') needed to reposition the economy in the rising phases of the cycle. While it is impossible to know precisely the elements of technology that will become manifest during this period, Hong Kong realizes that strategic investments in infrastructure have to be considered in the long-term context of expected technology, economic and social outcomes (Enright et al. 1997).

5.3.5 Futures as Scenarios

Scenario analysis has been used extensively to examine options or possibilities for future economic outcomes. It is a valuable—and some would argue an almost essential—tool of strategic planning under conditions of rapid change (Godet and Roubelat 1996; Godet 1990). It can be used to test and evaluate environmental, quality of life, transport, social, political and cultural issues, and as well to assess the impact of policy parameters on economic development processes. In Chap. 4 a case study was presented using scenario analysis in conjunction with the input-output analysis to test five economic development scenarios for the Brisbane-SEQ region in Australia. Scenario analysis takes time, and the techniques used to develop and test each scenario need to be considered carefully. Poor scenario analysis can lead to serious planning mistakes and may result in unnecessary expenditure on projects and infrastructure that may end up being underutilized or abandoned.

Scenario testing may also use linear programming techniques. For example, in 1994 the Queensland State Government in Australia evaluated three scenarios on the size and development of a 1200 km gas pipeline from the Western Gas Fields to service a mixture of large industrial projects in the Brisbane Gateway Ports area in the Brisbane-SEQ region (Brown 1994). The economic benefits of the three scenarios were evaluated using 12 variables relating to industry mix, cost, consumption rate of delivery of gas, electricity prices and demand for water. An optimum pipeline size to service the Gateway Ports Area was identified using a Monte Carlo simulation technique. That analysis provided a useful basis for examining the future long-term planning of energy to service the economic development in the Gateway Ports Area over the next 30 years.

5.3.6 Chaotic Futures and Games

Over half a century ago Van Neuman and Morgenson (1943) postulated that all systems are dynamic and can fluctuate from one form to another. A small event can lead to an exponential series of cascading events that might lead a system into chaos. A classic hypothetical example is a butterfly flapping its wings leading to a major weather event of catastrophic proportions. Eventually destabilized systems develop a more stable state–a kind of harmonic–before becoming destabilized again. Systems thus develop a continuum of stable and unstable states. More advanced theory has argued that determinants exist which operate to bring order to apparent chaotic conditions. Chaos theory (or fractal theory in the view of some) has been developed into a wide range of uses for predicting possible outcomes of events—for example, the spread of disease and congestion modeling.

Game theory has been used widely for solving business problems. It involves what are termed 'conflict of interest situations'. The strategic situation in game theory lies in the interaction between two or more persons (actors), each of whose actions are based upon the expectations of others over whom one has no control. The outcome is dependent upon the moves of participants. Game theory has been used extensively for predicting outcomes of conflict. It is now being used in scenario testing for a wide range of potential outcomes, including urban form. Game theory is based on using probability analysis of combinations of events to develop future outcomes. It is a useful planning tool for testing future negative and positive outcomes of change to urban systems.

Grabowski (1994) has used game theory concepts to evaluate differences between the arbitrary, short-run powers of the state and the long run, infrastructural power of the state on economic development in Japan. The evolution of infrastructural power is related to culture/ideology. Grabowski showed that successful economic development is closely linked to the infrastructural power of the state. The research findings explained why numerous short-term measures to stimulate the Japanese economy between 1991 and 1998 would and/or did fail.

5.4 Recent Developments in Economic Strategy Formulation

The growing importance of futures analysis within regional economic strategy development is being recognized widely by governments, business and development organizations. However, there remains considerable confusion about what is meant by strategy. What is the current status of strategy, and what is it that shapes future strategy? There is also confusion about the role of strategy and the role of planning, and whether they are the same or different things. It is thus important to trace the factors that have contributed to thinking about future strategy, and to examine recent developments in strategy in the areas of competitiveness, sustainability and collaborative competition—issues which are now widely held to be crucial considerations for regional economic development strategy formulation and planning in a wider context than just economic considerations.

5.4.1 Strategy as Competitive Advantage: Porter's Diamond Model

Globalization has fundamentally changed the way nations and regions compete for business in the contemporary world. As discussed in Chap. 1, it has shifted the focus of strategy from comparative advantage to competitive advantage. A problem for regions and communities is how to measure competitive advantage. Much of our thinking on *competitive advantage* has developed from management theory.

One of the leading theorists on competitiveness is Harvard management guru Michael Porter. He explores techniques for analyzing industries and their competitors, and how firms might develop strategies to gain competitive advantage (Porter

1980; Porter 1985). Strategy is seen to develop from an understanding of four forces driving industry competitiveness: potential entrants, buyers, substitute suppliers and industry competitors. This introduced the concept of internal and external environmental analysis. For Porter's model the 'environment' was exclusively the economic/business environment.

Porter realized that the study of firms and industries was not sufficient to explain competitive advantage. This led to further research on the competitiveness of global industries (Porter 1986) and later to research on the competitiveness of nations (Porter 1990).

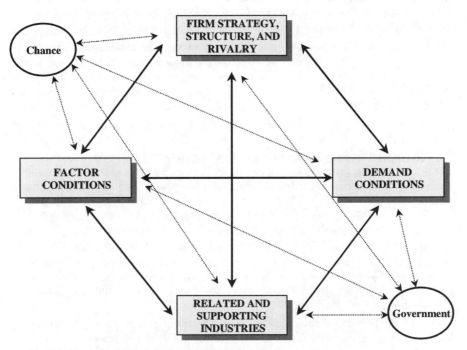

Fig. 5.5. Porter's diamond model of competitiveness
Source: Derived from Porter 1990

In the diamond model of competitiveness (see Fig. 5.5) Porter (1990) introduces four broad attributes that shape the environment in which local firms compete for business (Porter 1990, p. 71):

(a) *Factor conditions*, which include skilled labor, resources, technology and infrastructure necessary to create competition in a given industry or cluster.
(b) *Demand conditions*, which are the nature of home demand for industry products and services.
(c) *Related and supporting industries*, where the presence or absence of supplier and distributor industries in support of a sector industry or clusters will determine its competitiveness.

(d) *Firm strategy, structure and rivalry*, which relate to conditions in a nation governing how companies are created, organized and managed and the nature of domestic rivalry.

Porter (1990) states that:

… These determinants individually and as a system create the context in which a nation's firms are born and compete: the availability of resources and skills necessary for competitive advantage in an industry; the information that shapes where opportunities are perceived and the directions in which resources and skills are deployed; the goals of the owners, managers and employees that are involved in and carry out competition; and most importantly the pressures on firms to invest and innovate…The determinants in the 'diamond' and interaction amongst them create the forces that shape the likelihood, direct and speed of improvement and innovations by a nation's firms in an industry. (p. 173)

Porter also recognizes two other factors that significantly affect competitive advantage:

(a) the role of governments; and
(b) the role of chance.

Chance relates to events or occurrences that have little to do with the circumstances of the country, but may be influenced by national governments. Governments have significant roles in aiding competitive advantage through public policy that is favorable to investment and profit performance. Porter identified the importance of clustering competitive industries to create rivalry and stimulating innovation.

Porter's work on the competitiveness of nations has had an important influence on strategic thinking and analysis of business performance and economic development. While he was concerned with the competitiveness of nations *per se*, many of his case studies provide insight into the competitiveness of regions. Several attempts have been made to use Porter's analysis to identify the competitive advantage of regions, including in services (Daly and Roberts 1998) and education (Lewis 1993).

However, the lack of regional data and difficulties in comparing data between regions and countries gave arise to criticism of the Porter diamond model (Rugman 1991). There are also other difficulties with the model. Ohmae (1996) argues that regions are replacing nations as the engines of economic growth. And Carnoy (1993) points out that globalization means that firms and multinationals operating largely outside the realm of the state. He says that:

… can create competitive advantage in regions of outsourcing by entering into alliances with other regions to overcome competitive weaknesses or leveraging competitive strengths. This does not suggest that the competitiveness of the nation state is no longer relevant. It is. The nation state can influence significantly the environment in which multinational corporations make choices about the location and scale of investment and production (p. 45).

But despite its critics, Porter's diamond model nonetheless provides a useful framework for strategic thinking on regional and community economic development. The model can be used to help identify and analyze the interaction of those factors that underlie local competitiveness, as well as to help formulate strategy for regional economic and industry cluster development based on identified elements of competitive advantage. Porter's analysis can be adapted to enable regions and firms to identify best practices that can help facilitate competitive advantage. This has led to many economic development strategies focusing on operational effectiveness, but Porter does point to the danger of confusing or equating operational effectiveness with strategy. The focus on operational effectiveness has led firms and industries to concentrate on best practices in quality management, benchmarking, partnering, research and development, change management, technology and innovation to enhance their competitiveness. In regional economic development, there is a growing interest on applying best practice in:

(a) regional organizational management, marketing and promotion;
(b) industry cluster building;
(c) benchmarking economic performance; and
(d) developing smart infrastructure and core competencies.

Clones et al. (1988) illustrate how *benchmarks* may be used to compare economic performance and provide a basis for weaker regions to adopt best practices to enhance their economic performance and competitive position. However, the danger in this approach is that firms, organizations and regions will seek to improve their competitive advantage through the adoption of best practices, but ultimately this minimizes the differences between competitors, effectively neutralizing competitive advantage and leading to the contraction of profit margins and performance. Porter (1996) has looked at this convergence of best practice in terms of a *productivity frontier curve* (see Fig. 5.6). The frontier constitutes the sum of all existing best practices at any given time for an industry. This may be thought of as the maximum value that a company or region delivering products or services can create at a given cost using best practices. The best practice frontier is set by a small number of industries, but the majority of industries will fall well below the frontier line. In order to improve economic performance, regional firms or industries would need to adopt best practices relative to their cost/buyer value position, with the intent of positioning themselves on the curve to maximize their competitive advantage or invent new production process innovations.

Consider, for example, two competing regional industry clusters involved in tourism. Each has a similar buyer value/cost position and competes for tourists in the global market place. Industry cluster A in region Y is operating at best practice level on the productivity frontier. Its economic performance will be better than industry cluster B in region Z. The strategy for industry cluster B will be to improve its competitive position by adopting similar best practices to A. Industry cluster B has the potential to catch up to A. If it does, neither will be in a position of competitive advantage (see Fig. 5.6).

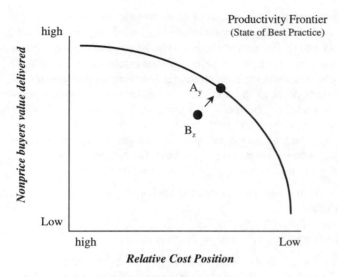

Fig. 5.6. Productivity frontiers of best practice
Source: Derived from Porter 1996

In this example, if all tourism industry clusters with similar cost/value positions to A and B decide to adopt best practices, there will be a convergence on the frontier, and competitive advantage will be neutral. This situation is beginning to emerge in some industries, due mainly to the dominance of multi-national corporations forcing local or global supplier firms in different geographic locations to adopt uniform best practices. As best practice becomes the benchmark by which firms and industry clusters operate, competitive advantage will narrow, and profitability margins globally will fall.

How do local industries and clusters gain competitive advantage when faced with a convergence of best practices? Porter (1996) suggests the answer to this is to recognize that operational effectiveness must not be considered as strategy. Strategy is what it takes to position an industry or industry cluster at a higher point along the best practices productivity frontier, or to extend the frontier. The *best practices productivity frontier* will constantly shift outwards as a result of new technologies and management approaches. However, the role of strategy becomes important in providing the thrust or strategic direction in which a region or an industry cluster must go to maintain its position on the new frontier or even extend the frontier by becoming a leader. The reality is that relatively few regions or industries will be in the position to do this.

A critical decision relating to strategy formulation is how a regional firm should or industry cluster shift its relative cost or value delivered position relative to a new frontier? That decision can have a significant impact on economic performance and the position of an industry on a new productivity frontier. Using the example of the tourism industry cluster, it could move in one of three directions on the new frontier (see Fig. 5.7).

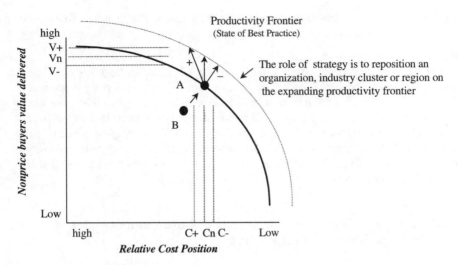

Fig. 5.7. Strategy leading to new productivity frontiers
Source: Derived from Porter 1996

If it moves in the same direction as the expansion of the new frontier, it will lose its cost and buyer value delivered position (V, C). This may result in the industry moving to a lower value-adding state or possibly exiting from the market. If it chooses to move vertically, its positions will remain neutral (V_n, C_n in Fig 5.7). If it chooses to increase its value position, it must move to a position of higher relative cost (V_+, C_+). This will probably mean moving to a higher value adding position. Whichever direction is selected depends on circumstances. Not all choices will be available, but the neutral and higher position should achieve a position of increased competitive advantage.

The *best practices productivity frontier concept* is important in formulating economic development strategy for a region. If regions do not develop strategy to stay on the frontier, then they will be in danger of loosing competitive advantage. One of the principle reasons Japan is now facing difficulties in regaining competitive advantage in many industries in which it was seen to earlier lead over the last couple of decades is that the rest of the world has caught up with Japanese best practices, while industries in Japan have not developed strategies to move to the next level of the productivity frontier. The expanding frontier indicates the need for strategy to continuously position and reposition industries on new frontiers that are constantly being redefined through process and sometimes product innovations by rapidly changing best practice. To do this, economic development strategy needs to concentrate on factors that create the new frontiers. According to Hamel and Prahalad (1994), these include:

(a) regional innovation;
(b) management and technology;
(c) resource stretching and leveraging; and

(d) core competency building, network building and building new economic or strategic architecture.

Some of these terms are discussed below.

The best practice productivity frontier concept has particular relevance to enhancing the competitiveness of *industry clusters* in regions, which are discussed in Chap. 6. The role of regional economic development strategy should give direction to regional industry on positioning and extending the threshold of best practice through enhancement of the factors described above. Regions need to fully understand their competitiveness and to know where their competitive position lies on the productivity frontier. The role of regional strategy should be to maintain the position on the frontier, or to advance it outward.

5.4.2 Strategy as Stretch and Leverage of Core Competencies Through Strategic Architecture

Hamel and Prahalad (1994) have developed an approach to thinking on regional development strategy and the future. They challenge the view that strategy is primarily concerned with fit, arguing that strategic fit does not necessarily lead to competitive advantage. Their view of strategy is that it is not enough to optimally position an organization within existing markets. Organizations need to challenge and pierce the fog of uncertainty and develop greater foresight into the whereabouts of tomorrow's markets and economic opportunities.

For firms and regions to be successful in the future, it is necessary for them to *unlearn* much of the *past*, and to *learn* a new strategy. This involves exploring the different frontiers and markets and understanding where a firm or region can create opportunities in future—without fully understanding what these opportunities might be. The future is not something that can be extrapolated or guessed (Prahalad 1996). To compete for the future, firms and regions need to build a unique *strategic architecture* that provides a blueprint for building those *core competencies* needed to dominate future markets. To gain competitive advantage, firms need to develop *competitive core competencies* that enable them to *leverage* and *stretch* resources to capture emerging economic opportunities as the future unfolds. The Hamel and Prahalad model incorporates four basic components of strategy building (see Fig. 5.8). Those are:

(a) identifying and developing core competencies;
(b) defining economic possibilities to strengthen or create new opportunities that build upon competitive advantage;
(c) defining strategic intent; and
(d) defining strategic architecture.

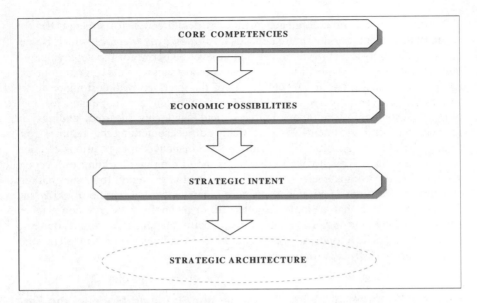

Fig. 5.8. Processes for strategic thinking
Source: Derived from Hamel and Prahalad 1994

These require further explanation:

Core Competencies. These are not resources, but rather technologies, skills, re-source applications and management that, when combined in certain ways, enable a firm or region to produce goods and services for export or domestic consumption competitively. Building *core competencies* requires an understanding of the interrelationship between factors creating the right mix of skills and resources that enable specific enterprises to develop in an organization or a region, and in addition it requires the provision of missing infrastructure to create a powerful array of competitive competencies to advance to the next stage of development.

Defining Economic Possibilities. This involves stretching and leveraging of re-sources, markets, infrastructure and core competencies to realize new opportunities that lie in areas referred to as *white spaces* (Hamel and Prahalad 1994, p. 252). Leveraging, redeploying and recombining core competencies and resources produces opportunities which may be thought of as falling between a firm's existing business units and leverage. In the context of a region, if it can stretch resources this will open up a rich and fertile ground for innovation and development of new business opportunities. It involves networking or chaining relationships within and between industry sectors that result in new or expanded business opportunities.

Strategic Intent. This describes the *path* for a firm or region to create it's economic future. Strategic intent goes well beyond a basic statement of mission or vision—rather, it conveys a sense of clear direction, a sense of discovery, and a sense of destiny.

Strategic Architecture. Strategic intent provides direction for strategy, but the capacity to compete for the future depends on *strategic architecture*, which Hamel and Prahalad (1994, p. 118) define as a high level blueprint for the deployment of new capabilities, the acquisition of new core competencies or the migration of existing competencies, and the reconfiguring of the interface with customers. It describes the process or recipe for developing and utilizing core competencies, mobilizing resources, developing markets, and achieving desired endings in realization of strategic intent. Strategic architecture is something that requires continual rebuilding to enable a firm or a region to maneuver into positions of competitive advantage. This involves continuous monitoring and building to leverage and stretch core competencies within an organization or region to create markets and service customer requirements. A core competence is a bundle of skills and technologies rather than a single discrete skill or technology. A core competency is not a resource, but rather a unique and competitive application of a resource to support business development or customer need. Developing core competencies is part of a *learning process* for organizations.

5.4.3 An Example: Applying the Hamel and Prahalad Model in the Far North Queensland Region

The Hamel and Prahalad model provides a useful basis for developing new strategic thinking for regional economic development. The key concepts contained within it have been applied in regional economic development in Australia, including drafting a strategy for the Brisbane Ports Area in the City of Brisbane in Australia (Maunsell and AHURI 1998). This strategy incorporates a statement of strategic intent and a description of strategic architecture to support the development of six industry clusters. The strategic architecture for the area focuses on high quality infrastructure, the creation of strategic alliances and industry networks, and a focus on building core competencies necessary to facilitate the growth of the six industry clusters. Core competencies targeted for development include:

(a) integrated business management;
(b) marketing capacity for export development amongst cluster industry stakeholders;
(c) collaborative marketing intelligence;
(d) research and development;
(e) government business partnerships;
(f) intermodal logistics capacity; and
(g) sustainable production and environmental management.

These elements of strategic architecture are those considered vital to enhance the capacity of the area to compete for business and investment.

In the Far North Queensland region of Australia, an economic possibilities matrix was developed to identify 'white space' opportunities (that is, opportunities that are not known but may represent future potential) by stretching and leveraging resources within and between industry sectors (Roberts and Stimson 1998). The technique involved deriving a matrix in which an informed group of regional business managers and government officers were asked to identify and evaluate opportunities for leveraging resources between industry sectors to create new business and investment (see Fig. 5.9).

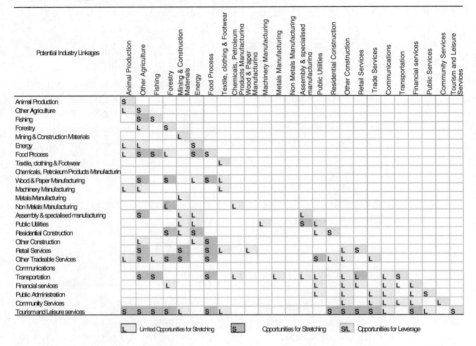

Fig. 5.9. Matrix of economic possibilities for the Far North Queensland region, Australia
Source: Roberts and Stimson 1998

The analysis, conducted in 1994identified over 65 'white space' opportunities for leveraging and stretching (AHURI 1995). That analysis led to the creation of several new developments by the primary industry sector with the tourism and business services sectors. The methodology used to apply the Hamel and Prahalad model is described in more detail in Chap. 7.

5.4.4 Strategy as Sustainable Development

Increasingly it is becoming recognized that finding sustainable solutions in regional economic development is paramount to ensure that future generations enjoy a good quality of life, social equity is improved, current levels of wealth and prosperity are maintained, and the desirable environmental outcomes are achieved.

Historically, the approach to sustainable development has tended to be to address it on a 'sector basis'. The initial concept of sustainability involved a focus on ecologically sustainable development. This involved governments developing measures to reduce the destruction of the natural habitats and the consumption of non-renewable resources. Later the incorporation of social and economic issues added new dimensions to the concept of sustainability. What emerged in the late 1980s was a tripartite model of sustainability often represented diagrammatically by three intersecting circles depicting the natural, social and economic environments (Fig. 5.10). Sustainability was to be achieved somehow by a convergence of views to achieve balanced development outcome, taking into consideration the economic, social and economic environments impacts of development.

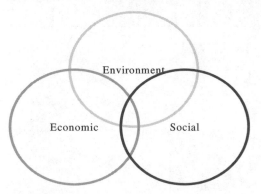

Fig. 5.10. Traditional model of sustainability

This tripartite model, however, presents conceptual difficulties on how to integrate the three dimensions of sustainability and create mechanisms that allows trade-offs over the use of social, economic and environmental resources. This led to concepts such as 'triple bottom line accounting' (TBLA) (Elkington 1997) being developed which tried to provide a basis for measuring the use of different kinds of capital for development and other purposes. The difficulty with the tripartite model is that the union of the three environmental domains was never well explained. The intersecting areas are in fact trading areas or zones where capital is exchanged and converted according to governance mechanisms defined by societal values, principles, customs, laws and rules. These determine how individuals and society are permitted to use natural (capital) resources and transform these into economic and physical capital. These values and rules vary from country to country.

If we consider sustainable development as a trade-off between environmental, social and economic values over the use of economic, social and natural capital, then it is conceptually possible to plot these values on axes to create what we might call the triangular plane of sustainability. Figure 5.11 shows how these values might be represented conceptually as the interplay between three value components: economic, environmental and social.

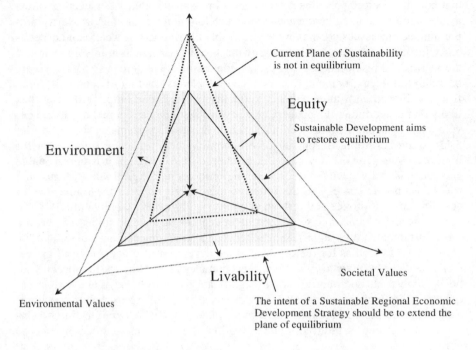

Current Plane of Sustainability
is not in equilibrium

Equity

Sustainable Development aims
to restore equilibrium

Environment

Societal Values

Livability

Environmental Values

The intent of a Sustainable Regional Economic
Development Strategy should be to extend the
plane of equilibrium

Fig. 5.11. Framework for sustainable development

All societies, governments, businesses and communities use these values to make choices or decisions about many things that determine the functioning of society. The play-offs between these value factors define levels of environmental sustainability, equity and livability acceptable to society and the community.

In many societies, it has been common for economic values to hold a higher place than societal and environmental values. The thrust for economic growth has been driven by the belief that economic prosperity is gained by utilizing and transforming natural resources to create wealth that ultimately improves the social well-being of society. This is the fundamental basis of capitalism. The realization that economic development has not achieved these goals, that resources are not always unlimited, and that equity and social justice are important to civil society, is challenging many beliefs and assumptions we have about economic development processes. It is forcing us to look towards more sustainable approaches to economic development in the future. The Asian economic crisis of 1997, the collapse of the previous Soviet Union economies in the late 1980s and the bursting of the ICT bubble in 2001 are three examples of events of world significance that demonstrate the need for more sustainable approaches to both national and regional economic development.

The net result of past economic development practice has been that the current plane formed by the intersection of these value systems, as represented in Fig. 5.11 is not in equilibrium. Environmental resources are being consumed at a rate that may increasingly become problematic from a sustainability perspective and at a cost that may jeopardize the achievement of long-term economic prosperity. Important societal issues concerning the equitable distribution of wealth and accessibility to finite resources has led to growing social dissatisfaction in some parts of the world. The result has been that some communities are demanding greater care be given to the environment, that a more equitable distribution of economic benefits be achieved and improvements to quality of life be accomplished. It is thus likely that in the future the primary objective of sustainable regional economic development will be to restore equilibrium between value systems as shown in Fig. 5.11. Some constraint on pursuit of economic values, particularly the quest for high economic growth, may be necessary in the short term, but it is unreasonable and unrealistic for societies to lower long-term economic aspirations. Thus, the role of economic strategy in the future is more likely to push the framework of equilibrium to improve the threshold of equity, sustainability and livability.

The need for economic development in regions and communities to become more sustainable poses significant challenges for governments, businesses and the community, as well as for regional economic development organizations. Few societies will be willing to give up the economic gains of the past, especially in newly industrialized countries where significant gains in poverty alleviation and quality of life have been achieved. However, unreasonable or excessive consumption of environmental resources, as demonstrated in the past, could lead to declining economic returns, reduced social aspirations, and in a worst-case scenario civil unrest.

The need to achieve sustainable economic development is giving rise to a greater emphasis on integrated approaches to strategy. The 1991 Resource Management Act in New Zealand represents an early attempt by national government to seriously address sustainable economic development planning at both the national and regional or local levels. That Act introduced integrated holistic planning. More recent changes to laws in the States of Queensland and New South Wales in Australia require all planning schemes to adopt integrated strategic planning at the local government level and for economic development strategies to incorporate principles for sustainable development. Integrated approaches to strategy require more extensive consultation, and accepting the reality that environmental and societal values are legitimate constraints to the pursuit of economic development goals. In short, there is a need to set strategies that aim to restore equilibrium and advance the framework of sustainable development.

In addition to emphasizing the development of high-level economic capabilities in a region, the principles of environmentally sustainable development (ESD) are increasingly being incorporated into regional economic strategy. The impetus comes from the global principles of ESD established under United Nations Agenda 21 adopted at the Earth Summit conference in Rio de Janeiro in 1992.

The United Nations Habitat II Agenda in 1996 and the Greenhouse Gas Agreement in 1997 are international policy documents that are also influencing economic decision-making and investment for the future. While governments have been slow to respond to and implement global agreements, continued environmental degradation ultimately will tend to move all governments, corporate business and communities toward greater conformity with these agreements.

5.4.5 Strategy as Structured Chaos

Earlier in this chapter typologies of future states were discussed and it was argued that strategy will become more concerned with managing probable futures, noting that probable futures lie in a domain between chaos and planned futures. Brown and Eisenhardt (1998, p. 11) argue that business strategy is moving close to a state that they refer to as the edge of chaos, "a natural state between order and chaos, a grand combination between structure and surprise". They argue that to achieve successful strategy it is necessary to learn to manage the unpredictable and uncontrollable, to accept inefficiencies as part of the learning process, and to be proactive, continuous and diverse to capture new opportunities the future will offer. Strategy is not just about focus and position, but also about ensuring a firm or region is robust and diverse, making a moves to ensure some will be brilliant enough to ensure success and survival in an environment which will become increasingly dynamic and competitive.

Brown and Eisenhardt have identified four models of competitive strategy that have relevance for regional economic development (see Table 5.1).

Table 5.1. Four concepts of strategy

	Five forces model	Core competencies model	Game theory model	Competing on the edge model
Assumptions	Stable industry structure	Firm and regions as a bundle of competencies	Industry in regions viewed as dynamic oligopoly	Industries in rapid unpredictable change
Goal	Defensible position	Sustainable advantage	Temporary advantage	Continuous flow of advantage
Performance Driver	Regional industry structure	Unique regional competencies	Right moves	Ability to change
Strategy	Pick an industry; Pick a strategic position, fit the organization	Create a vision; build and exploit competencies to realize vision	Make the right competitive and collaborative moves	Gain the 'edges', time, pace, shape semi-coherent strategic directions
Success	Profits	Long-term dominance	Short-term win	Continual reinvention

Source: Adapted from Brown and Eisenhardt 1998, p. 8

The first column in the table lists the assumptions, goals, performance driver, strategic intent and success benchmarks for strategy. This 'five forces model' is aligned with Porter's ideas on competitive strategy discussed previously. The core competencies model is aligned to Hamel and Prahalad's view of strategy. The 'game theory' model has embedded within it the ideas of Hammer (1997) and Moore (1996) with strategy being offensive and defensive depending on the situation. The fourth concept of strategy -competing on the edge—is based on the view that a firm needs to constantly organize for change and that semi-coherent strategy direction should emerge as a result: in other words, it is about combining two parts of strategy by simultaneously or concurrently addressing the questions of where to go and how to get there. A 'semi-coherent strategic direction' is fundamentally different from what traditionally is called 'strategy'. Traditional strategy is long-term focused and coherent. Semi-coherent strategy involves identifying and capturing short-term opportunities on the edge of the future. 'Competing on the edge' is a perspective of strategy that is different to that normally used by organizations, firms and governments. It tells us that strategy is not a uniform process, and that strategy must be more closely aligned and exposed to the future states discussed earlier in this chapter.

For some regions—especially those associated with advanced technology and innovation—change will become more rapid, and it will produce operating environments less stable or more chaotic. Traditional approaches to strategy may no longer be applicable in these regions as the restructuring and reinvention processes are so rapid that long-term strategy is unlikely to remain relevant for long. The transformations taking place in the economies of regions such as Silicon Valley in California, the Research Triangle region in North Carolina, Austin Texas and the Washington D.C. National Capital region in the United States suggests that regional economic development strategy needs to be more dynamic and semi- coherent. In rural areas and in some process manufacturing regions there is an especially acute need for strategies built around the five forces and aimed at building core competencies.

5.4.6 Collaboration as Strategy

The belief that strategy primarily is concerned with competition has been under challenge (Moore 1994). As discussed earlier in this chapter, the pursuit of best practice leads to a convergence of competitive differences, which suggests that firms, industries and regions should be looking for new ways to gain competitive advantage.

One means of gaining competitive advantage has been for businesses and public agencies to improve internal efficiencies by outsourcing or privatizing. This practice is leading to greater specialization—especially in small and medium size component and producer service industries. However, outsourcing leads to supplier companies working for competitor clients. Information is shared, reducing competitive position, and thereby forcing organizations along a path of continuous innovation (Kash 1989). For a firm or an organization to be openly competitive, it would need to collaborate with suppliers and distributors that provide components and services needed to develop and sell new products and services. In some situations, collaboration with a support organization will be the difference in winning or losing a competitive contract.

However, collaboration is also evolving along other paths. There is a growing realization that no organization has the resources to develop and trade new products or services in isolation. Even if it did, it is unlikely to be in the position of controlling the distribution, marketing and retail networks required to get products and services to consumers and other markets. Firms are becoming increasingly dependent upon networks, partnerships and alliances to do business. This is leading to the emergence of strategy as collaboration through strategic alliances, networks and partnerships (Yves and Hamel 1998; Cunnington 1991; Segil 1998). Strategy as collaboration leads to opportunities to stretch and leverage resources, technology, and infrastructure in ways not previously conceived. It enables business, government and communities to capture and create new opportunities for economic development, and to reduce risk to expand economic development frontiers that single or pioneering firms or organizations would be unlikely to attempt. Collaborative strategy can reduce risks, improve efficiencies, and help conserve resources and advance innovation frontiers in a way single firms or industries can never achieve.

Collaboration as a strategy requires one important element to transform the benefits of networks, alliances and partnerships into development outcomes: this element is catalysts. Catalysts are persons, mechanisms or instruments that facilitate or trigger linkages and transactions between business, government and community so as to:

(a) identify potential business investment contracts and infrastructure;
(b) create networks and clusters of core competencies attractive to business;
(c) draw together people, resources and technology; and
(d) facilitate investment through a range of finance, land and business packages.

The involvement of catalysts allows greater creativity in drawing upon available resources in a region than would be possible through a single development organization or the individual efforts of business enterprises or government. In the context of regional economic development, this catalytic element is about leadership, an issue on which we focus in Chap. 8.

An example of a collaborative strategy for economic development is the planning for a new large urban development to the north of Brisbane within Australia's third largest metropolitan region. The framework for the strategy is shown in Fig. 5.12. The economic strategy was developed in conjunction with a master plan to create an 'edge city'. The intent of the strategy is to facilitate rapid development of value adding enterprises through building virtual strategic architecture pulling together collaborative networks, alliances and partnerships, and business and community catalysts to create new investments and infrastructure, management services, community development organizations, projects and programs which will make the new edge city a highly competitive and pleasant environment for business, employment and living.

Management or Commercial Governance Framework

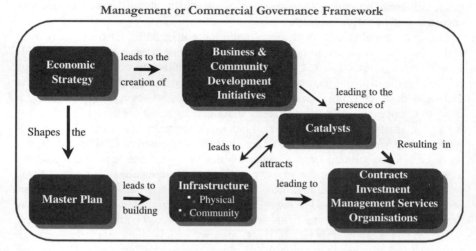

Fig. 5.12. Framework for a collaborative economy for a new edge city in Brisbane

5.5 A New Synthetic Strategic Framework for Planning

So far in this chapter different approaches to strategy which have influenced, and are likely to continue to influence strategic thinking and planning for regional economic development have been described. Below a new synthetic approach to strategy which incorporates many of the concepts discussed is presented. That approach involves the 14 stages set out in Fig. 5.13 that elaborates on the process introduced in Fig. 1.6 in Chap. 1.

5.5.1 Establishing a Platform for Change

For an economic development strategy to be successful it is essential that the planning process begin by establishing a platform for change (stage 1 in Fig. 5.13). If a region or local community is not supportive of change in its economic position, then the preparation of an economic development strategy will be a far less fruitful exercise, as there will be little commitment to implement any strategy developed. Economically successful regions or local communities typically are those which have embraced the need for change, and have methodically set about building a *platform for change*. The process involves regional or community leaders mobilizing support for change—using a range of media, community and organizational support groups to educate the community and members of organizations of the benefits of managed change—as opposed to having change imposed upon the community, which often results in defensive or reactionary strategies. With this approach, once a platform for change has been established, a steering committee needs to be formed comprising business, community, government and other stakeholder interests to develop a process to prepare an economic development strategy and plan.

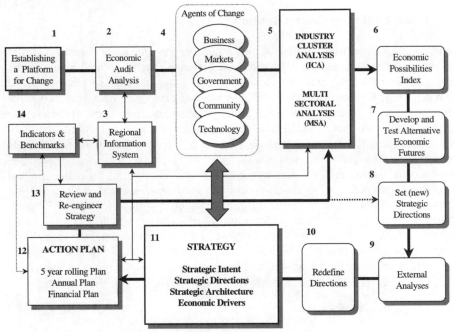

Fig. 5.13. A new framework for regional economic development strategy planning

5.5.2 Economic Audit

The first priority of a strategy steering committee is to initiate the preparation of an economic audit (stage 2 in Fig. 5.13). The audit involves the methodical collection of regional, national and international data to provide information to investors, business governments and other organizations concerned with economic development in the region. Typical approaches used to undertake a regional or local community economic audit were discussed in Chap. 2. The audit should incorporate the collection and assessment of at least the following information (stage 3 in Fig. 5.13):

(a) *economic data*, including gross regional product, trade data, production data, capital base;
(b) *labor markets*, including skills base, education levels, special competencies;
(c) *resource base*, including natural resources, other economic and environmental resources;
(d) *infrastructure*, including type, capacity, quality and flexibility;
(e) *technology*, including the range, level of advanced and new technologies used in production processes; and

(f) *social capital*, including business networks and community organizations.

5.5.3 Regional Information System

There is no single 'best' format for preparing an economic audit. The availability of resources and access to information will determine the extent and nature of the audit. When collecting data it is useful to assemble this in an electronic format and standardize it for use in national and international comparative analysis. To make comparisons of local or national data, it may be necessary to transpose information into an internationally accepted standard. Standard Industry Classification (SIC or now NAICs) is used universally by international trade organizations to enable international comparisons. However, care should be taken in converting cross-national data to the SIC/NAIC. Techniques for collecting and analyzing data collected in economic audits have been described in Chaps. 2 and 3, and then should be used at this stage. The outcome of this process should be to build an interactive regional information system (stage 3 in Fig. 5.13), which might already exist, but which typically will be of an inadequate nature and will require substantial resources and effort to redesign in order to conduct analyses required to evaluate the existing nature and performance of the regional economy.

5.5.4 Agents of Change

The next stage in the process (stage 4 in Fig. 5.13) involves the identification of actors who are agents of change in a region's economy. This can be assisted through the identification and assessment of institutional capacity, capability and preparedness discussed in Chap. 2. Identifying actors who play a role as agents of change in a regional economy is an important first stage of strategy formulation. Those actors may include: community, business and organizational leaders; scientists and technologists; educators; environmentalists; entrepreneurs; and media personnel. Mobilizing the support of local and external change actors helps to involve these people who will derive benefit from the strategy. It is important that the agents of change be encouraged to become involved in industry cluster and reference groups. Those people and organizations will have a key role in facilitating strategy development and implementation in a later stage of the process. Typically there are six main types of change agents:

(a) Business agents: Businesses are an important agent of change, as they create new investment, develop new products and services, and introduce new technologies that change production, service and distribution systems in a region or community.

(b) Market agents: Markets operate both internally and externally to a region. Changing consumer demands and patterns, product cycles, demand for information, and distribution systems change the way economies operate over time, in both the wider society context as well as the local regional or community context.

(c) Government agents: Government at all levels has a role in regulating, supporting and planning for regional economic development. Changes in government policy affect the way a regional economy operates and competes, both nationally and internationally.

(d) Community agents: Regions and communities comprise a wide range of interest groups and organizations, whose beliefs and values impact upon change. These need to be identified and co-opted to the process.

(e) Technology agents: As technology product cycle times decline, the impact of technology as a change mechanism in production and services will become more rapid. Changing technology is the most rapid agent of change upon economic systems.

(f) Information agents: With the rapid growth of the global economy, information availability has grown rapidly, resulting in massive changes in the way social and economic decisions are made and the way a region or community is managed.

5.5.5 Industry Cluster Analysis (ICA) and Multi-Sector Analysis (MSA)

As shown in Fig. 5.13, stage 5 of the process involves two specific analytical techniques which identify those *industry clusters* which drive a region, or which might potentially evolve to drive its development in the future. This involves obtaining expert assessment of industry sector performance and potential in order to stretch and leverage to enhance industry cluster development. This is achieved through using a qualitative technique called *multi-sector analysis*. These methodologies are discussed in detail in Chaps. 6 and 7, but a brief explanation of them is necessary at this stage.

Industry Cluster Analysis (ICA). This involves the identification of interrelated industrial sectors that drive the basis of the economy, or have the potential to develop new economic activities to create a different and/or more diverse economy in future. Industry cluster analysis involves: identifying core bundles of industry sectors or clusters; analyzing the supplier distribution network of core industry clusters in a region; identifying strategic infrastructure supporting existing or potential future industry clusters; and defining mechanisms for enhancing leadership, developing networks, alliances and partnerships within industry clusters.

Multi-Sector Analysis (MSA). This technique, developed by Roberts and Stimson (1998) involves understanding the sectoral strengths and weaknesses of a region's economy in terms of its core competencies, strategic infrastructure, marketing intelligence and risks. The analysis is used to reveal factors that create competitive advantage, and to identify elements that may need to be enhanced as part of economic strategy. The results derived from MSA enable regions to concentrate resources in priority areas that will lead to increased economic performance.

5.5.6 Economic Possibilities Index

The next stage of the process (stage 6 in Fig. 5.13) is to undertake an *economic possibilities analysis* (EPA). This identifies opportunities and capabilities to stretch existing industries, or leveraging them to create new economic possibilities. Using EPA, an *index of economic possibilities* can be prepared to show industry sectors or clusters with the greatest opportunity to improve economic performance in a region or community in future. This index is based on the most competitive elements of the economy viewed from the strength of its core competencies, strategic architecture and a future industry perspective. Figure 5.14 shows an index of economic possibilities developed for an economic development strategy study for Far North Queensland in Australia. How this is done and applied is described in Chap. 7.

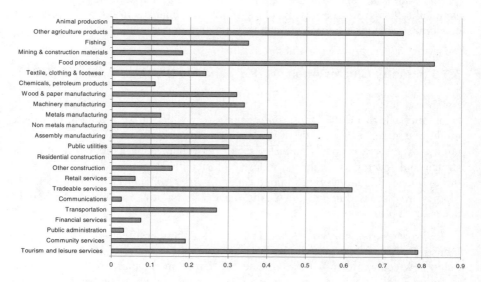

Fig. 5.14. Economic possibilities index for Far North Queensland

5.5.7 Scenario Analysis

At this point it is necessary to bring together stages 4, 5 and 6 in Fig. 5.13 in order to undertake stage 7, which involves the formulation of a number of options for the future development of a region. Not all of these will be realistic, given available resources, population, and core competencies, strategic architecture and scale industry factors present in a region. Regional values, legislation, physical and conservation measures will narrow further the options. The process of arriving at different strategies to be evaluated can take time, depending on the size of a region, options available and weighting of values that will apply to the evaluation processes. An illustration of scenario options testing using input-output analysis is provided in the case study of the Brisbane-SEQ region in Australia as outlined in Chap. 4.

There exist a range of qualitative and quantitative techniques available to evaluate scenarios. Some of these are: cost/benefit analysis; internal rate of return (IRR) analysis; I/O analysis; Monte Carlo simulation; and linear programming. Qualitative techniques for analysis include: the goals achievement matrix (Hill 1973); Delphi; and other heuristic methods as discussed earlier in this chapter. It is important to realize that the outcome of scenario testing is heavily dependent upon assumptions made about the future. Generally speaking, the greater the number of assumptions built into the modeling or testing process, the greater the number of potential outcomes. Also, assumptions will have a greater impact on the results of scenario testing than others. Thus sensitivity testing of scenarios is advisable.

5.5.8 Setting Strategic Directions

Scenario analysis provides a valuable basis for selecting strategy options. Once a final option or combination of options has been decided upon, strategic directions to realize *strategic intent* should be prepared (stage 8 in Fig. 5.13). Strategic intent will describe what is to be achieved by the planning process. *Strategic directions* describe that *path* for strategies to achieve *strategic intent*. Strategic directions will provide options, accommodate changes, channel resources and determine priorities and a framework for decision making by management. Strategic intent can be likened to designing and packaging a product or idea in location A to be assembled or realized in location B. For strategic intent to be realized, enabling mechanisms are needed for the journey along the strategic path. Strategic directions therefore will need to:

(a) plot the course for the journey;
(b) determine the mediums, pace and options for transportation;
(c) provide alternative routes and contingencies;
(d) place the navigation beacons in advance for guidance in adverse conditions ahead; and
(e) allow for some transformations in the design of ideas along the way.

Strategic intent, and strategic directions for economic development strategies, should state both *where* a region or community intends to go and *how* it intends to get there. The strategic intent may be to develop specific core business activities and competencies that will enable a region to diversify or become more specialized. The strategic directions to achieve this may be to develop key strategic alliances, partnerships and resources, or to develop key smart infrastructure. Details of the planning strategy are developed once strategic directions are set, and these will prescribe in detail the *path* to achieve strategic intent. A major weakness in most strategic planning for regional economic development is that few strategies describe or set strategic directions. Often this leads to strategies becoming a series of disjointed ideas and actions that are not guided along a path to achieve strategic outcomes.

5.5.9 External Analysis

A major problem with strategic planning processes is that most of the analysis and planning are undertaken internally to an organization or a region or community. Often it is the case that external investors, customers, observers and clients are excluded from strategic planning processes. Globalization is fundamentally changing decision-making processes, such that most major investments and often up to or in excess of one-half of local investment will be determined by decision-makers residing outside a region. The key to successful strategy is to have it evaluated externally by the customers, investors and other representatives with interests or associations within the region.

External analysis and evaluation (stage 9 in Fig. 5.13) thus helps to eliminate mismatches on critical issues related to perceived competitiveness, relevance and focus of strategy and investment attractiveness. If an economic development strategy does not match the expectations, knowledge and values of customers and investors, the prospects of failure will be high. External analysis has a second useful purpose; it enables some competitive advantages or disadvantages that external customers and investors may not be aware of to be highlighted. Where advantages are identified by external evaluation, these should be incorporated strategies for marketing and promotion.

External analysis can be undertaken at three stages in the strategy formulation process in Fig. 5.13: the multi-sector analysis and the industry cluster analysis in stage 5; the EPA analysis in stage 6; and following the setting of strategic directions in stage 8. Stage 8 is the most critical. The analysis is usually undertaken by providing external evaluators (customers, agents, investors, public policy advisors) with the results of the cluster analysis, MSA and EPA analyses. They are asked for verification of those findings, to test the strategic directions. The external evaluators may be asked their opinion on whether these are realistic and achievable. While some perverse results inevitably will occur from this evaluation process, the overall benefit will be to give confidence to strategy formulation, and to reduce the risk that strategies will miss the mark. It assists also in broadening the ownership of the economic development strategy and can help by mobilizing external support for its implementation. The external evaluation process is a first step in marketing strategy. This is different to most regional economic development strategies, which usually end up having to be sold once the strategy formulation process is completed.

5.5.10 Redefine Directions

The external analysis process often leads to the setting of new directions to ensure strategies are more closely aligned to markets and customers (stage 10 in Fig. 5.13). It might even be necessary to rethink the entire approach to strategy but in most situations this is unlikely. It is probably not possible for strategy to completely match the region's expectations with those of its firms and their trading partners. The value systems of investors and developers are often different or at odds with local enterprises and with the aspirations of the local community. This is frequently the case with multi-nationals seeking to invest in regions. Also, sustainable development policies are not likely to be universally supported, especially if business margins become squeezed or threatened. Economic strategy should, therefore, recognize these polarities, but should seek also to minimize these where possible. This may mean, for example, excluding from a strategic plan the promotion of industrial activities which have potential for development, but which face community or other opposition for one reason or another.

5.5.11 The Strategic Plan

This brings us to the stage of *developing the strategic plan* (stage 11 in Fig. 5.13) for a region. There exists no universal format for the preparation of strategic plans for regional economic development. However, there are some guidelines which can be applied in deciding the content and structure of a strategic plan document for economic development. A strategic plan document should normally include:

(a) *Statement of strategic intent*: This conveys the purpose of the strategic plan. Strategic intent will describe the main thrusts, philosophies, expected outcomes and organization of the plan, and the framework to manage implementation.

(b) *Planning context*: Most strategic plans commence with a detailed description of background information relating to a region and a summary of the analysis used to prepare the plan. In many strategic plans, much of this information is superfluous to reader requirements and should be compiled in supplementary reports or appendices. The planning context should give a succinct statement of details of a region's economy, its organization and competitive position, and strategic planning issues addressed by the plan. It should describe the futures environment in which the plan will operate and the implications of these on the region and the success of the plan.

(c) *Strategic directions*: These describe the main paths for implementation of the plan. Strategic directions determine such matters as the key future drivers of economic development–such as: the key industries, clusters and markets to develop; and the strategic architecture needed to facilitate the implementation of the strategy and manage strategic outcomes.

(d) *The strategic architecture*: This provides the framework and mechanisms for plan implementation. It describes: the core competencies to develop, leverage and stretch the regional economy; the key elements of strategic infrastructure needed to support key economic drivers and the resources and risks to be managed in and outside an economy; how to create business catalysts to stimulate investment; and the networks and strategic alliances to be established.

The key future drivers of economic development are then established, identifying those industry clusters or enterprises and markets that form the basic economy supporting regional development. Primarily these are likely to be the exporting industries; but they may also involve import substitution industries. The focus of development of key industry drivers is on both industry cluster development and on marketing intelligence.

5.5.12 Action Planning

Action planning provides detailed programs for implementing strategy (stage 12 in Fig. 5.13). *Action plans* are separate from a strategic plan, as by nature they are prescriptive. There are three key elements of an action plan:

(a) *Roll-over plan*: Having a three to five year roll over plan is a relatively new component of action planning. They provide an instrument for continuity for multi-agency funding for projects and programs that run over several budget cycles. Previously action plans have tended to operate on annual cycles, but this causes serious problems in project implementation when funds are cut or an agency withdraws funding for a multi agency project. Organizations in developed nations are moving to five-year rollover plans to support strategic planning processes. In former socialist nations five year planning at the central and regional level was practiced for many years, but they lacked flexibility.

(b) *Annual action plans*: These provide details of funds, resources, responsibilities and scheduling for annual projects and programs supported by strategic plans. It has been common practice for development organizations to use committees, departments, and single agencies to implement action plans. Implementation of action plans for economic development is now moving more towards team, network or partnership approaches to plan management. There are several reasons for this, the primary one relating to outsourcing and the need for government to secure multi-agency or partnership funding from the private sector.

(c) *Financial plan*: Seldom will the resources for completing projects and programs of roll over and annual action plans be fully committed. Some projects will require public and private investment capital to be raised in financial markets. A financial plan should detail the levels and structuring of capital and recurrent funding for projects and programs included in roll over and annual action plans, the expected returns or recovery rates on capital investment and outlays, and any special conditions attached to funding. In some nations state and enterprise zone tax breaks are given for research and development investment and for employment creation. These need to be specified in the financial plan supporting the economic development strategy. A financial plan can be used to benchmark economic performance or set performance indicators.

All action plans need to be reviewed annually, and should be undertaken in parallel with public sector and business budget cycles where possible. It is important to note that these plans are all in support of strategy implementation and must have the flexibility needed to adjust as conditions change. This is in contract to the socialist 5-year planning cycles of the past.

5.5.13 Review and Re-engineering of Strategy and the Provision of Indicators and Benchmarks

Regional economic development strategies should not be static documents. For strategy to remain relevant it must be reviewed and re-engineered at regular intervals of not less than three to five years. *Benchmarking* and *monitoring* economic performance through the use of appropriate *indicators* is an important element of the strategy review process. Benchmarks help define where a region is on the best practice productivity frontier. The extent to which a strategic plan fails to meets its targets will determine how far back in the plan preparation process a review must go. If a plan fails badly, a complete review may be necessary. Experience tends to indicate that many economic development plans after five years need to be completely re-engineered. These are the crucial ongoing phases of the new framework for regional economic development planning and incorporate stages 13 and 14 in Fig. 5.13.

There are a number of examples of alternative approaches to strategic planning for economic development which incorporate the concepts outlined above. For example, Adams and Parr (1997) provide ten case studies of best practice in metropolitan level regional development strategy in the United States and the reader is recommended to look at these cases.

5.5.14 An Example: Applying the Framework to the Far North Queensland Region

The framework for regional economic development strategy planning outlined above has been applied through a project to help develop the Cairns Regional Economic Development Strategy in the Far North Queensland region of Australia. Industry cluster analysis and multi-sector analysis were used to identify the potential to develop new industries within existing clusters by leveraging, and capitalizing on the region's core competencies—particularly in tourism—through its geographic proximity to Asia. The focus of the strategy is two-fold: firstly to facilitate the development of industry clusters in the service and value-adding primary production and processes sectors; and secondly, to improve the region's strategic architecture to develop new opportunities for business and investment. The conceptual framework for the strategy is shown in Fig. 5.15, and while fifteen industry clusters were identified as having potential for development in the strategy, only the main ones are shown in the figure. An active program of facilitating the development of these is now in place. The specific outcomes of the strategy at the end of 1999 were:

(a) a fully developed regional information system;
(b) 15 industry clusters involved in joint marketing and development;
(c) new leveraged industries involving tourism, culture, education horticulture and tropical foods; and

(d) a regional development organization owned and funded by all regional indus-
    try clusters.

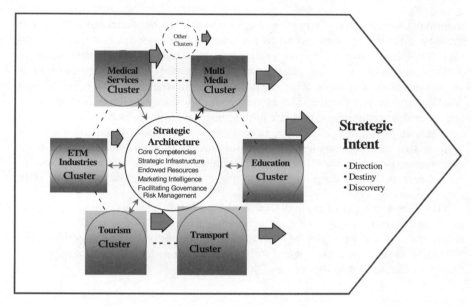

Fig. 5.15. The conceptual framework for the Far North Queensland regional economic
development strategy

5.6 The Future Direction of Strategy for Regional Economic Development

5.6.1 Status of Strategy

From the discussion in this chapter it is evident that strategy for regional economic
development planning is undergoing significant evolution. Conventional approaches to strategy formulation that were discussed in Chap. 2—based on data
analysis, the use of strategic planning methods such as the SWOT methodology,
and extrapolating futures from current trends—have been re-evaluated and have
been evolving into these new approaches which emphasize strategy formulation
building on core competencies and assessing risk in striving to position a region to
be competitive in an uncertain environment. Globalization and rapid economic,
technological and social change are creating a new agenda and landscape for regional economic development planning. While the future direction of the evolution of strategic planning for regional economic development is unclear, a paradigm shift is occurring in futures analysis which suggests we will move away
from predicting and targeting futures to an approach involving anticipating and

managing futures. This will continue to have a profound impact on thinking about strategy for regional economic development in the future.

In the past it has been common for strategic planning for regional economic development to have failed. There has been a failure of planners, and in particular of land use planners, to understand economic planning (Calavita and Caves 1994). Analysis in regional economic development has tended to become somewhat too focused on measuring performance, rather than on the factors that contribute to competitiveness of regions. This requires that greater attention to be paid to non-market as well as to market factors that contribute to the competitiveness of regions, as was suggested in the late 1980s by Byars (1987). Increasingly it is the sensitivities to variations in intangible factors—such as social capital (Bolton 1998), core competencies (Hamel 1996) and value factors (Information Design Associates 1997)—that are determining the path of regional economic futures.

5.6.2 Principles for Future Strategy

What, then, might be the significant principles for future economic development strategy thinking?

By way of an analogy, consider the following. The architecture of a building has a significant influence on the form and function of activities that can take place in that building. Good design is judged by the degree to which a building meets a range of needs by different users, at different times and across different ages. Architecture sets parameters on peoples' movement in and use of space, as well as affecting across the qualities of experience and the sense of association with a place. These are enduring as long as the architecture can adapt to the needs of time, space and changing demand. Strategic architecture for future regional economic development may be thought of in much the same way. The building of strategic architecture for regional economic development requires a wide range of supporting infrastructure, resources, technology and information to enable a modern regional economy to function and compete in the global economy. In the old economy, much of the architecture was fixed and designed for a more defined future. In the new economy, the architecture supporting regional economic development needs to be much more flexible, strategically located in time and place, and intelligent in a sense of creating—rather than waiting for—opportunities to happen in a region. The building of this new form of architecture requires innovative approaches to strategy and analyses of the future.

The conclusion drawn from the analysis of futures and strategies provided in this chapter is the need for broad principles as a framework for strategy formulation. Some of these principles are now discussed as a guide.

Strategy as a Framework for Developing Strategic Thinking. Just as no man or woman is an island, no region is a fortress. Globalization has fundamentally changed the rules of economic engagement for the future. With the gradual removal of tariffs, the free flow of capital and information, and the ease of travel between nations, the economic infrastructures of regions are becoming more internationalized, more integrated and more competitive. As suggested in Chap. 1, for regions to develop in the future, economic development strategy thinking must become focused on the dual principle of 'Think Global: Act Local', and 'Act Global: Think Local'. This implies that changing mindsets will be vital to unlocking the potential of regions. The greatest impediment to regional economic development options in the past has been the incarceration of strategic thinking in its application to strategy.

Strategy as a Framework for Dynamic Change. Strategy in the future will not operate in a steady state environment. In the concept of 'future states' presented earlier in this chapter, it was argued that regions must begin to take steps to anticipate and create the future in an environment of dynamic change. The tendency for regional economic strategy to follow the linear paths for success laid down by others leads to a position of catch-up and the loss of competitive advantage. By developing strategy that creates futures from a position of competitive advantage, regions are not left playing catch-up. An important principle for regional economic strategy, therefore, is that it needs to be revolutionary (Hamel 1996)—not in the sense of promoting radical directional or structural change, but in the sense of developing highly innovative approaches that stretch and leverage strategic architecture—and that it maintain a focus on perpetual re-invention of strategy to position for competitive advantage. That suggests strategy is not something which is absolute, but rather it is something that steers a region along economic paths which are constantly evolving while capturing opportunities for new business, investment and trade development.

Strategy as Convergence and Divergence. The tendency for strategy to encourage diversification often leads to the broadening of the economic base of a region, but it can also over-stretch resources and undermine the competitiveness of an economy. As a principle, a regional economic development strategy should converge on the development of core business and competencies, and on the positioning and provision of strategic infrastructure and resources that give a region competitive advantage and/or the region to become divergent in areas that capturing economic opportunities that form an extension or leverage between core business activities. That maximizes the efficiency and effectiveness of strategic architecture in a region, and will help ensure that a strategy for economic development is based on competitive advantage and the minimization of risk.

Strategy as Collaboration. Globalization, economic uncertainty and the striving for best practice are forcing business and organizations to find new ways to become more competitive. As economic margins are predicted to continue tightening in the future, collaboration—through strategic alliances, partnerships and networking—will become the thrust and trust of future strategy. An important principle for regional economic development strategy in the future will be the focus on collaborative strategy that attempts to maximize strategic advantage rather than compromise competitive advantage. The principle of 'one and one make three' forms the tenet for the principle of strategy as collaboration.

Strategy as a Focus on Clusters as Engines of Economic Growth. Globalization is leading to the spatial agglomeration and specialization of industries in clusters, each with highly developed networks of suppliers and distributors. The role of strategic architecture in a competitive environment is to support the development of industry clusters. The principle of economic strategy should be to facilitate industrial clusters as the engines of economic growth. This will be taken up in Chap. 6.

5.7 Summary

This chapter has provided an overview of emerging new ways of thinking, of analyzing, and of formulating regional economic development strategy for the future. The chapter began with a discussion about understanding the future, followed by techniques we can use to analyze futures. The discussion then focused on new aspects of strategy, including: strategy as competitive advantage; stretch and leverage of strategic architecture; sustainable development; structure chaos; and collaboration with application to regional economic development. A proposed framework for the process of regional economic development strategy formulation incorporating elements of new strategic thought was introduced. The chapter finally looked at future directions and principles to incorporate into economic strategy. That discussion is intended to provide the rationale for regional economic development planners and practitioners to embrace a new paradigm for approaching the formulation of a regional economic development strategy.

There are no set rules for formulating economic strategy that will ensure a region's successful path to the future. However, there are emerging best practices which may be selectively applied and adapted to suit the situation of a particular region. The lessons learned from the failure of economic strategy is that regions must carefully analyze their competitive position, anticipate the road ahead, define strategic intent, and then formulate strategies and create paths to the future. Two new tools to assist regional economic development analysis and futures formulation are of specific importance—industry cluster analysis (CIA) and multi-sector analysis (MSA). Those tools are discussed in detail in Chaps. 6 and 7. Then Chap. 9 discusses the application of geographical information systems (GIS) based methodologies to develop decision support systems–incorporating visualization techniques – that can be used to assist impact assessment and scenario valuation in regional economic development.

6 Industry Clusters and Industry Cluster Analysis

6.1 A New Context

Early chapters in this book discussed how, during the three decades that followed the end of World War II, governments of all persuasion had embarked upon programs to develop national industries to create employment and to achieve greater self-sufficiency in the production of domestic goods and services. Large heavy industry and assembly towns emerged under national industry plans. Many of those industries had strong horizontal and vertically integrated systems of production orientated to the manufacture of total or fully assembled products. Interaction between industry sectors was limited, and there was significant duplication of research, service provision and resource consumption. National industries were also protected by tariffs and monopoly provisions, leading to inefficiencies, reduced competitiveness and declining innovation. By the 1980s, national restructuring policies, globalization and new production technology began challenging traditional production systems leading to emerging global corporations looking at ways to improve efficiencies and competitiveness.

Industry clusters, involving the concentrated production in specialized commodities and merchandise products or services, have played a leading role in the development of cities, nations and trade for over three millennia. Thus there is nothing new in the concept of industry clusters. However, the industry clusters of the past are very different from the clusters that drive regional economies today. Many elements of old industry clusters still survive despite many of their core elements either disappearing or relocating to lower cost production centers. Old industry clusters were firmly entrenched in manufacturing. The new industry clusters—especially in developed economies—combine services with technology and/or high value added production processes.

Undoubtedly there has been a growing interest by governments, international development agencies and business in the value of, facilitating the growth of industry clusters to support regional and local community economic development (Jacobs and deJong 1992; Anderson 1994; Enright 1995; Porter 1998; Bergman and Feser 1999a). The successful development of industry clusters requires careful analysis into the way clusters function, the smart infrastructure needed to support their development, and the processes required to mobilize support to facilitate cluster development.

This chapter examines factors that have led to the changing nature of industry clusters. The concept of industry clusters as a new strategy for regional economic development is outlined, and it is demonstrated how new industry clusters are often the key drivers of metropolitan regional economies in the post-industrial era of the information age. The chapter proceeds to discuss how industry clusters are identified and analyzed, and what the lessons are for regions seeking to undertake cluster-based development. Case studies are presented showing different applications of *industry cluster analysis* (ICA).

6.2 Globalization, Industry Structure and Localization

In earlier chapters, it was discussed how, since the 1970s, transformations have been taking place globally in the economic development and structure of nations, cities and regions, with the shift from the old industrial (Fordist) to the new post-industrial (post-Fordist) era of the services and information economy. The international economy is being driven increasingly by financial and information markets that support production processes that are intricately linked to each other in different parts of the world. Multi-national corporations orchestrate synchronized production systems in regions of the world that have competitive advantage in the production and distribution of goods and services (Porter 1990). Information technology continues its rapid penetration into every sphere of national and local economies helping to create new products and integrate markets (Carnoy et al. 1993). At the same time, production systems are requiring new skills and strategies to cope with dramatically accelerating change driven by globalization.

These massive changes to production, trade, information and employment are driving firms and nations to put in place measures to improve their competitiveness. Paradoxically, as the global economy gets larger the component parts of the economy are getting smaller. Naisbitt (1994) shows how "we are making business units smaller and smaller so we can more efficiently [compete and] globalize our economies". Economic competitiveness now has a significant impact on investment location decisions (Porter 1998), and this can be enhanced through the strategic alliances and supply linkages between firms, and by the development of industry clusters, and through the agglomeration of related activities in particular regions or locations within cities.

6.2.1 An Old Issue Considered in a New Context? Modern Theory of Industrial Districts

It may be claimed that industry clusters are groups of industries that are highly interdependent in that they buy and sell from each other, and that their products are functionally interrelated, so that their components (sectors or industries) are usually geographically concentrated in specific regions or in specific parts of states or metropolitan regions. But this notion is not new.

Stigler (1951) provided the first analytical treatment of what determines the division of labor within and across firms. He suggested that for a firm producing a particular specialized manufactured input in-house where level of cost minimizing production is greater than that required to satisfy the firm's own needs, then it is reasonable to suggest that the firm may spin-off the inputs of production, this being converted into external spatial division of labor with the original producer having reduced costs and the number of independent producers having increased. Later Scott (1986, 1988, 1992) drew on the notion of markets of hierarchies proposed by Williams (1975) to analyze the division of labor within and between firms as a series of transaction costs, with economies of scope being present when it is more cost effective for the firm to outsource particular functions to move from vertical integration to vertical disintegration, with firms seeking proximity with their key suppliers and other service providers. The assertion is that heightened global competition makes predicting market conditions much more difficult for firms, thus limiting the ability to vertically integrate with standardized production markets. Instead they seek to achieve greater flexibility by contracting out, developing strategic alliances for or as supply chains. All of this encourages proximity between suppliers and buyers, as Scott (1992) shows in an analysis of industry structure in Los Angeles and Gertler (1988) in an analysis of firms in the United Kingdom. Thus, as Sweeney and Feser (1998) note, the modern theory of industrial districts emphasizes spatial proximity for small firms. Humphrey (1995) writes:

...the basic principle is that clusters of predominantly small firms can gain economies of scale and scope and increased flexibility through specialization and inter-firm cooperation. If they cluster they can be as competitive or more competitive than small firms (p. 1).

This is in line with new growth theory (discussed in Chap. 1), where the emphasis is on business externalities and agglomeration economies, and spill-over effects. Researchers such as Jacobs (1969), Lucas (1988) and Glaeser (1994) have drawn attention to the direct role of cities and regions in spurring national and global economic growth and advancement, particularly through the interaction and mutual learning that takes place between firms in spatial proximity through spill-over effects. Empirical studies have tended to either (a) focus on agglomeration in industrial districts of small firms because of technologically related factors, whereby functionally related firms tend to co-locate in space (see, for example the early studies of Czamanski 1964, 1976), or (b) on second-order clustering patterns of different categories of firms where there is no relationship between plant size and clustering (see, for example Barff 1987). In a study of plant size and clustering of manufacturing activity in North Carolina in the United States, Sweeney and Feser (1998) used statistical spatial analytic techniques to measure the degree of clustering dispersal among different size classes of manufacturing establishments, finding that clustering tendencies were not a progressive function of size per se, but rather as being restricted to particular size classes, with most clustering occurring among establishments in the 20 to 30 worker range, especially within small densities. This confirms the established notion emphasizing agglomeration tendencies of groups of small and medium size firms occur primarily on the basis of flexibility rather than internal economies of scale. However, those authors also note that environmental and land use control and other institutional factors, such as the location of North Carolina's research triangle play an important influence (pp. 60–61). They also propose the use of detailed input-output accounts to examine spatial clustering among groups of industries.

It is worth bearing in mind this historic context for the analysis of agglomeration in industrial districts in considering the contemporary interest in the concept of industry clusters.

6.2.2 New Dimensions of Regional Economic Development

The current phase of global economic restructuring and competitiveness has changed five important dimensions of regional economic development which are said to influence the development of industry clusters. These are discussed below.

Geographic Scale. Globalization is resulting in the weakening of the nation state with a renewed focus on regions as the centers of economic growth (Ohmae 1996). Regions, and more particularly larger metropolitan regions, dominate the growth of employment, investment decisions and distribution networks in a global market place. Traditionally local government jurisdictions have been the geographic scale for focusing development. Today, however, key metropolitan industries are located throughout a region with their linkages spanning the entire area. Increasingly it is the metropolitan region that is the geographic unit of analysis at which competitive economic activities take place, and where prosperity is generated. Regions are not defined by political boundaries, but spread across local and international jurisdictions to encompass the broadest definition of contiguous economic activity. For example, the Shenzen region in the Pearl River Delta areas in Southern China is an integral part of the Hong Kong economy. Most businesses do not confine their economic activity to a specific jurisdictional boundary, and when the form of their organizational dependence is the industry cluster, important economic inter-relationships are even more likely to spread across jurisdictional boundaries (Schriner 1995).

Industrial Organization. Economic strategies are focused primarily on the company industry sector level. Such strategies fail to recognize the increasing importance of inter-organizational dependencies that exist between industries and companies across industries, regions and nations. Globalization and restructuring in national economies has resulted in the outsourcing of production and services. As developed economies open up to international competition, national industries are not able to compete on price or achieve economies of scale or maintain quality standards, and subsequently restructure or move production offshore. From these restructured industries emerge new hybrid businesses that begin producing and exporting components and services into global markets. This leads to a growing network of suppliers and distributors and the formation of new industry clusters.

Economic Inputs. The types and location of input factors crucial to the formation, expansion and attraction of industry are changing. There is a stronger emphasis on networks and clusters, on value adding factors related to efficiency, on technology applications, on skills placement, and on leadership. Instantaneous communication, computer aided design (CAD), and microchip technology have become an integral part of economic production and information dissemination processes. We are moving inevitably towards the telematic age, where business and person transactions are instantaneous and not geographically confined. Networks begin the processes that lead to strategic alliances, partnerships and joint ventures and other types of contracts that form the basis of modern economic development in the global economy. This is being enhanced through e-commerce. Few firms are capable of dominating or controlling markets and production processes in the global economy. Even firms like Microsoft and Boeing work in collaborative competition to develop new products and services. Collaborative competition fosters innovation, leading to continual processes of innovation to maintain competitiveness in the global economy.

Sustainability. Regional economic performance is no longer being viewed in purely economic outcomes. There is little value to communities if economic development does not improve the quality of life, maintain safe and pleasant living environments or resolve social disparities within communities. There is increasing evidence suggesting regions that have high environmental quality through economic policies that promote sustainable development are more attractive to business investment, generate more jobs and have a better standard of living and amenities that those regions that are indifferent to it.

Governance. Globalization is also bringing about major changes to governance systems. There is emerging a 'borderless society' with unrestricted movement of information, travel and currency between countries. Greater levels of transparency and standardization are occurring in government and business. Governments have less control and influence over economic development and investment decision-making than was the case in the past. Government's role is moving towards facilitating and managing development processes that fit the global forces shaping the changes and patterns of investment and production. This situation poses difficulties for governments, especially for those evolving from centrally planned to market oriented economies. Local communities too are demanding greater empowerment, forcing governments to delegate and execute their public responsibilities in a more consultative and efficient manner (Robertson 1992). New partnerships between government, business and communities are emerging to execute many of the functions and responsibilities undertaken by government agencies. Communities and business are having a more direct role in the formulation of strategy and economic development processes. These trends are a paradox for government. On the one hand governments must shape public policy and public investment in services and infrastructure to facilitate the growth of competitive business, but on the other hand they must empower communities and business to mobilize resources to provide and manage much of the infrastructure and services needed to support sustainable economic development. What is emerging is a transformation in governance processes that have been incumbent for over 100 years.

The factors described above are substantially changing the way industries develop in regional and metropolitan economies. Governments and business that do not respond to these changes risk losing market position and new economic development opportunities that are created by open markets. It makes little difference to global corporations and financial markets where industries and businesses are located, provided they have the capacity to deliver services, commodities and products in a timely, efficient, predictable and cost-effective manner, taking into consideration acceptable risk.

6.3 Industry Clusters: A Contemporary Strategy

6.3.1 The Concept

The classic concept of the *regional export base*, consisting of *clusters of economic activities* that draw in revenue from outside the region, has even greater relevance today with the expansion in regional and international trade. Industry clusters have always existed. However, the historic forms of industry clusters involving national sector industries, with strong vertically and horizontally linked firms, are rapidly disintegrating. Work in the United States on the concept of *power clusters* (for example, by Stough et al.1995; Waits 2000) has focused on support for critical masses of clusters or agglomerated industries which have a major role in developing new export industries. Industrial sectors in the core of a cluster for the most part produce for the market outside the local region of concentration and are thus export-base industries. Of course there are different types of industry clusters including, but not limited to, traditional industry clusters (the dominant industry or group of related industries in a region), as well as new emergent or propulsive clusters to service-based clusters.

Industry clusters have formed the industrial organizational basis for much regional economic planning and development since the late 20th century, and it is a widely held view that they will be a dominant concern in strategy formulation for regional economic development as we move into the 21st century, especially in the promotion of new technology industries.

A useful definition of industry clusters is given in a report by the Information Design Associates and ICF/Kaiser International Inc. (1997) to the United States Department of Commerce:

...Industry clusters are agglomerations of competing and collaborating industries in a region networked into horizontal and vertical relationships involving strong common buyer-supplier linkages, and relying on a shared formulation of specialized economic institutions. Because they are built around core export-oriented firms, industry clusters bring new wealth into a region and help drive the region's economic growth (p.3).

Figure 6.1 provides an example of a typical industry cluster in this case based on electronics, showing the structural linkages between supply, export, distribution firms, and manufacture, research and development, wholesale, marketing, transportation, infrastructure, and finance functions. The internationalization of national economies has removed many inefficient industries forming the old cluster chains of the industrial era, and in some cases removed the core industry itself.

Non-local, offshore or intra-national suppliers or distributors are replacing many elements of the old industry chains of production and services. Contemporary globalization in the post-industrial era has seen all economies experience significant increases in non-local linkages with suppliers and distributors outside their region in trade in materials and products, in information and in services. Even in the nations of the developing world key elements of industry sector production chains—especially services—must be imported. The concept of totally integrated production centers is no longer realistic or efficient in the global economy.

Research in the United States found that, by the early 1990s, 54 per cent of American employment was attached to just 18 industry clusters (U.S. Department of Housing and Urban Development 1996). Most large cities have four or five industry clusters. The new industry clusters are much more complex than old industry clusters.

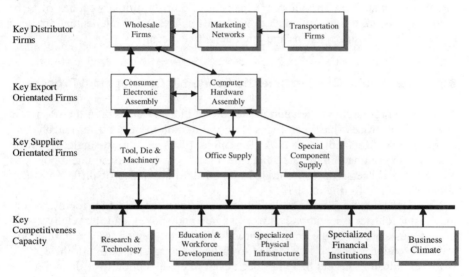

Fig. 6.1. Structure of typical electronic industry cluster
Source: Derived from Information Design Associates and ICF/Kaiser International Inc 1997, p. 2

6.3.2 Attributes of Clusters

There are three significant attributes associated with contemporary notions of industry clusters:

(a) *Shared end-markets*. This typically includes companies that produce similar goods for several markets, and have similar relationships with customers, with other industries, wholesale retail trade sectors or directly with consumers. End customers can be separated into buyers that sell their products to wholesalers or intermediaries.

(b) *Strong buyer-supplier linkages*. Clusters tend to have important relationships with suppliers. This characteristic has become more important as industries have decreased their degree of vertical integration and out-sourced inputs into production.

(c) *Shared technology and know-how*. Clusters are likely to share technologies, information and skills in their work. This is particularly true on knowledge-based industries where a close association of skilled workers helps to create a synergy of new ideas and foster innovation. Shared occupational characteristics are important as an indicator of technology utilization as a large part of all technology transfer takes place through the hiring of skilled workers. In regions such as California's Silicon Valley in the San Francisco Bay Area, the mobility of workers between member companies in an industrial cluster is one of the primary modes of technology diffusion, despite attempts to hold staff and valuable intellectual property through high industry pensions and other attractive incentives.

6.3.3. The Role of Clusters and the Economic Development Process

Much of the research on the appropriate role of industry clusters in the economic development process has been focused on the definition of a cluster (Jacobs and de Man 1996; Roelandt et al 1997; Rosenfeld 1997). While no single definition has come to dominate in the literature, the role of economic development policy in response to clusters has concentrated around the importance of differing policy approaches for differing clusters.

The types of inter-firm linkages, number and scope of industries represented in the cluster, cluster orientation (network or hub-and spoke), and the relative position of cluster firms in the value chain will differ from region to region, thus policy intervention will also likely need to differ on a regional basis. In the applied work on industry clusters there has been even more dissention over the definition and role of clusters, since many of the studies carried out at the state or regional level (more so in the United States than in Europe) have been politically motivated and seen as value-added more as a marketing tool than a policy informing tool.

Held (1996) has identified four of the common pitfalls of cluster studies that have been carried out in the U.S. in government attempts to employ cluster analysis for policy. These include:

(a) *Prescribe but do not identify*. In this case the rush to come up with policy prescriptions overrides the understanding of what firms and institutions within the region are integral parts of a cluster, and more importantly there is little or no effort to understand how these firms and institutions work in conjunction to promote growth within the region. Industries included in the cluster usually have no measurable linkages, and will be identified by names with more marketing power than systemic usefulness—for example, 'Aerospace Cluster', 'Biotechnology Cluster', or 'Telecommunications Cluster'.

(b) *Identify but do not prescribe*. This purely quantitative approach is usually comprised of a list of industries (or SICs) that may have some loose connection since they are components of higher level SICs and have been identified as industries of concentration (with high location quotients). Little attention is given to the important components of the cluster, such as shared labor pools, strong buyer-seller relations, firm structure and supporting institutions.

(c) *Say cluster but think industry*. The policy prescriptions in these studies usually follows closely those that are associated with particular industries, typically at a much broader level and are not usually very meaningful at the regional or local level for the promotion of clusters or groups of related industries. Such studies generally address larger single industry concerns of a dominant industry within a region with an indirect message such as 'what is good for General Motors is good for the United States'.

(d) *Only the 'classic' cluster will do*. Research guided solely by the conventional tools of regional analysis—such as input-output modeling to identify buyer-seller relationships, and export-base theory to describe the propulsive industries within a region—typically will miss the importance of informal institutions and conventions that are present within a region, and that are critical to the emergence of a cluster. Emerging clusters often will be missed in the typical analysis of vertically integrated clusters since there are no buyer-seller relationships established. However, they do benefit from a specialized regional labor force.

6.3.4 Methodology

There is no standard methodology for the analysis of industry clusters. Researchers and practitioners using the concept have utilized and explained a variety of approaches, both quantitative and qualitative, with the more useful ones utilizing both approaches. The quantitative approaches typically analyze industrial sector data using methods that range from measures of industry size and change—for example, employment, wage level, establishments and related dynamics—to measures of inter-industry linkage levels—for example, by using input-output models. Qualitative analysis—using techniques such as interviews, focus groups and surveys—is needed, however, to learn about the structure of supply chains and to describe supporting hard and soft infrastructure. The *multi-sector analysis* (MSA) technique discussed in Chap. 7 is an example of such an approach. What follows in this chapter we describe a basic approach to how industry clusters might be built, and outline both quantitative and qualitative approaches to measuring and analyzing industry clusters in regions.

6.4 A Basic Approach to Building Industry Clusters

The focus upon governments and business facilitating the creation and development of industry clusters had become a recognized best practice in regional economic development by the late 1990s. Key success factors which have been identified as crucial to building industry clusters are:

(a) strategic leadership and infrastructure;
(b) networking;
(c) market intelligence;
(d) capacity building; and
(e) maximizing the use of endowed resources.

But industry cluster building is also a *learning process* involving the identification, development and acquiring of specific skills, knowledge, personnel, technology, resources, infrastructure, markets and management that will support the growth of a specific cluster. Collectively these factors create a kind of critical or strategic architecture that will support industry cluster development. All clusters require a strategic architecture. However, the strategic architecture to support, say, an electronics industry cluster will be very different to that for a chemical industry.

Best practice experience tends to suggest that regions seeking to build industry clusters need to spend considerable time working out the core competencies, strategic infrastructure, and marketing intelligence systems needed to facilitate the development of competitive industry clusters. This involves putting in place a *cluster learning process*. Industry clusters are not just things that can be created or manufactured quickly; they need to be learned. It takes time—often many years—to build a successful industry cluster. The process requires nurturing from inception to maturity and metamorphosis from time to time. Part of that process involves industries learning new ways to collaborate competitively. The idea that industries may have to competitively collaborate is incomprehensible to many businesses, except when it is realized that businesses share highly overlapping if not common distributor and supplier networks, and that collaboration through clustering achieves economies of scale, rapid rates of technology transfer and efficiencies through resource leveraging.

6.4.1 Identifying Industry Clusters

Enhancing regional economic performance now involves the development of industry clusters. Building industry clusters involves the four-stage process shown in Fig. 6.2.

Fig. 6.2. The stages of analysis in building regional industry clusters
Source: Derived by the authors from ideas in Information Design Associates and
ICF/Kaiser International, Inc. 1997

It was noted earlier how researchers and practitioners have identified industry clusters using both quantitative and qualitative techniques. A cluster is a significant core group of industries in a region or locality that are linked through a complex network of inter-related suppliers and distributors. Industry clusters may be a local or regional phenomenon. Shift share, input-output models, and statistical methods of analysis may be used to analyze patterns of employment, industry establishment or production characteristics and industry interaction potentials undertaken geographically in a city. Those patterns may then be compared with national or international competitor average factors used for the analysis—for example, a location or city that has employment characteristics for an industry that is more than 1.5 the standard deviations from a national average could be considered a significant industry cluster. From that analysis it may be possible to describe the key types and characteristics of industry clusters within a region. The aviation industry cluster in Seattle in the United States is an example of a global cluster. London in the United Kingdom is an example of a global financial and insurance cluster. Bangalore in India is an example of a global software industry cluster.

6.4.2 Identifying Industry Clusters through Qualitative Assessment: The Example of Industry Clusters in Ho Chi Minh City

When statistical data is neither available nor reliable, qualitative techniques can be used. These might involve Delphi, focus and discussion groups to identify industry clusters within a region. Most businesses involved in national or international trade will be aware of the structure of suppliers and distributors within the region, and the magnitude of sales or purchases made between core industries and the supplier network. By using qualitative and interpretive techniques in the absence of reliable data, the best possible picture of key clusters in an economy can be established. This is then used to start the next stage of the cluster building process.

Figure 6.3 shows how a pictorial representation of a cluster can be provided. It gives a conceptual representation of the magnitude of clusters for Ho Chi Minh City (Saigon) in Vietnam, based on an analysis of concentrations of employment in industry sectors.

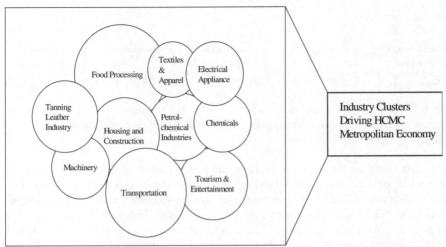

Fig. 6.3. Identifying the industry clusters of the region: the case of Ho Chi Minh City, Vietnam

Source: Roberts and Lindfield 2000

6.4.3 Identifying Market Potential of Local Clusters through Analytical Techniques: The Example of South-East Los Angeles

Where reliable data are available on regional economic structure, it is possible to use quantitative methods, including techniques such as those discussed in Chaps. 3 and 4, to identify industry clusters in a region. An important step in the industry cluster building process is to analyze the relative strengths of each cluster in terms of market growth potential and competitiveness within a metropolitan, regional, national or international market place.

Figure 6.4 shows how the competitive position of industry clusters in a regional economy may be plotted and described by using concentric circles to scale the relative importance to competitor clusters and markets. The example here is for the southeast sub-region of Los Angeles, California. Here the size of the circles indicates the cluster size, and the clusters are positioned on two axes which relate to the level of employment growth (vertical axis) and a measure of the relative concentration of employment in the cluster vis à vis that for the industry at the national level. From this type of analysis important questions can be asked about how to reposition or to maintain stability of an industry cluster relative to competitor clusters and markets. In the example in Fig. 6.4, the benchmark is the data for the United States.

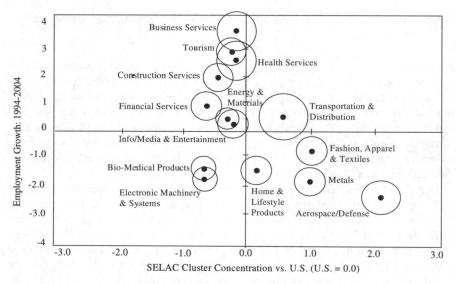

Fig. 6.4. Cluster industry opportunities in the south-east sub-region of Los Angeles
Note: SELAC Clusters Growth/Share Matrix (circles indicate cluster size)
Source: Information Design Associates and ICF/Kaiser International Inc. 1997, p. 36

6.4.4 Small Clusters and Growth Potential: Examining Changing Markets

While an industry cluster may hold a weak competitive position and have limited potential for growth, this does not mean that it should be neglected or not encouraged to grow. Many small clusters provide highly specialized services in a global marketplace, continue to experience growth, but lose relative market share because of the rate at which a market is growing. This is true of software industry clusters. Many manufacturing industries often experience this situation during restructuring.

Market potential analysis thus becomes an important basis for strategy later in the industry cluster building process as shown previously in Fig. 6.2. To be competitive, industries need to understand the comparative advantage a cluster might have, and in addition will need to know how these advantages might be leveraged or stretched to develop new economic possibilities. Market potential analysis requires a careful evaluation of competitor products and markets, especially their growth potential and marketing strengths. Failure to analyze competitor position may well result in the loss of competitive position, leading to lagging or under performance by an industry cluster seeking to trade in the global marketplace.

6.4.5 Analyzing Suppliers Needed to Help Core Industries be More Productive

As shown in Fig. 6.2, the third step in the industry cluster building process involves a detailed investigation of the linkages, core industry suppliers, and distributors that support an industry cluster. The volume, type and value of these linkages should be investigated, together with common user suppliers and distributors for different business units forming the industry cluster. Common user suppliers often service more than one cluster. The absence of a key local or co-industry supplier may create an opportunity for import substitution suppliers—although they would need to be competitive.

Careful attention should be given to seasonal and market factors associated with suppliers and distributors as well as the propensity to out-source externally from local markets for services. Attention also should to be given to product differentiation and service delivery—in particular specifications, quality assurance standards or conditions applied to delivery of products and services used by core industries that form a cluster. Figure 6.5 illustrates the relationship between the export and supplier element of the computer industry cluster.

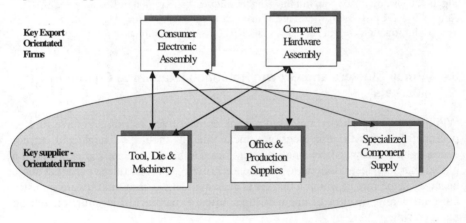

Fig. 6.5. Analyzing supplies needed to help core industries be more productive: the case of a computer industry cluster
Source: Derived from Information Design Associates and ICF Kaiser International Inc 1997

6.4.6 Identifying the Economic Foundations that Clusters Need to Evolve

Figure 6.2 showed the fourth step involved in an industry cluster approach to regional economic development. It concerns the identification of the form of the economic foundation needed to support the development of industry clusters. There are six issues that need to consider as shown in Fig. 6.6.

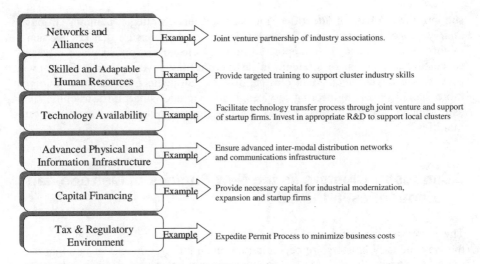

Fig. 6.6. The economic foundations of clusters
Source: The authors, using the ideas in Information Design Associates and ICF Kaiser
International, Inc. 1997

These are:

(a) *The availability of skilled and adaptable human resources*—for example, the
availability or multi-skilled, multilingual, and contract labor.

(b) T*echnology availability,* which is important for improving production efficiency. The wide use of CAD, CAM and GIS systems to assist manufacturing and
analytical processes is important to improve the efficiencies of production processes and the delivery of value-added producer services.

(c) *Advanced physical and information infrastructure*—for example, fiber optic
cable systems, data processing centers, efficient transportation systems, waste
management services, education training and community facilities will have a significant impact upon equipment and human performance in supporting the development of industry clusters and attracting new industries to regions.

(d) *Access to financial capital*—for example, equity, venture capital, debt-
financing, etc, are important economic functions needed to support investment in
industry clusters.

(e) *The taxation and regulatory environment* have a significant impact on the cost
of business and economic performance. Regions that offer incentives—for example in the form of taxation incentives for research and development, flexible building codes, sound policies on environmental performance, and support for cleaner
production—will provide the kind of economic foundations that foster industry
cluster development.

A critical component in this stage of industry cluster development is *strategic
planning*. Strategic planning will help to identify strategic infrastructure needed to

support the growth of identified industry clusters. Strategic planning through *multi-sectoral infrastructure planning* (MSIP) will help ensure that key elements of infrastructure needed to support an industry cluster are put in place. The absence of one element of strategic infrastructure may severely constrain or even prevent the development of a cluster growing. Strategic infrastructure involves more than physical engineering services; it also includes smart infrastructure, such as advanced research centers, marketing intelligence systems and learning communities.

6.5 Industry Clusters as the New Engines of Metropolitan Economies in the US

The increasing importance of metropolitan city regions and their economies and the way industry clusters are becoming the main building blocks of national economic prosperity through the wealth they are generating, was recognized in a report by the U.S. Department of Housing and Urban Development (HUD) in the mid 1990s. It shows how, after 25 years of change during the transition to the new information-based economy, during which the processes of globalization and technology innovation have changed the dynamics of metropolitan regions, the United States has become better positioned for economic growth and prosperity. The report by HUD analyzed 114 of the metropolitan regions in the United States. It found that nearly 80 per cent of the population and almost 90 per cent of job growth is located in metropolitan regions. Some of these have prospered as centers of the new knowledge-intensive economy, growing through services—business, professional, financial, health—and spawning whole new manufacturing industry sectors—from computers and electronics to telecommunications and multimedia. Other regions have become more competitive by transforming older manufacturing industry sectors, such as automobiles or apparel, into new, more productive, technology-driven industry clusters (US Department of Housing and Urban Development 1996). It is the focus on industry clusters to explain this growth that is significant in differentiating among metropolitan regions.

6.5.1 Diversification with Complementarity

As the U.S. economy has shifted from an old industrial economy to the new post-industrial economy, the dominant pattern of industrial organization has changed. Whereas before the old economy gave rise to many undiversified regions and 'company towns', the new economy requires:

(a) *greater specialization* of skills, technology and services at the advanced end of manufacturing. United States firms are now more highly competitive in the global marketplace;

(b) *greater flexibility* as businesses are taking steps to become more flexible in order to improve efficiency and respond more rapidly to market signals; and

(c) *greater diversification* creating a diverse group of producer and supplier businesses within a region that complement each other working on specialized aspects of a broad industry cluster.

The new economy is demanding tighter connections between related businesses, as well as between public and private institutions. These connections are
fostered by being located in close geographic proximity within metropolitan regions. Groups of complementary firms and industries seem to come together and
locate in the same place, often within metropolitan regions form industry clusters.
The industries in a particular cluster share common markets and suppliers, and
primarily produce for export, either to other parts of the U.S., for example, between metropolitan regions, or globally.

6.5.2 The HUD Approach

In its report, HUD approached cluster identification initially at a national level.
Using data derived from regional input-output analysis, those industries that had
similar patterns of sales to final demand categories, or had similar markets, were
grouped. More specifically, the 183 commodity (or industry) definitions were related to the 139 categories of final demand using a factor analysis, which resulted
in the assignment of 84 industries to 21 distinct clusters. 'Similar, in this context,
means that the correlation between the market distribution of a given industry and
the customer market of industries within a given cluster, was higher than its correlation with any other cluster' (US Department of Housing and Urban Development 1996, p. 12.2). The HUD report goes on to explain that:

...Part of the cluster concept involves close coordinating relationships between suppliers
and final producers. For this reason, 21 clusters were added to certain industries that comprised a relatively large share of output of the entire cluster. To accomplish this, the I-O table was collapsed to show the purchases from each industry, of the 21 clusters. If an industry's sales to the cluster accounted for more than 2 per cent of the cluster's total output,
then it was added to the cluster as a 'dedicated supplier industry' (p. 12.2).

A total of 21 clusters were defined using vertical linkages evident in national
data. These definitions are then applied to metropolitan regions using a rationing
process based on the noted tendency for most concentrated metropolitan clusters
to capture a high percentage of national cluster employment. Specifically, the rationing process includes the following process and criteria:

(a) Metropolitan regions were ranked according to their concentration ratios (similar to cluster quotients). For each cluster category, a metropolitan region was considered as specializing in that cluster if its concentration ratio was greater than
1:10, or if its concentration ratio was greater than that of the region that cumula-
tively accounted (in a descending order) for 50 per cent of national cluster employment.

(b) If a cluster employed more than 250,000 it was designated as a metropolitan
regional cluster.

(c) Unique metropolitan clusters were determined on a case-by-case basis where regions had an industrial constituency that was not consistent with the cluster patterns evident in the national data. This was usually achieved by linking the region's individual industries that had a concentration ratio greater than 1:10.

Implicit in this approach by HUD is the assumption that clusters should be identified working backwards from an end product or service. However, a weakness of this approach is that it appears to ignore the linkages not represented in the input-output table while it further assumes that all co-operation and collaboration can be traced through the output of final goods and services.

6.5.3 The Fastest Growing Clusters

The HUD report claims that economic growth and prosperity in the metropolitan regions of the United States is being generated by the 18 industry clusters, which comprise 54 per cent of all employment in the United States and which span the breadth of the United States economy, from business and consumer services to manufacturing to natural resource extraction (US Department of Housing and Urban Development 1996).

The fastest job-growing industry clusters are mostly in the services industries, tourism and entertainment, health services, business and professional services, financial services, housing and construction, and transportation and trade services. The most productive industry clusters—measured by their rate of increase in value-added per employee- are in manufacturing or resource extraction, and include: electronics and communications, apparel and textiles, industrial machinery, consumer goods, natural resources, agriculture and food processing, medical products, and transportation equipment. The shift from the old to the new metropolitan economy may be seen most strikingly in the growth of high technology manufacturing and advanced information services industries in every industry cluster. Some examples are:

(a) *electronics and communications* - semiconductors, computer hardware and cellular telephone services;
(b) *materials supplies* - fiber optics;
(c) *medical products* - optical devices, specialty pharmaceuticals;
(d) *industrial machinery* - laser technology, photonic equipment, factory automation systems, electronic machinery;
(e) *transportation and trade services* - overnight delivery services, freight transport arrangements;
(f) *business and professional services* - computer software, laboratory services, management services; and
(g) *financial services* - new financial instruments, brokerage services.

The HUD report showed how businesses related through producer-supplier relationships operate as an industry cluster. The economic activity of these clusters takes place throughout a metropolitan region, cutting across political and jurisdictional boundaries. Thus, in the new economy it no longer makes sense to distinguish cities from suburbs in economic terms. However, in that new pattern of metropolitan economic activity, some central cities (which are large administrative entities that often incorporate significant tracts of suburb) have key roles to serve as:

(a) centers of innovation, advanced services, information and technology;
(b) centers of education, research and health care;
(c) centers of culture, recreation, sports, entertainment, conventions, and tourism;
(d) centers of transportation and trade; and
(e) market and workforce centers.

6.5.4 Types of Metropolitan Regions

The transition to the new economy is occurring in different ways and at different paces in the metropolitan regions of the United States. Some regions are already well advanced into the information-and technology-intensive, globally oriented growth of manufacturing and services industry clusters. Those metropolitan regions range from smaller urban areas in the South and West—experiencing new prosperity on the basis of electronics, advanced materials or other high technology manufacturing, research and services—to the nation's largest urban regions that are global competitors in business and professional, financial, transportation and trade services. Other regions are transforming more slowly as they seek to identify their competitive advantage and target appropriate market opportunities.

To explain some of the differences in current patterns of economic change, the HUD report identified four types of metropolitan regions (US Department of Housing and Urban Development 1996):

(a) Booming regions, which were experiencing the most rapid job growth, fuelled by the emergence in these regions of industry clusters that are driving the new economy—for example, semiconductors in Austin, banking in Charlotte, telecommunications in Atlanta.

(b) Mega regions, which were the very large, highly urbanized regions facing greater complexity and diversity in their economies and populations, and are distinguished by the presence of some of the highest value segments of the world's leading industry clusters—for example, advanced financial products and multimedia in New York, international trade and transportation services in Los Angeles, information technology and bio-technology in Washington DC, and international banking services in Miami.

(c) Revitalizing regions, which were strong economic competitors sustaining economic performance as they manage the transition from traditional manufacturing-based economies to advanced manufacturing and more knowledge-intensive services—for example, Detroit, Portland, Jacksonville, Akron—and other revitalizing regions share traditions of good government, strong civic leadership and relative cohesion within their communities.

(d) Transitioning regions, which were those still struggling to emerge from the old economy, and often are leading manufacturing centers that are beginning to find new opportunities in revived manufacturing clusters and advanced services.

6.5.5 Successful Strategies

The HUD report highlights how regional leaders can successfully implement effective strategies to make the transition from the old to the new economy. The most successful strategies seemed to focus on the key industry clusters of a region, with coordinated metropolitan leadership, targeting being important, with resources that enhance the global competitiveness of the industry clusters.

Ten broad elements of economic strategy which regions were using to support industry cluster development and improve regional competitiveness and prosperity were identified (US Department of Housing and Urban Development 1996). These are:

(a) transportation and infrastructure;
(b) research and technology;
(c) trade promotion and the market development;
(d) tax and regulatory policy;
(e) education and workforce development;
(f) financing;
(g) environmental preservation and restoration;
(h) quality of life;
(i) economic and community revitalization; and
(j) business development and attraction.

During the 1990s federal policy in the United States increasingly seemed to be recognizing the critical importance of metropolitan economic strategy in securing national and local economic prosperity, placing metropolitan economic growth within the mainstream of national economic policy making. The federal government was helping support development of metropolitan economic strategies by:

(a) identifying the key industry clusters in each of the 317 metropolitan regions

(b) developing a competitiveness index to measure and rank regional economic competitiveness and help region's identify their strengths and weaknesses by benchmarking their performance.

There is a role for federal government to support the implementation of locally-determined metropolitan economic strategies by tailoring federal investments to support regional industry cluster growth, along the lines of the ten broad elements of economic strategy listed above.

6.6 Quantitative Analysis to Identify and Measure Regional Industry Clusters

There have been a large number of studies at the state and sub-state regional (local) levels aimed at identifying and measuring industry clusters. Those studies have used approaches ranging from guessing or inferring from experience the clusters that exist to relatively sophisticated analysis of both quantitative and qualitative data. Despite the diverse and large set of industry cluster studies (see Stough et al. 2000; Bergman and Feser 2000; Steiner 2001; and Karlsson et al. 2005) there has been little systematic development of quantitative methodologies specifically designed to identify and measure such clusters in regional economies and that would produce replicable results. However, recently such an attempt at methodological development has been undertaken in a study by Stough et al. (2000) which sets out to identify and measure the strength of concentration of industry clusters in the state of Virginia in the United States. Two separate investigations were conducted–one at the sub-state regional level and the other at the state level. Two of the techniques developed and applied to the state level analysis are presented first in section 6.6.1. The sub-state regional analysis methods and some results are presented in section 6.6.2. Later, Yang and Stough (2005) developed and applied a cluster analysis method for decomposing regional input-output data into clusters whereby the clusters are systematically identified and replicable. The technique used by Yang and Stough is presented and assessed in section 6.6.3.

6.6.1 State Level Analysis

Three types of clusters at the state level of aggregation were defined by Stough et al (2000): resource, manufacturing or service based. Clusters are derived by examining each two-digit SIC sector to determine which other two-digit SIC sectors were highly related to the sector being considered. While this is a somewhat superficial way of defining clusters it does provide a quick first approximation of the important clusters in a state. To undertake this analysis an input-output table for the Commonwealth of Virginia was developed using the IMPLAN methodology developed by the United States Forest Service. A cluster is defined by combining all highly interdependent two-digit sectors with the sector under consideration. In all, four resource-based, five manufacturing and eleven service-based clusters were identified, and the potential of high growth clusters explored.

The basic assessment process for this approach is based upon 15 economic performance measures. These are:

(a) employment;
(b) employment change;
(c) annual average wage;
(d) rate-of-change in the annual average wage;
(e) establishments;
(f) rate of change in the number of establishments;

(g) wage level relative to the national industry wage level;
(h) rate of change in relative wage;
(i) inter-industry dependency;
(j) productivity;
(k) rate of change in productivity;
(l) contribution to gross state product;
(m) rate of change in contribution to gross state production;
(n) location quotient; and
(o) change in location quotient 1992 to 1998.

Coordinating and conducting an analysis using 15 different measures is complicated and can be quite confusing to interpret. Thus to facilitate the analysis, 'spider diagrams' were created with 15 spokes, one for each measure as illustrated for the transport equipment manufacturing sector in Fig. 6.7 and for the professional services sector in Fig. 6.8.

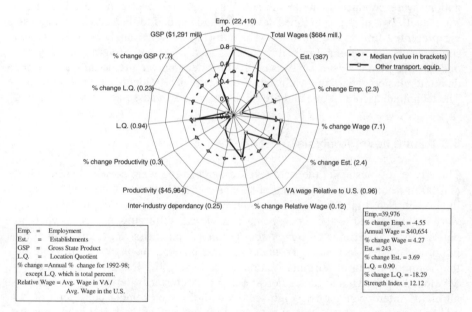

Fig. 6.7. Spider diagram of the transportation equipment manufacturing cluster in Virginia, 1998
Source: Mason Enterprise Center, School of Public Policy, George Mason University, Fairfax, VA; unpublished data and analysis results

These spider diagrams present a large amount of information on each of the industrial clusters. While interpretation of the strength of a cluster is facilitated by each of the 15 data points on the spokes of the diagrams, the amount of data can still be confusing. However, the shape of the diagram and the amount of area encompassed are rough but good indicators of the strength of the cluster both in terms of size and growth. Thus, spider diagrams that are full in shape (that is not characterized with sharp edges and points) and that fill up sizeable amounts of the total space (area) available are larger and generally more propulsive clusters; for example, see Fig. 6.8 representing the professional services cluster.

A *cluster strength index value* appears on the diagram for each cluster (Figs. 6.7 and 6.8). These values are computed as the ratio of the area inside the spider diagram to the total area available, multiplied by 100. Thus, the larger the index value, the larger and more growth oriented the cluster and vice versa.

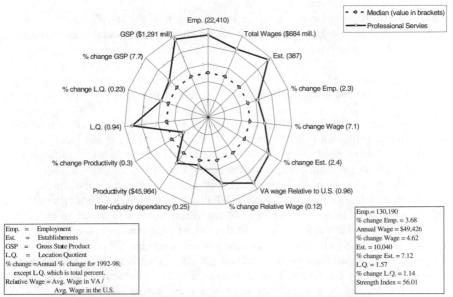

Fig. 6.8. Spider diagram of the professional services cluster in Virginia, 1998
Source: Mason Enterprise Center, School of Public Policy, George Mason University, Fairfax, VA; unpublished data and analysis results

For example, in Fig. 6.7 the strength index for the transportation equipment manufacturing cluster is 12.12, indicating that it is a relatively weak cluster despite being relatively large, compared to the professional services with a strength index of 56.01 in Fig. 6.8. This is further confirmed by comparing the regularity of the shapes of the diagrams, with the spider diagram for professional services (Fig. 6.8) being more regular than for transportation equipment (Fig. 6.7). Finally, four somewhat interrelated size or scale variables are located at the top of the diagrams. These are employment, total wages, number of establishments and contribution to gross state product. For large clusters, the upper part of the diagram will tend to encompass a large amount of the area available. Another measure of the overall strength of the clusters can be computed by counting the number of spokes for which the values are greater than the median value for all Virginia clusters. Dividing this value by 15 (the number of spokes) and multiplying by 100 would be another way to create a cluster strength index.

This measure has not, however, been computed as it is likely highly correlated with the original cluster strength index and, moreover, the reader can easily compute it if desired. Each of the 15 variables is measured in terms of its own metric, which means that few measures have comparable metrics.

Consequently, it becomes necessary to create standardized scores to ensure metric commonality for the spider diagrams. This is accomplished by ranking each 2-digit SIC sector from the smallest to the largest and assigning a score of 0.0 to the smallest and 1.0 to the largest and to scores in-between according to the rank of the sector then aggregating these scores across all 15 variables. This helps to interpret the full performance of a given sector.

However, it does require care in the interpretation of the relative wage measure as this variable is referenced (or base-lined) nationally, not locally, as in the case of the other 14 measures.

6.6.2 Sub-State Regional Analysis

Cluster analyses were also conducted by Stough et al. (2000) for each of nine substate regional economies in Virginia. The objectives of these analyses were to identify propulsive clusters that could then be used as the focal or guiding elements of science and technology investments in the state.

The nine regions were defined in terms of major urban cores and the jurisdictions in their sphere of market influence. Methods exist for systematically defining such urban market areas, for example, Reilly's law of retail gravitation and other related measures. However, in their study, expert judgment was used to identify and define the sub-state regional economies' geographies. Clusters for the substate regions were derived from 3 and 4-digit SIC sector data. While this made it possible to identify clusters at a more disaggregated level it also created problems of inclusion especially when highly interdependent sectors with very small scale or importance properties were added to a cluster. The methodology used is now described and then illustrated in a case study analysis of one of the clusters for the Shenandoah Valley economic region in Virginia.

The first stage of the analysis identifies the emergent or propulsive sectors of the regional economy using a *propulsiveness index* (PI). The PI value is defined as a weighted combination of a sector's employment size (50 per cent), relative wage (30 per cent) and employment change (20 per cent) over a specific time period (all weightings are applied to standardized values on these variables). This weighting or allocation was made for the current study on Virginia's region by the authors of the study based on their knowledge of the state's economy and the goal to identify propulsive clusters. It was made in collaboration with state technology and economic development officials in Virginia. For making applications in other regions, or for the identification of non-propulsive clusters, a different weighting scheme would possibly be more appropriate. Ideally, time periods for such studies would span a recent growth period. The PI values should be computed for all industrial sectors in the regional economy at least at the 2-digit or at a finer level. Given that wage and employment change are included in the index, propulsive industries will tend to have the highest PI values. Thus, sectors with the highest PI values are defined as a region's propulsive industries

The second stage of the analysis uses input-output modeling to measure interdependence among propulsive industries and other sectors. An input-output model provides measures of interdependency among all sectors in a regional economy. Of particular interest are interdependencies among propulsive industries themselves and with their supporting industries. When the largest propulsive industries (sectors) are highly interrelated, only a few (one or two) clusters should be expected. Where the level of interrelation among propulsive industries is less concentrated there will tend to be more clusters. When these have been identified, the industrial structure of a cluster(s) is defined. Thus, with the aid of input-output analysis and the PI values initial propulsive clusters are systematically defined and can be separated by others using this same analytical routine.

The Shenandoah Region stretches more than one hundred miles along Route I-81 that runs parallel to the Shenandoah River and the Blue Ridge Mountains. Given that the region is quite elongated in a north-south direction and somewhat impermeable in an east-west direction suggests that it has more than one labor pool. This may explain why the analysis for this region yielded four industrial clusters rather than a more limited number. Below the results for the first cluster identified, the Food Products Cluster, are presented to further illustrate the method.

The food products cluster employs 11,605 with a location quotient of 4.1 that has been increasing. The average annual wage for this cluster is US$25,047 that is 105 percent of the regional average and 93 percent of the statewide industry average. Employment increased at an annual rate of 5.5 percent. In short, it is a fairly propulsive cluster. Its sectoral composition is heavily dominated by meat and poultry products (64 percent of cluster employment). At the same time, it exhibits rich diversity with a number of smaller industry components including beverages, sugar confectionary products and dairy products. The full list of sectors appears in Fig. 6.9, along with employment share.

Supporting industries for this cluster employ 25,673 and have an average location quotient of 1.0 (Fig. 6.10). Employment had been increasing in all of the 8 support industries with business services the greatest at 10.2 percent per year and wholesale trade in durable goods the least at 1.7 percent. The average annual earnings are all above the regional average except business services which is at 63 percent of the regional average. However, only chemicals and allied products and trucking and warehousing are above the statewide industry average earnings. The largest supporting industries are special trade contractors (27 percent of support industry employment), business services (26 percent), wholesale trade in durable goods (15 percent), trucking and warehousing (14 percent) and chemicals (13 percent) as illustrated in Fig. 6.10.

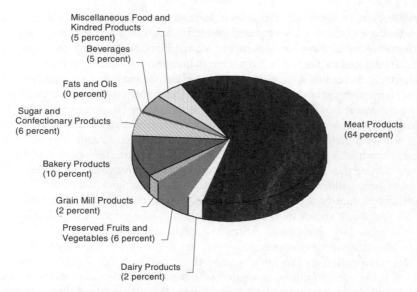

Fig. 6.9. Composition of Shenandoah Valley cluster 1 (employment, 1999).The total cluster industry employment is 11,605 for the year 1999
Source: Stough et al. 2000

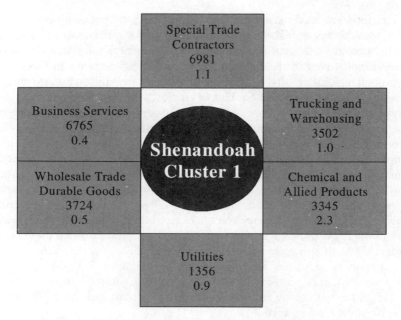

Fig. 6.10. Composition of support industry employment in Shenandoah Valley cluster 1, 1999. The total cluster support industry employment is 25,673 for the year 1999
Source: Stough et al. 2000

The food products cluster is relatively propulsive given the nature of the Shenandoah Valley regional economy as are its supporting industries. In short, it is a relatively dynamic set of industries with relatively high wages. However, the average earnings in the chemicals and allied products sector at US$60,207 is nearly twice as high as for any other supporting industry and for the dominant cluster itself. Despite employment growth, three of the supporting industries' location quotients have been decreasing, indicating that while these industries are growing they are doing so at a slower rate than the national average. One of these industries, business services, has a wage level that is far below the regional average and even further below the statewide industry average. Despite these qualifying observations, the food products cluster exhibits propulsiveness.

In addition to the methodology presented here the methods used to illustrate the cluster elements and the supporting industries are of interest. Such diagrams show the investigator what sector components of a cluster have the greatest weight or importance. At the same time one might argue that some small sectors in a cluster are propulsive and therefore should be given greater weight than indicated by the Location Quotient. This can easily be addressed by weighting either the Location Quotient or the interaction measure by a sectoral growth measure (for example, employment or earnings change); or by adding a third growth or change dimension to the diagram).

6.6.3 A Reliable Quantitative Cluster Analytical Methodology

One of the largest problems in cluster analysis research is the lack of a standard methodology for identifying the basic clusters researchers and thus policy makers and planners will work with. The method presented here goes a long way toward providing such a methodology. The methodology is first presented and then applied to the Baltimore, Maryland (US) case.

First, industry sector location quotients are used to identify economic base industry sectors and input-output tables to identify their major suppliers. This is an important step because it enables provision of a set of sectors that are part of the economic base and their major suppliers. These sectors are then retained for the analysis. A matrix showing input-output relations is then formed.

Second, sectors are divided into groups using Ward's (1963) clustering method which aims to form partitions P_k, , P_{k-1}, ..., P_1, in a fashion that minimizes the loss of variance associated with each grouping. Ward defines information loss in terms of an error sum-of-squares (ESS) criterion (MD*Tech-Method 2003). ESS is defined as:

$$ESS = \sum_{k=1}^{K} \sum_{x_i \in C_k} \sum_{j=1}^{P} (x_{ij} - \overline{x}_{kj})^2 \qquad (6.1)$$

With the cluster mean

$$\bar{x}_{kj} = \frac{1}{n_k} \Sigma x_i \in C_k \, x_{ij}$$ (6.2)

where x_{ij} denotes the value for the i_{th} sector in the j_{th} cluster, k is the total number of clusters at each stage of the Ward cluster analytic technique, and n_j is the number of individuals in the j_{th} cluster. Ward's minimum variance approach enables the identification of compact clusters and in this way also differs from other approaches as single linkage and complete linkage methods.

The application of Ward's minimum variance method to data for Baltimore, Maryland, provides an example for demonstrating the method and to show some of the graphical illustration techniques that help the analyst interpret the results of the analysis.

Baltimore is located on the east coast of the U.S. near the head of the Chesapeake Bay, a large embayment that ends at Norfolk, Virginia and begins at the mouth of the Susquehanna River in the northern part of Maryland. It is an old industrial urban region that has had a large thriving port both historically and more recently in the age of containerization. Its economy therefore is a mix of older manufacturing based industry as a legacy effect of its role as an industrial city and a growing services sector that is more linked to the contemporary economy.

The result of the location quotient analysis produced 245 industry sectors at the 3 and 4 digit level that had location quotients greater than 1.05 which was the economic base cut-off adopted for this case. The cluster analysis focuses on these 245 industrial sectors.

Figure 6.11 presents the 'cluster dendrogram' for the Baltimore analysis which is the array of the sectors on the horizontal axis and the ESS value on the vertical axis as all of the sectors are added sequentially. The dendrogram is organized from left to right in terms of the sectors that first minimize the percentage change in ESS value and then continue until all 245 sectors have been added to the analysis. Values increase from left to right as more sectors are added until a break point (large jump in the value for ESS) occurs. As the dendrogram values unfold, the percentage increase tends to be gradual and the analysis technique continues to merge more and more sectors. When the ESS value increase is relatively large (compared to previous value increases) a relatively large information loss has occurred with the addition of the last sector. When this occurs the analysis selects the previous step as the cut-off. This is illustrated in Fig. 6.12, which shows a relatively large increase of ESS at the 131st step (marked with a star). Thus it is determined that the merging process should be stopped at the 131st step. Nine clusters were identified using this threshold. The clusters are not analyzed further here because the purpose of this analysis was to illustrate how to identify clusters in a relatively unambiguous way and in a way that was generally replicable. This method satisfies both criteria.

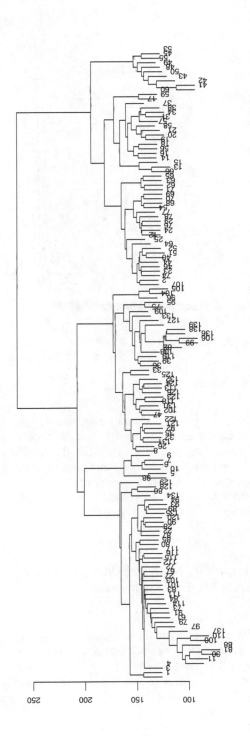

Fig. 6.11. Clustering Dendrogram

Note: The numbers shown in the dendrogram are not the IMPLAN codes for the sectors. They are the order numbers for the IMPLAN sectors when the 139 sectors are arranged using their IMPLAN codes

Source: Yang and Stough 2005

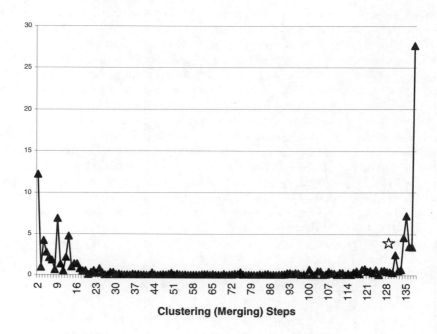

Clustering (Merging) Steps

Fig. 6.12. Baltimore cluster- Increase of ESS (percent)
Source: Yang and Stough 2005

One aspect of the Yang and Stough (2005) analysis is important to note. The goal of this research was to develop ways to unambiguously identify functional clusters, the nine industry sectors that were identified, and to also identify the spatial distribution of the establishments that composed each of the functional clusters, that is spatial clustering. While there is not sufficient time to go into this latter interesting question and analysis, the reader interested in methods for measuring spatial clusters are referred to Chap. 9 (Sect. 9.2) focusing on GIS related methods, where this analysis is presented.

6.6.4 Merging Quantitative and Interpretive Analysis: Using Industrial Cluster Analysis to Inform State Technology Investment Decisions

The Virginia studies by Stough et al. (2000) were aimed ultimately at using industry cluster analysis at both the state and sub-state regional levels to inform the state technology and investment process. The goal was not just to advise how technology could be used to accelerate the larger urban regional areas, but also the more diffuse and rural parts of the state.

What follows summarizes the results of the analysis for the Shenandoah Valley region where the topography varies from mountainous to broad and agriculturally productive valleys; for example, the Shenandoah Valley region and the population is quite dispersed and loosely organized around several small cities or towns with the maximum size of about 25,000. The dominant industry cluster for the Shenandoah region was found to be food processing and related agricultural products. On the ground this translates into the presence of a number of hog and chicken growing and processing facilities. Following the initial analysis a group of executives from the processing facilities and other experts on the local economy were convened to discuss the analysis in an effort to validate it and to explore ways that technology investments in the state might contribute to improved competitiveness in general and to stretching and deepening the food processing cluster in particular. One outcome from the discussion was the validation of the quantitative analysis. Another was the revelation that three of the food processing facilities had outsourced their logistics totally to a firm located in the Mid West of the United States. This led to an observation that many firms in the manufacturing sector in other parts of the state and the country were achieving improved competitiveness by outsourcing the technologically advanced logistics function. A proposal developed that while the state of Virginia housed a large and impressive information technology industry, the kind of technical skills that are required to support advanced logistics, none of this capability was supporting a local state supplier of these services. Despite the fact that many Virginia companies would continue to outsource logistics none of those services were being provided by Virginia companies. The upshot of this meeting was a technology investment oriented recommendation that Virginia attempt to either grow a logistics support industry out of its existing information technology industry cluster or attract a major provider of logistics services to the state.

This example is offered for several reasons. First, it shows the results of the application of a quantitative analysis that identified and defined an industry cluster at the sub-state regional level. Second, it shows why a combined methodology of quantitative analysis and interpretive analysis of expert views is more powerful than using only one of the approaches. Third, it shows how industry cluster analysis can be used to advise development policy and technology investment decisions. Finally, it shows the importance, even in more remote areas, of the impact of processes such as globalization (logistical integration and outsourcing is a global trend), technology development, heightened competitiveness and decomposition and outsourcing in product if not service oriented firms.

6.6.5 Quantitative Analysis of Industry Clusters at the National and Global Level

The Virginia examples discussed above focus mostly on using measures of interdependence to define clusters. Another approach that has been used at the national level by Bergman and Feser (2000) uses principal components analysis of the

United States input-output table to identify some 20 or more industry clusters–including, for example, metal working; motor vehicles; chemicals/ pharmaceuticals; electronic/electrical; food; wood/paper; and construction materials. These clusters, as with those identified in the Virginia studies, are termed value-chain clusters because the methodology combines highly interdependent sectors; that is, those that buy and sell from each other and in the process add value to the product as it moves up or down the value delivery chain. Some of the Bergman-Feser national industry clusters consist of as few as four sectors and some, for example, motor vehicles, contain more than 100 sectors.

The national value-chain clusters are important for two somewhat related reasons. First, Bergman et al. (2001) argued that these 20 or so national level industrial clusters in the United States arise in all advanced economies (further discussed in Bergman and Feser 1999b) and, therefore, can be thought of as clusters that exist not only in large advanced economies but also across such trading partners. Second, as Bergman et al. (1997) argue, state and sub-state regional industrial cluster analysis should be related to broad cross-national clusters for appropriate interpretation. In short, state and regional level analyses run the risk of being misunderstood or misapplied if they are not presented in a way that shows how they fit into the broader more globally predominant value-chain clusters. While this is a reasonable argument, it is also important to note that the value chains that are used to define these clusters are themselves in a continual state of flux and therefore investigators should take care not to assume that they are stable, long term benchmarks for analyses at lower levels of scale.

6.6.6 An Example of Quantitative Cluster Analysis in Europe: The Case of Styria

Styria is one of nine provinces of Austria and is located in the south central part of that nation. Graz is the capital city. This region experienced marked economic decline and de-industrialization that heavily impacted the region until the late 1980s. Over the following decade Styria moved from last among Austria's provinces on many economic indicators to first or near first on the same indicators some ten years later. Much of this growth was in manufacturing and technology intensive industry. As Steiner (2001, p. 280) states: "Styria underwent a process of development from the province with the most aggravating problems (especially in terms of labor market indicators) to the one with the highest level of job creation."

Steiner (2001) applies five different methodological approaches to industry cluster identification and evaluation in an effort to find a basis for a new policy orientation that could offer guidance for sustainability. These approaches range from the more familiar one of examining input-output linkages to the role of learning and cooperation.

To begin, Steiner first applied cluster analysis to input-output coefficients to identify functional clusters economic clusters. Then, via correlation analysis of pairs of employment distributions by sector across the nine provinces, he identified spatially concentrated groups of industry sectors (clusters). Comparing the functional clusters with the spatially based clusters showed that there was considerable dissimilarity in the clusters, contrary to theory that suggests that there is a strong tendency for functionally related industry sectors to locate proximally; for example, see the spatial analysis of technology based industrial sectors in the previous part of Steiner (2001). One reason for this may be that the recent transformation of the economy may be so recent that stable supplier chains and relations have not yet developed. However, the input-output based approach shows the existence of an old economy cluster (iron, steel, foundry and mining) that has both functional and spatial proximity. Further, supporting the supposition that spatial-mismatch in this case is arising from the newness of the dynamic parts of the regional economy.

Steiner (2001) also used the methodology of Porter (1990) to identify what are called competitive clusters. Here a diversity of factors such as international market shares, foreign trade specialization, degree of international division of labor and export distance are used to identify competitive clusters. The results of this analysis showed that Styria has a significant presence of Austria's highly competitive sectors. As is often the case, however, these clusters represent only a small portion of Styria's economy. In contrast, next patents were used as the basis of identifying technology based clusters. In all there were five industry clusters that had the same pattern of patents for all of Austria. These included: electro/electronics and telecommunications; transport and traffic; construction and housing; sport articles; and pharmaceuticals and chemicals. Styria was strongly represented in two of these clusters: transport and traffic; and electro/electronics and telecommunications. That analysis shows, however, that only small niches in these clusters and a few firms are responsible for Styria's performance.

Another interesting attribute of Steiner's (2001) study is the examination of the role of learning and cooperative organizational behavior both within and across organizations; that is, the role of networks and networking. Both of these processes were found to be critical to the formation of propulsive and fast growing clusters like those found in Styria and responsible for the most part for its recent economic success. These results, combined with knowledge about the clusters, inform proposed technology policy for the region. The policy prescriptions include those to enhance cooperative behavior among firms in or tied to a cluster and to increase information absorption capacity and diffusion. For the latter proposal, a new coordinated information initiative and the formation of demonstration centers are proposed. The analysis also resulted in a proposal to provide innovation assistants for small and medium sized enterprises. That proposal bears resemblance to the large number of technology business incubators and accelerators that have evolved in the United States over the last several years in an effort to address the same goals.

This is an interesting case study for several reasons. First, it addresses the issue of functional and spatial proximity of clusters and provides a methodology for examining this issue but at a higher level of scale or resolution compared to the analysis by Stough et al. (2001) discussed earlier in Sect. 6.6. Second, it offers a relatively broad gauged combination of methods for identifying competitive clusters. Third, a methodology using patents and patent structure by industry is developed and applied. This provides a way to identify technically propulsive clusters. Fourth, the three cluster analysis methodologies applied to Styria enabled an analysis of the role of learning and cooperation in cluster dynamics and thus regional economic development dynamics. Finally, the different methods of cluster analysis and related processes provided a framework for making informed policy recommendations.

6.7 The Process for a Region Undertaking Cluster Industry Development

A report on cluster-based economic development as a key to regional competitiveness by Information Design Associates and ICF/Kaiser International Inc (1997), which was undertaken for the United States Department of Commerce, looked at 17 case studies of cluster-based development initiatives (see Table 6.1). It found that when regions had been successful in nurturing industry clusters, typically they had developed "high quality, economic institutions, responsive to the specialized needs of existing and emerging clusters in the region" (p. 4). That study stressed how this occurred through strong civic leadership and collaboration creating a regional culture which is termed *collaborative advantage*, in which public-private collaboration is a hallmark of the ability of a regional economy to support its industries and how it adapts to economic cycles and the changing market requirements of industry clusters.

The study suggested that:

…an economic development approach based on understanding industry clusters in meeting their economic infrastructure needs can assist economic development leaders in identifying their industries that are key to the region's economic future and in developing the information and civic collaboration that is essential to achieving the region's economic future and in developing the information and civic collaboration that is essential to achieving the region's economic development goals (p. 4).

It is important to recognize that each region will have its own specific set of characteristics and factors that are driving economic change; thus it is important for economic development leaders to seek an approach to cluster-based development that is adaptable to the political, economic and social circumstances of their region. The experience from the review of the 17 case studies in the United States is that a cluster framework for economic development can be a valuable tool for effective economic change in a region because it:

(a)   is market driven, considering both demand and supply;
(b)   is inclusive of large and small business, of supplies, and of supporting economic institutions;
(c)   seeks collaborative solutions;

Table 6.1. Overview of cluster-based development initiatives in 17 regions in the US

Region	Economic challenge	Cluster development strategy
State of Arizona	Rapid population-driven growth, but concentrated in lower wage sectors and highly vulnerable to cyclical changes in economic conditions.	Grow emerging, technology-based industries such as Optics, Software and Environmental Services.
Austin, Texas	Low growth economy dependent on non-market sectors (i.e. state government and public university).	Build Electronics and Communications industry cluster with diversified strength in R&D, manufacturing and related business services.
State of California	Recession aggravated by defense industry downsizing and environmental quality problems.	Launch Advanced Transportation industries building on aerospace, software, and environmental engineering talent released by defense industry.
St. Louis, Missouri	Downsizing of defense-related federal facilities and contractors combined with longer-term decline in heavy manufacturing industries.	Target 'critical technology' competencies in the region to build new, high value-added clusters in Biotech/Agriculture, Environment/Energy, and Telecommunications.
Jacksonville, Florida	Population-driven growth combined with downsizing of military facilities and environmental quality problems.	Create manufacturing and higher-value service sector jobs by developing Health Services and Medical Products Manufacturing industry cluster with supporting business
State of Ohio	Reductions in federal spending on defense and space research.	Strengthen and adapt supply-side inputs of aerospace research and production to serve federal and commercial markets.
Southwestern Pennsylvania	Long-term decline of heavy manufacturing industry.	Reorient regional economic infrastructure to support growth of high technology industries (e.g. biotech) and advanced manufacturing).
State of Oregon	After recession of the 1980s the state established a state lottery for economic development.	Establish "Oregon Shines" to form 25 industry partnerships in 14 industries, each focus upon industry actions, such as training, research, capital, marketing.
Camino Real (El Paso, Chihuahua, Mexico-NM)	Low income, slow growth economy facing new competitive pressures, particularly due to NAFTA reductions in trade barriers.	Increase cross-border industry and government cooperation to remove constraints on regional economic development.

Table 6.1 (cont.)

South Eastern Los Angeles, California	Decline of federal defense spending and downsizing of related federal facilities and contractors.	Take advantage of growing foreign trade and planned construction of the Alameda Corridor to create business and job opportunities in Trade and Transportation, and link to surrounding clusters.
Venture County, California	Downsizing of defense-related contractor firms, with economic conditions worsened by the Northridge earthquake.	Link technology competencies of defense firms to commercial applications for the region's health, environmental, media, agribusiness clusters.
State of Florida	Economic growth driven by tourism and retirement income with weak indigenous industry development.	Help emerging business services and technology industries grow higher value-added manufacturing clusters through an improved statewide economic infrastructure system.
East Tennessee (Knoxville to Oak Ridge)	Downsizing of national laboratory and weakened competitive position of traditional manufacturing sectors.	Leverage technology assets of the national lab to diversify and strengthen traditional and emerging clusters in East Tennessee economy.
State of Washington	Decline of federal defense spending leading to cuts in aerospace industry.	Create "flexible manufacturing networks" to improve collaboration among small and medium enterprise to pursue opportunities in non-military markets.
Silicon Valley, California	Defense cutbacks, increased global competition and regional business climate difficulties.	Implement collaborative initiatives to improve training, communications, finance infrastructure for the region's advanced technology clusters.
State of Connecticut	Recession combined with downsizing at federal defense contractor firms and high business costs.	Strengthen existing clusters by actions to add-value and accelerate diversification through providing advantages to emerging clusters (optoelectronics).

Source: Information Design Associates and ICF/Kaiser International Inc 1997 p. 6

(d) is strategic, helping stakeholders create a vision for the next generation of their regional economy; and

(e) creates value through vertical linkages that enhance exporting industries reinvent new local suppliers or business services that enhance productivity and generate local employment. (Information Design Associates and ICF/Kaiser International Inc. 1997, pp. 5–7).

6.7.1 Checklist of Questions for a Region

Information Design Associates and ICF/Kaiser International Inc (1997) provides a useful checklist of diagnostic questions a region or local community might ask to determine whether or not an industry cluster approach to development is appropriate (Table 6.2). These are important questions both for professionals and for civic leaders to ask in order for them to know if launching a cluster-based strategy initiative will be the 'right' action to take for the future of their regional economy.

Table 6.2. Diagnostic questions for regions or local communities to ask to help determine if a cluster approach is right

Diagnostic questions	Cluster approach right/yes/no/perhaps
A. *The right economic scale:*	
A1. Are you thinking regionally?	Yes
A.2. Are you a small community?	No, need to link with nearest region.
A.3. Are you a sub-region?	Yes, helps build identity within broader region.
B. *The right economic challenge:*	
B.1. Are you responding to economic restructuring?	Yes
B.2. Are you thinking about improving the economic inputs used by different industries?	Yes
B.3. Are you covered with a specific project or investment?	No, but can help in market focused linkages development.
C. *The right economic focus:*	
C.1. Are you thinking about your region's cluster 'portfolio' and regional vision?	Yes
C.2. Are you thinking about one industry?	Perhaps, but one industry approach may miss opportunities.
C.3. Are you thinking about one company?	No.
D. *The right leadership and strategy process:*	
D.1. Do you have a tradition of leaders caring about the region's economy?	Yes if so.
D.2. Is there an organization that can help bring stakeholders together?	Yes, this is crucial.
D.3. Are you ready to use an inclusive and collaborative change process to engage industry and institutions?	Yes if so.
E. *The right capacity to take action:*	
E.1. Is there a tradition of working with the surrounding or nearest metropolitan region?	Yes if so—interdependencies are crucial
E.2. Is there a tradition in the region of assessing and facing economic challenges?	Yes if so.

Table 6.2 (cont.)

E.3.	Does the region have the technical capabilities and financial resources to undertake a sufficiently thorough diagnosis of economic structure and conditions?	Yes if so; if not, and can't get it, then no.
E.4.	What is the history of communication and cooperative action among public and private organizations in the region?	Yes if so.
E.5.	Is the region prepared to respond to challenges by changing or working in new ways?	Yes if so.

Source: Derived from Information Design Associates and ICF/Kaiser International Inc 1997, pp. 11–12

6.7.2 Lessons Learned from Successful Cluster-Based Strategy

In the review of case studies of cluster-based economic development as a key to regional competitiveness conducted for the United States Department of Commerce, Information Design Associates and ICF/Kaiser International Inc (1997) pointed out that

Although the practice of cluster-based economic development is evolving, there is a growing body of experience from which lessons can be drawn about how to develop and carry out a winning strategy (p. 16).

The report highlighted the following recommendations:

(a) *Recruit highly committed leadership*; including 'champions' who may be from business, the private sector, or institutions like universities. This is identified as civic entrepreneurship. Public-private collaboration is seen to be vital.

(b) *Have a strategy to ensure adequate resources, throughout the process*. Cluster-based initiatives require the commitment of much time, energy and financial resources, which might be tapped from a variety of sources, including private firms, major employers, trade associations, chambers of commerce, retired professionals, academic institutions, governments, and non-profit foundations.

(c) *Choose the right geographic level of focus*. Cluster strategy initiatives are more effective at a regional rather than at a statewide level because the economic linkages of industry clusters are regional by definition. Also, mobilizing resources and achieving collaboration is more likely to be effective at the regional level through personal interaction physical proximity factors. This is not to say that a state level leader, such as a governor, may play a catalytic role.

(d) *Find tools to sustain momentum between stages*. It is common for the energy levels of leaders and participants to wane after six to twelve months, thus maintaining commitment and momentum becomes crucial in the process of building industry clusters in longer-term. The use of 'participatory diagnostic' reports to supplement economic analysis, with expert opinion drawn from industry and other actors, keeps the key players in touch and interested in the project, rather than allowing the entire diagnosis to be completed by expert consultants and for university researchers. Also, it is important to have a strategy for continuous media involvement to help sustain momentum.

(e) *Engage potential implementing institutions from the earliest stages of the process.* Ultimately the success of a cluster-based development initiative will depend on whether or not members of a region take specific actions required to implement strategies for economic change. Experience shows that more action is likely to occur if the responsibility for implementation is given to an organization whose mission and resources are closely aligned to the objectives of the action initiative.

6.7.3 The Four Stages of Cluster Initiatives

The regional cluster development process has been conceptualized as a four-stage process (The Information Design Associates and ICF/Kaiser International Inc., 1997). These are:

(a) mobilization;
(b) diagnosis;
(c) collaborative strategy; and
(d) implementation.

Table 6.3 provides a summary overview of the key lessons for undertaking each of the four stages of the process that are derived from the 17 case study examples of cluster-based development initiatives in the United States that were listed earlier in Table 6.1.

Table 6.3. Key lessons for undertaking the four stages of the regional cluster development process

Stage 1:	*Mobilization: educating stakeholders about economic challenges and opportunities, developing leadership, and building stewardship*
Lesson M.1:	Use economic challenges as windows of opportunity to bring stakeholders together.
Lesson M.2:	Kick-start mobilization by creating or identifying an organization dedicated to the initiative's goals.
Lesson M.3:	Cultivate broad private and public sector participation and early 'bring-in'.
Lesson M.4:	Cultivate responsible stakeholders and 'champions'.
Stage 2:	*Diagnosis: understanding the region's economic portfolio and recognizing regional capacities*
Lesson D.1:	Provide a neutral independent analyst/resource who can 'tell it like it is'.
Lesson D.2:	Involve and build momentum for the initiative across the community through a participatory diagnostic process.

Table 6.3 (cont.)

Lesson D.3:	Use cluster analytic techniques to identify regional strengths and strategic opportunities and benchmarking the region's performance.
Stage 3:	*Collaborative strategy: bringing clusters together to discover common needs, committing to collaboration, and building cross-cutting agendas for action.*
Lesson C.1:	Hold events that confront and engage participants from key industries and institutions in the region.
Lesson C.2:	Create highly inclusive cluster 'working groups'.
Lesson C.3:	Select working group leaders who are committed and can recruit high profile participants reflecting small and large companies in the region.
Lesson C.4:	Create accountability mechanisms and progress milestones throughout the process.
Lesson C.5:	Develop concrete action plans for cluster-specific initiatives as bridges to implementation.
Lesson C.6:	Establish the market viability for each initiative to ensure the necessary support from stakeholders for implementation.
Stage 4:	*Implementation: taking responsibility for implementation and building action teams*
Lesson I.1:	Create a management stewardship group by identifying a new or existing organization to spearhead and oversee the initiative.
Lesson I.2:	Use the cluster framework to facilitate supply and demand side connections.
Lesson I.3:	Identify sources for ongoing funding commensurate with the type and scale of action initiatives.
Lesson I.4:	Sustain sources of new leadership.
Lesson I.5:	Build a monitoring system to track activities and communicate outcomes.

Source: Derived from Information Design Associates and ICF/Kaiser International Inc. 1997

6.8 Conclusions

As cities, regions and communities around the world become more closely linked through trade, technology and communications, there will be increasing competition for jobs, investment and development. For a region or community to complete in a global market place, it will need to learn how to interface with other competitive and potential trading regions nationally and internationally. Successful places are now those which demonstrate the ability to be creative, flexible, innovative and responsive to changing environmental circumstances. Global changes are demanding quick local responses. Both business and local communities will be required to learn to think globally and act locally—as well as doing the opposite.

Several times in this book it is noted how it seems that regions—particularly large city regions—are replacing nations as the principle competitive instrument for trade and economic development. Across the world there is a growing regional specialization and concentration or clustering of industries in response to increasing competition, outsourcing and corporate downsizing as a result of national economic reforms and globalization. Industry clusters have become an important instrument for building economic capacity for regions to compete in the global market place. Regions that have evolved highly specialized industry clusters are showing consistent performance, both in their economic development and through improvements to their quality of life.

Most of the examples that have been used in the chapter are from the United States and are used to illustrate the important impact industrial cluster analysis is having on economic development planning and practice. This should not be taken as an indication that it is only a North American development. Steiner (1998) provides a number of detailed studies that have been conducted in Europe. And a study by Roberts and Lindfield (2000) on the identification of industrial clusters in Ho Chi Minh City in Viet Nam is a good example of a methodology for situations where archival information does not exist or is suspect for the identification and examination of clusters. Steiner (2001) provides an analysis in which he identifies value-chain clusters in Austria and then uses the analysis to advise and inform sub-national regional development policy. Finally, Bergman et al. (2001) offer an interesting study that using industrial cluster and contingent-valuation methodologies to examine the costs and benefits to be achieved from investment in alternative freight transfer centers in Austria and surrounding regions. That is the only study that the authors have identified using industrial cluster analysis to investigate and inform a policy area other than direct economic development.

The building of industry clusters is a progressive and learning process. Winning the confidence of businesses and public agencies to share information, collaborate and operate as a cohesive industry cluster may take many years to develop. Overcoming those difficulties requires a strong commitment by government and industry champions to provide the leadership, the vision and wear-with-all to make industry clusters happen.

7 Multi-Sector Analysis: Approaches to Assessing Regional Competitiveness and Risk

7.1 The Need for a New Technique

In Chaps. 3 and 4 it was shown how the application of quantitative techniques of regional analysis using secondary data can provide us with an understanding of both change over time in the performance of a regional economy and the interdependencies between sectors of the economy. That may include the use of projections to test potential regional development futures. Quantitative techniques are certainly very powerful analytic tools, but they do have limitations in enabling us to confidently predict regional futures (Mintzberg 1994). Qualitative techniques also may be used to help explain the *raison d'être* for economic development patterns occurring in a region and to improve our capacity to speculate on a range of economic possibilities or outcomes for the future based largely upon the collective experience, wisdom and judgment of actors in regional economies.

A challenge in regional economic analysis and planning strategy is how to develop further more refined and new techniques that bridge quantitative and qualitative methods of analysis to help give us a better perspective on the factors that contribute to the creation and maintenance of competitive regional futures and which help us to assess the risks regions face or may face which warrant consideration in formulating regional development planning strategies and policy and program interventions.

The discussion in Chap. 6 showed how industry clusters are now regarded as one of the primary engines of economic development of regions, and in Chap. 5 we showed how, in the global economy, the success of regional development depends upon the competitiveness of firms, industry cluster development (Porter 1991a), and the provision of strategic architecture (Hamel and Prahalad 1994). However, most of the tools commonly used to evaluate the competitiveness of regional economies tend to look at an economy either as a macro entity or as specific industry sectors or large firms that operate somewhat independently of each other. The reality is that every regional economy is driven by a maze of interlinking, multi-sectoral factors and flows, some of which are highly specialized and may play a key role in defining the future structure and form of an economy. Using input-output analysis to help derive industry clusters is one of the tools available to regional economic analysts that provides a good understanding of the relationships or linkages that occur between industries in a region. While intersectoral financial and other flows are an important indicator of competitiveness, they do not *per se* explain the complex factors, or deep structural elements, that contribute to the making of competitive advantage in firms, industry sectors and clusters in a region.

Economic base theory and related research (Richardson 1973) tells us that it is often a small number of firms and businesses that play a pivotal role in defining the economic structure, performance and development of a regional economy. Many of those industries comprised in networks or clusters (Cook et al. 1994; Saxenian 1995; US Department of Housing and Urban Development 1996), and in Chap. 6 it was shown how industry clusters comprise core industries, supplier/distributor networks and basic economic infrastructure which support their development and operation in a competitive market place (International Design & Associates & ICF Kaiser International Inc. 1997; Williams 1996). However, the competitive elements which support firms and enhance the development of industry clusters may differ significantly between industry sectors. And indeed they will likely vary somewhat between regions. Thus, to maximize the economic development potential of a region it is important to know what factors in what industries contribute to competitive advantage, to measure the strength and weaknesses of these, and to identify the relationship and inter-dependence of factors in supporting single or cross-cluster industry development (Aacker 1998). In view of the increasing levels of uncertainty in a rapidly changing world in the contemporary era of globalization and the increasing uncertainty and competitiveness of business environments, it is also necessary to develop more rigorous methods of assessing regional risk as we know that it is not just business but also regions that experience profound and rapid changes in their competitive advantage. To do this, analytical techniques are required to measure and compare characteristics of and the relationships between industries or clusters. Techniques that involve some form of *multi-sector analysis* (MSA) are needed, as proposed by Roberts and Stimson (1998).

This chapter outlines the development and application of MSA as a tool to help address this challenge, and in particular it provides a guide for regional economic development practitioners in formulating strategy which is based on a methodology that is explicitly designed to assess and appraise regional competitiveness and risk. MSA is an eclectic tool developed from different qualitative and quantitative analytical techniques and schools of thought. It begins by discussing the analytical techniques that collectively provide elements of the framework for the development of the MSA methodology. This is followed by a discussion of key aspects of regional competitiveness and risk that may be analyzed by MSA. Case studies are used to demonstrate actual applications of MSA in practice, in both developed and developing nations. An important use of MSA is to enhance the industry cluster-based strategy approach to regional economic development, which was discussed in Chap. 6.

7.2 Theoretical Basis for Multi-Sector Analysis (MSA)

7.2.1 Early Development of MSA

Roberts and Stimson (1998) initially proposed MSA as a qualitative tool to assist in measuring the competitiveness of factors contributing to the development of regions. The early application of MSA was in one of Australia's fastest growing regions, the Cairns-Far North Queensland region. The primary objective of developing the technique was to provide evidence to support the development of an economic strategy for that region - undertaken in the mid-1990s - that sought to capitalize on its competitive advantages and take due account of the risks the region currently faced and potentially might need to face (Roberts 2003). This early work provided a basis for further development of the technique for analyzing the performance of, and for developing a strategy for the Washington D.C. region in the United States (Stough 2001), and also for assessing the infrastructure needs of Ho Chi Minh City in Vietnam (Roberts and Lindfield 2000). The theoretical basis of MSA and the operational approach for assessing regional competitiveness and risk are discussed in the sections that follow.

7.2.2 Multi-Sector SWOT Analysis

The widespread application of SWOT analysis as a tool in strategic planning for regional economic development was discussed in Chap. 2. If we take the framework used to conduct a SWOT analysis (refer back to Fig. 2.8) and apply it on an industry sector basis, that analysis might reveal patterns or clustering of factors that affect the influence the competitiveness of a region's economy. Some factors will be common to several industry sectors. Other factors may be strong in one industry sector but overall week for an economy. The subtle differences in SWOT factors between industry sectors may help identify opportunities or strengths in one industry sector that can be leveraged with other industries to improve the competitiveness of an economy as a whole. But a problem with general SWOT analyses is that they tend to be all encompassing. Consequently, many of the micro-factors in different industry sectors that could be utilized and developed to improve the competitiveness of other industry sectors or clusters go unnoticed or are cancelled out in a traditional SWOT analysis. It is the cumulative effect of factors contributing to the competitiveness of specific industries or clusters that are responsible for giving a region its competitive advantage.

To undertake a sectoral SWOT analysis, it is necessary to develop a set of evaluation criteria ($Ec_a, ...Ec_m$) and to compare these across different industry sectors, ($Is_1...Is_n$) for the four SWOT elements—strengths, weaknesses, opportunities and threats. This can be done using a basic analytical grid. By comparing the results on an industry sector basis, the critical elements of strengths and weaknesses and of opportunities and threats that might affect the economic performance in a region can be identified. Appropriate strategies may then be developed to address negative factors—or to reinforce positive factors—on an industry-by-industry sector basis. Unfortunately, multi-sector SWOT analysis is time consuming, and for this reason it is seldom used in regional development and economic planning analysis. It has a further limitation in that data is not in a qualitative format that is easily analyzed, although computer programs—such as NUDIST—can be of assistance. Structural analysis (Godet 1991) overcomes many of the problems of multi-sector SWOT analysis.

7.2.3 Matrix Theory

Matrix theory is used extensively in mathematics and economics to represent often complex sets of information, and it is used often to simplify notations when dealing with a large number of simultaneous equations. It was adapted and applied extensively by Leontief (1953) and Isard (1960) in the field of regional science. Saint-Paul and Teniere-Buchot (1974) pioneered the application of matrix theory for qualitative analysis. Matrices are used for a wide range of analytical and presentation purposes using both qualitative and quantitative data. MSA uses several matrix applications to develop a number of indices.

The primary application of MSA is the development of an index or a set of indices enabling comparisons between industry sectors on a range of criteria. These can be derived from the assessments made of industry sector performance against the evaluation criteria or of the importance of the evaluation criteria for the competitiveness of the industry sectors. Those evaluation scores may be represented in matrix form as shown in Table 7.1. The advantage of such matrix is that it can be used to provide a common basis for interregional comparisons of factors contributing to the competitiveness and development of economies. The basis of developing an index of industry sector performance is derived by summing down the columns in the matrix, while an index of evaluation criteria importance is derived by summing across the rows of the matrix.

As we will see later in the chapter, this type of matrix, which is the basis of the MSA approach developed by Roberts and Stimson (1998), may be used to compile a range of indices that are of use to evaluate the competitiveness of a region's industry sectors and the importance of a set of evaluation criteria with respect to those industries and to assess the risks that might impact the region and the exposure of industries to those risk factors.

It is worth pointing out that multi-variate analytic techniques, such as factor analysis and structural analysis, could be applied to the data in such a matrix to derive dimensions and identify the factors that contribute to the competitiveness of a region's economy.

Table 7.1. The basic matrix format used for MSA

Evaluation criteria	Industry sectors					
	Is_1	Is_2	Is_n	$\sum$	Maximum (where $\sum s_n=3$)	Index of evaluation criteria importance
Ec_a	0	1	0	1	15	0.07
Ec_b	1	3	0	4	15	0.27
Ec_c	0	0	2	2	15	0.14
Ec_d	2	4	3	9	15	0.60
Ec_m	2	4	6	12	15	0.80
$\sum$	5	12	11			
Max (where $\sum c_m=5$)	25	25	25			
Index of industry sector performance	0.20	0.48	0.44			

7.2.4 Structural Analysis

The MSA technique developed by Roberts and Stimson (1998) has its genesis in two areas of research: regional input-output matrix theory; and structural analysis (Saint-Paul and Teniere-Buchot 1974; Lefebvre 1982; Godet 1994). Already in Chap. 4 an extensive discussion of input-output analysis has been provided, with the typical inter-sectoral interaction input-output technique illustrated. Structural analysis is concerned with the application and manipulation of qualitative sets of data. Structural analysis enables analysts to describe a system using a matrix, which interconnects all the system components. The method permits the identification and analysis of the relationships of the main variables that give structure to the system under investigation. Godet (1991) describes the framework for structural analysis as:

...A system consisting of a set of interrelated elements. The system structure, i.e. the network of relationships between these elements is essential to an understanding of its devolution, because the structure maintains certain permanence. The aim of structural analysis is to highlight the structure of the relationship between the qualitative variables—quantifiable or not—which characterize the system under study (for instance, a company (region) and its strategic environment) (p.83).

Godet (1991) makes use of structural analysis to evaluate the underlying causes of unemployment in France, and to use it to evaluate future scenarios, such as for airport planning and marketing analysis. For a detailed description of MIMMAC (Cross-impact Matrix-Multiplication Applied to Classification) and MACTOR (Matrix Alliance and Conflicts, Tactics, Objectives and Recommendations) techniques see Godet (1991). MINMAC is used to investigate the strength or interconnectedness of factors comprising a matrix. The matrix must have the same vertical and horizontal variables. Structural analysis has been used to evaluate the determinants of employment and unemployment in France in the early 1980s. Figure 7.1 shows the matrix developed by French researchers Barrand and Guigou, as reported in Godet (1991). They looked at the relationship of 41 exogenous and endogenous variables affecting employment using qualitative assessment—ranging from very strong (vs), strong (s), average (a), weak (w) to very weak (vw), and using a category described as 'potential'. The matrix is developed using Delphi or focus group techniques, and is refined several times until a consensus is reached on the importance of different relationships. From such analysis a picture can be developed of variables having the strongest determinants on other variables of employment and unemployment. Godet (1991, p. 89) suggests that only about 25 percent of the matrix is filled.

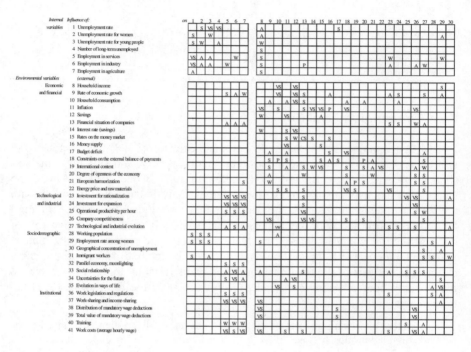

Fig. 7.1. A structural analysis matrix for evaluating interrelationship between employment and unemployment in France
Legend: *VS*=very strong, *S*=strong, *A*=average, *W*=weak, *VW*=very weak, *P*=potential
Source: Godet 1991

MACTOR is another structural analysis technique used to evaluate convergence and divergence of variables between two sets of criteria, that is, actors involved in regional development and their support or opposition to different economic objectives.

Structural analysis has several advantages in regional analysis. It can:

(a) stimulate thinking about the way regional economic systems operate;
(b) be used to evaluate strategic choices and planning scenarios; and

(c) help with communications and discussions on determining specific economic development options.

In practice, structural analysis is being used in two main ways:

(a) in decision-making research on the key variables which bear on the future dimensions and structure of systems; and
(b) in forecasting.

7.2.5 Analytical Framework for MSA

MSA brings together key elements of the analytical techniques described above. The principle tool for MSA is a matrix to measure and analyze different variables across industry sectors that comprise a regional economy (see Fig. 7.2).

Sector/Criteria	Animal Production	Other Agriculture	Fishing & Forestry	Mining & Construction Materials	Food Process	Textile, Clothing & Footwear	Chemicals, Petroleum Products Manufacturing	Wood & Paper Manufacturing	Machinery Manufacturing	Metals Manufacturing	Non Metals Manufacturing	Assembly & Specialised Manufacturing	Public Utilities	Construction	Trade Services	Communications	Transportation	Financial Services	Public Administration	Community Services	Tourism & Leisure Services
Criterion A																					
Criterion B																					
Criterion C																					
Criterion D																					
Criterion E																					
Criterion F																					
Criterion G																					

$\frac{\Sigma\ \text{(Criterion Sector Score)}}{\Sigma\ \text{(Max score sectors)}}$

$\frac{\Sigma\ \text{(Sector Score)}}{\Sigma\ \text{Max Score Criteria}}$

Fig. 7.2. Matrix framework used for MSA

Using this basic matrix structure, it is possible to develop a range of tools for regional analysis applying the MSA framework. The number of industry sectors selected for the MSA depends upon the level of detail desired. It is useful if the number of industry sectors selected were to match the industry sectors used for producing I/O tables derived from input-output analysis if these are available. That enables us to make comparisons between I/O tables and MSA matrices. For example, using the standard industry classifications available in Australia, selecting around ten to twelve two-digit or around 40 or so four-digit industry sectors is likely to demonstrate good results. MSA can also be used to evaluate industry clusters, although significant research is required to map industry clusters before conducting an analysis. Its application to industry sectors is easiest.

7.2.6 Methodological Issues in Assigning Weights to Matrix Data

A significant methodological issue with MSA is the weighting of the factors or criteria that are used to assess regional competitiveness or risk. Not every factor measured by the technique carries the same weight across all industries. For example, the importance of research and development (R&D) in supporting regional economic development and competitiveness will vary significantly between the industry sectors and regions depending on the nature of the economy. In most cases there are significant differences in the importance of factors between sectors. In goals and achievement matrix theory (Hill 1973) different weights are applied to factors used in the evaluation, depending on the importance, priority and values attached to different factors. In MSA, in some cases we apply weights, in recognition of the different values or importance placed on factors measured across industries sectors. Using a Likert scale (Likert 1961) respondents to questionnaires or experts working in a group (see Sect. 7.2.7) are asked to record on a scale (for example, one to five) the importance of a factor to an industry. By averaging scores for different factors in each industry, a weighting factor can be generated for every factor by industry sector. The weight is multiplied by raw scores to give a weighted score. This technique removes the intensity of skewness that was found to exist in the pilot work to develop MSA (Roberts and Stimson 1998).

7.2.7 Multi Criteria Analysis

Multi Criteria Analysis (MCA) offers policy makers an alternative when progress towards multiple objectives cannot be measured by a single criterion (that is, monetary values). MCA can be used as a decision support tool to evaluate the optimal outcomes of a project, taking into consideration a range of predetermined criteria or variables. Nijkamp et al. (1990) have used MCA to evaluate planning scenarios for major infrastructure projects based on different criteria. This work evaluated a range of weighting techniques for different applications of MCA. MCA is used extensively in environmental decision making processes (Munasinghe 1994) where a range of often conflicting criteria are required to be taken into consideration as part of an overall decision making process. One of the simplest forms of MCA is Goals Achievements Matrix Analysis (GAMA). It was used to evaluate alternative project outcomes based on different criteria for transport planning by Hill (1973). In its more advanced form, MCA involves complex linear programming. A field of research is devoted to MCA, with many publications in the Journal of Multi-Criteria Analysis.

MCA has been successfully used to evaluate optimal land use planning outcomes for energy and for allocation of community support grants (Enache 1995). It has been applied to energy savings of different residential land use styles of living (Matsuhashi 1997). Munasinghe (1994, p. 15) argues that MCA clarifies the most important attributes or goals, eliminating many irrelevant options, and makes final trade-off process more transparent and more focused, while also providing the decision makers with more flexibility of choice. He refers to the value of MCA by the World Bank in the evaluation of more than 60 projects each year.

The choice of weights applied to different variables can significantly alter outcomes. Another issue is the complexity of the mathematics, and the way the modeling process handles the dynamics of change. Despite these short falls, MCA is a useful decision support tool for economic analysis, and forms an important platform for the development of MSA analysis.

Figure 7.3 shows the basic matrix used for MCA. The option criteria score X_{ij} is a measurement of the strength of a relationship between each option i_n and each criterion j_m is evaluated for all cells (i, j) in the matrix.

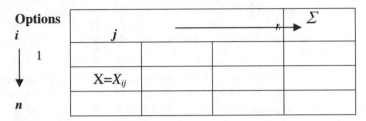

Fig. 7.3. Basic matrix used for MSA

The scores for each X_{ij} in a row are summed horizontally to give an option score:

$$\text{Option Score (Oi)} = \sum_{1}^{m} X_{ij} \qquad (7.1)$$

The highest score is the best or worst option depending on the measurement scale.

It is unusual for all criteria being evaluated to carry the same weight. For example, life-threatening issues are likely to carry a higher rating in a risk analysis. To accommodate the importance of different criteria, weights ω_{ij} are applied to each criteria. The scores of the weight X_{ij} values are calculated as follows:

$$\text{Weighted Option Score } (\omega Oi) = \sum_{1}^{m} \omega_{j} * X_{ij} \qquad (7.2)$$

This basic technique has been used extensively for prospect evaluation and decision making needs. Advanced applications of MCA can become mathematically complex, especially when criteria are given ranges. Optimization techniques are required to determine the best range of combinations for each option evaluated.

Techniques such as this can be used to evaluate regional competitiveness factors for industries in a regional economy. If the competitive factors (i) are listed in rows (i = 1 …n) and industries in columns (j) (j = 1…m) then values can be obtained for each (X_{ij}). The (X_{ij}) represents the strength or weakness of relationships between the two sets of variables being evaluated. If the X_{ij} values are added across the page (summing over j), the strength of different competitive attributes can be identified. If the column scores are added (summing over i) the industries with the strongest competitive attributes can be identified. The matrix provides a picture of the relationship between competitive attributes and industries. This matrix technique forms the basis of MSA.

7.3 Using MSA for Regional Strategy

7.3.1 The Strategic Planning Framework

MSA can play a valuable role in assisting the formulation of regional economic development strategy. In Chap. 5 we discussed a new *path setting process* for strategic planning for regional development (refer back to Fig. 5.12 in Chap. 5). MSA can be applied to address the following aspects of regional performance and development potential:

(a) The objective of MSA is to identify and evaluate factors that contribute to the underlying competitiveness of regional economies, such as core competencies, strategic infrastructure and risk management (addressing stage 2 in Fig. 5.12 in Chap. 5). The role of strategy is to capitalize upon these advantages by stretching and leveraging resources, competencies, and strategic infrastructure within and across industry sectors and clusters (Hamel and Prahalad 1993).

(b) The second objective is to identify new opportunities and markets for regional economic development by addressing stages 7 and 8 in Fig. 5.12 in Chap. 5. That involves considering potential trade development or cross-sector industry development. What follows is a series of discussions that show how the MSA technique can be used in a series of stages to appraise these aspects of regional performance and development potential in formulating strategy.

In applying the MSA approach it is necessary to use qualitative research methodologies involving panels of experts or focus groups to undertake the qualitative assessment of industry sector performance and risk factors assessment. The processes involved are discussed later in Sect. 7.5 of the chapter. The sections that follow discuss in detail the elements of regional competitiveness and risk assessment that might be addressed in the MSA approach.

7.3.2 Elements of Regional Competitiveness and Risk in MSA

Competitiveness is a function of many factors. Regional competitiveness (RC) can be conceptualized as a function of three key elements:

(a) core competencies (C_C);
(b) strategic infrastructure or architecture (I_S); and
(c) risk management (R_M).

Thus,

$$RC = f\ (C_C,\ I_S,\ R_M) \tag{7.3}$$

As discussed in Chap. 5, the notion of *core competencies* (C_C) is derived from Hamel and Prahalad (1994). In the context of regional economic development, they refer to specific applications or unique ways regions use resources, technology, skills, infrastructure, etc., for competitive advantage.

Strategic infrastructure (I_S) is the hard, soft and smart infrastructure which supports value adding activities. Not all infrastructure is strategic. Strategic infrastructure may include endowed resources—such as natural, fiscal, technological, human and other forms of capital with which a region is endowed—enabling a region it to compete for investment, development and trade. Strategic infrastructure also includes elements of physical infrastructure that facilitate production, transportation and trade export or valuing adding sectors.

Risk management (R_M) refers to elements of risk, natural, human, market, etc., that impact upon the performance of a region's economy. The management of risk significantly affects the competitiveness of a regional economy. The ability to manage risk is a significant competitive advantage.

MSA attempts to measure, in a systematic way, a region's competitiveness with respect to these elements using the method described above. Those elements of the MSA technique are discussed below.

7.3.3 Industry Sector and Core Competency Indices

Assessing regional core competencies requires the development of two indices. These are:

(a) an index measuring *industry sector competitiveness*; and
(d) an index measuring *regional core competence*.

These are derived through an assessment of industry sector performance against sets of core competency criteria, which may relate to regional competitiveness and/or attractiveness factors. Using the MSA matrix framework discussed above, a quantitative *index of industry sector competence*, is generated for the region by summing the column scores and dividing by the maximum score (refer back to Table 7.1). That enables different regions (for example, a region and its designated competitors) to be compared on a equal basis provided the same criteria are used for analysis. Similarly, an *index of core competencies* may be generated by summing the rows and applying the same technique to define the index. This analysis graphically brings into focus those features of a region's economy that are critical in supporting development; and it provides a basis for driving the economy in the future. It highlights also those weaknesses that need to be addressed if the region is to be more competitive.

There are no clear guidelines for developing a list of the core competencies that underlie the competitiveness of a region's economy. However, to develop a list of factors to use in MSA to derive an index of core competencies, reference can be made to the World Economic Forum's (WEF) Competitiveness Index (IMD-WEF) which is compiled annually by this services-based economic think-tank to analyze the competitive performance of nations) and specifically to those factors that represent competitive core competencies applicable to regional economic development and competitiveness. The WEF has developed over 380 factors of competitiveness, listed in 8 categories (see Fig. 7.4). These are measured annually to report on the competitiveness of national economies.

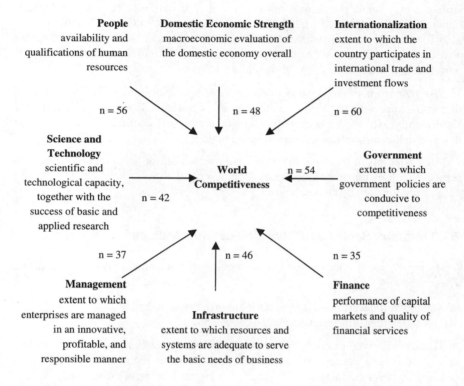

People
availability and
qualifications of human
resources

n = 56

Domestic Economic Strength
macroeconomic evaluation of
the domestic economy overall

n = 48

Internationalization
extent to which the
country participates in
international trade and
investment flows

n = 60

**Science and
Technology**
scientific and
technological capacity,
together with the
success of basic and
applied research

n = 42

**World
Competitiveness**

n = 54

Government
extent to which
government policies are
conducive to
competitiveness

n = 37

Management
extent to which
enterprises are managed
in an innovative,
profitable, and
responsible manner

n = 46

Infrastructure
extent to which resources and
systems are adequate to serve
the basic needs of business

n = 35

Finance
performance of capital
markets and quality of
financial services

Fig. 7.4. The competitive factors framework used by the World Economic Forum
Legend: *n* = number of criteria per factor
Source: Derived from IMD-WEF 1995

The IMD-WEF framework has been used to develop a competitive ranking index for the Gladstone region economy in Queensland, Australia (Kasper, et al. 1992). While that research was hampered by inadequate regional data, nonetheless it represents an interesting approach using national economies as benchmarks against which to compare a region's economic performance. Despite weaknesses, the Gladstone case study provides a valuable contribution to understanding and measuring factors contributing to the competitiveness of regions. A further application of the WEF framework is pilot work on MSA in Far North Queensland region in Australia, where composite WEF factors were used to evaluate 34 core competencies across 29 industry sectors (Roberts and Stimson 1998).

7.3.4 Strategic Infrastructure

Strategic infrastructure is well recognized as playing a critical role in the economic development and competitiveness of regions. Even minor weaknesses in strategic infrastructure can have a significant impact on the competitiveness of industries and organizations involved in economic and trade development. Using the WEF factors referred to above, it is possible to identify strategic infrastructure factors that contribute to the economic competitiveness, and to evaluate and compare these for different industry sectors. From this analysis, two further indices can be developed:

(a) an index measuring *industry strategic infrastructure competitiveness*, and
(b) an index measuring *factor strategic infrastructure competitiveness*.

These indices show the strongest elements of strategic infrastructure supporting regional development processes. The latter index is important in developing marketing strategies for regions to promote endowed factors that give a region a particular competitive advantage, and also for identifying infrastructure investment priorities. The identification of key strategic infrastructure factors is particularly useful for investors seeking locations for industries that can capitalize upon these advantages.

7.3.5 Risk Assessment and Management

Another important element of the MSA involves assessing regional risk, which is crucial in strategic planning, as well as for regional management and investment decision-making (Mason and Harrison 1995). While economic computer generated equilibrium (ECGE) models may be used to evaluate risk affecting national economies, most models are not developed sufficiently to enable risk to be assessed at a regional level.

How might regions assess risk and develop strategies to manage regional risk? There is no single answer to this complex issue which is becoming more and more important to address. Indeed it is difficult to precisely define what is meant by risk in the context of regional economic development and performance, and it is not easy to reach agreement on what specific criteria might constitute risk factors in a region and for a region. Nonetheless, there is an increasing recognition of the importance of incorporating risk assessment within regional development strategies.

The Australian and New Zealand Risk Management Standards AS/NZS 4360:1999 and AS/NZS 4360:2004 provide the most useful guide on risk analysis and strategy preparation. Those standards define risk as:

...the chance of something happening that will have an impact on objectives. It is measured in terms of a combination of the consequences of an event and their likelihood.

We would suggest that reasonably good general definition that might be applied to most types of risk is the following:

...Risk is a real, perceived or fabricated event or activity that has the potential to cause uncertainty, harm or disruption to economic, natural, or social systems.

This definition recognizes that risks are both real and perceived phenomena. They can also be fabricated and based on rumour. Factors or occurrences which cause risks range from life threatening situations to natural and environmental disasters, fluctuations in energy prices and exchange rates, civil unrest, competition, changes in consumer behaviour, beliefs and superstitions, and gossip.

There are two aspects of risk that are important to risk management. These are the likelihood and consequences or impact of risks. The AS/NZS 4360:2004 standards use a classification system to categorize risks according to a scale related to likelihood and consequence. This is shown in Table 7.2.

Table 7.2. The AS/NZS 4360:2004 risk scale

Likelihood label	Consequences Label				
	I	II	III	IV	V
A	Medium	High	High	Very high	Very high
B	Medium	Medium	High	High	Very high
C	Low	Medium	High	High	High
D	Low	Low	Medium	Medium	High
E	Low	Low	Medium	Medium	High

Note: The relationship between consequence and likelihood will differ for each application; the level of risk assigned to each cell needs to reflect this.
Source: Australian and New Zealand Risk Management Standards 2004

To develop strategies for regional risk management, it is useful to categorize risks. Seven broad categories might be considered important in assessing risk and its management in the context of regional economic development and strategy planning. These are:

Economic risk, which relates to the impact of global markets, trade factors, inflation, transportation and communication affecting goods and services.

Production risk, which relates to access to resources, profits, and production costs, such as labour disruptions, changes in material and energy prices affecting production, and corruption.

Governance risk, which relates to sovereign risk, government instability and loss of control over economic development processes by government.

Environmental risk, which relates to resource depletion, pollution, disease, natural and man-made disasters, and quality of life.

Societal or social risk, which relates to public liability claims against businesses and community attitudes towards development and pressure groups.

Technological risk, which relates to risks associated with the applications of technologies used in production processes.

Behavioural risk, which relates to the behavioural characteristics of people. The level of trust, sense of security and attitudes to work affect the performance and efficiency of firms, organisations and industry sectors.

Each of these seven broad categories of risk carries a range of specific risks factors that have the potential to impact upon a regional economy in some way.

Risk factors do not apply uniformly to every sector of a regional economy, and some risks will have knock-on or kaleidoscope effects. The 1997 economic crisis in Asia is an example of a sudden change in the macro-economic environment, affecting production, leading to changes in governance and civic disruption in regions in Indonesia. The more internationally exposed a regional economy, the greater will be the level of potential impact and knock-on effects that may occur from events occurring internally or externally to the region. Increasingly, however, it will be exogenous factors that regions must learn to manage if they are to remain competitive. That does not mean that the management endogenous factors are not an important consideration. Indeed, it is endogenous factors that are critical to the development of regional capital assets, which create competitiveness in the first place.

However, developing strategies to manage risk, is difficult. It requires regions to understand what risk factors need managing and to then develop appropriate instruments to manage these. Regional risk management requires a portfolio approach to the development of strategies that manage both *exogenous* and *endogenous* risk.

7.3.6 Exogenous Risk Management

Exogenous risks are particularly important to assess. Four approaches may be applied to reduce exogenous regional risk. These are: collective hedging; industry protection support; regional information systems; and networks and alliances.

Collective Hedging. Collective hedging involves industries or industry associations working with financial organizations to lock in advance prices for goods and possibly services–such as tourism–using futures and options. Those instruments provide greater surety on income to producers and may permit producers to adjust or diversity production in advance of the next trading cycle. Prices can be locked in for both exchange rate fluctuations and market prices. This strategy is particularly useful for resource and primary production regions.

Industry Protection Support. In some regions the loss of a large important regional industry may lead to the collapse of a large network of industries attached to it. In cases where an important core regional industry is linked to knowledge based industries that have significant long-term value-adding potential, short term government assistance may be necessary to tie an industry over a short period in time for a new metamorphosis to occur to replace old technology. In Australia this has been done in the Illawarra region south of Sydney for the steel town of Wool-longong. Over a five-year period the technology of the mills and a wide range of new technology industries were created as part of a defensive strategy to maintain and rebuild the competency base of the region. Defensive strategies of this type need to be considered very carefully, as they will involve public money that may have to be forgone to develop other sectors of a regional economy.

Regional Information Systems. The development of an accessible regional information system can reduce risk to producers and traders in regions by warning them in advance of exogenous changes in technology, markets, agreements and industry changes that are likely to have an adverse impact on local industry. Because of the lead times often associated with change, the more advanced warning that can be given to industry, the more time there will be to respond to events once they begin to impact upon a regional economy and its specific industries. An example of a well developed supporting regional information system is provided by the Center for Regional Analysis, at George Mason University in Fairfax, Virginia for the U.S. National Capital Region. The various continuously updated monitoring products of the region's economy appear at its web site *http://www.cra-gmu.org/.*

Networks and Strategic Alliances. Strategic alliances play a critical role in business development and trade in the global economy. Alliances can assist regions to maintain access to markets. Alliances can also maintain effective blocking of potential competitors into well-established markets. For alliances to work successfully, these need to be created through networks with other regions on an industry basis if they are to be effective. In New Zealand the establishment of the Berry Fruit Network, is an example of a strategic alliance that has resulted in significant expansion of office services, and the provision of a protection mechanism from import competition.

7.3.7 Endogenous Risk Management

The management of *endogenous risk* may be improved by adopting strategies focusing on developing clusters, enhancing local partnerships and networks, improving consultation processes.

Clustering. Industry clustering is an important strategy for supporting economic development. It leads to the more efficient use of slack institution and business resources and collective effort to develop research and markets. Clustering can also be used as a risk management tool. The same resources used to support development can also be used in the defence of regional business and competitiveness. By pooling resources, industries can strategically work towards offsetting risks, such as collective hedging and discounting from competitors. An example is the Far North Queensland region of Australia, where the fishing industry cluster has imposed a levy on the industry to pay for research into marketing and sustainable resource management, which is a major risk to the industry.

Enhancing Local Partnerships and Networks. The shift of economic competitive theory from competitive to collaborative advantage is also part of an overall strategy to reduce risk. Industries partnerships and networks enable business and other organizations to take a more collective approach to risk management. Regional networks, combined with information systems give early warnings of events, enabling quicker and more collective and coordinated responses there. However, partnerships reduce financial risks can also be an impediment to economic growth if organizational or regional fit are not compatible (Segil 1996).

Improving Consultation Processes. Improving consultation processes with communities can significantly overcome hostilities, parochialism and xenophobia in regions. The social risk factors can significantly reduce returns as well as add to development costs for new industries in regions. Public benefit from major projects are financial risk issues that will become more significant in the future, as local communities seek to extract greater public benefit from international investment by major organizations. For example, in Australia the Century Zinc Project on the Gulf of Carpentaria coast regions in Queensland could produce about nine per cent of the world demand for zinc. However, this project was placed in jeopardy for three years by the failure to embrace a more community consultative process during its development. For large development projects, project risk might be significantly reduced by adopting improved community consultation processes.

7.3.8 Measuring Regional Risk

Risk is largely perceptual, thus qualitative techniques can be employed, again through survey and Delphi methods using expert panels or focus groups drawn from industry sectors and/or selected organizations in the region.

As discussed above, it is important to stress that ideally risk assessments should be both *internal* (endogenous) and *external* (exogenous) to the region, as most regions are relatively highiy dependent on actors and agents external to a region for some if not a large amount of the decision-making that occurs with respect to investments in many of the economic activities—as well as in some aspects of public-financed infrastructure in the region. Thus we strongly recommend that the panels or focus groups comprise industry sector experts and investment decision makers .both internal and external to the region. In this way we may scale the perceived risks for each industry sector against categories of risk.

We would suggest also that the seven risk categories discussed earlier in Sect. 7.3.5 might be used, with a list of specified risk criteria being identified within each of those broad categories. Those criteria would then be used by the panels or focus groups to assess the risk criteria for the industry sectors.

Again using the MSA matrix framework, the members of the panels or focus groups would be asked to evaluate all of the risk criteria for each of the industry sectors in order to generate assessments of both the perceived *likelihood* (L) and the perceived *impact* (I) of different risk events. Once more the technique involves

296 7 Multi-Sector Analysis

a score system of perception using a Likert scale of, say, one to five where, for example:

     5 = very strong or very significant (total loss of assets or business)
     4 = Strong or significant (prolonged shut/slowdown with setback to profitability)
     3 = moderate (reduced output or performance for several months)
     2 = weak or discernable (inconvenience to business for short period)
     1 =insignificant (minor inconvenience).

For both the L and the I risk assessments, an *economic risk analysis matrix* may be generated. From the matrix, two further indices may be derived (using the column and the row summation procedure discussed earlier in Sect. 7.2.3):

(a) an *industry risk index*, derived by summing the weighted scores in the column; and

(b) a *risk factor index*, derived by summing the weights in the matrix rows.

However, the raw scores for risk impact developed in the matrix used to derive those indices assume that the perceived *impacts* or *likelihoods* fall equally across the region's industry sectors. But in reality, this would not be true, as some sectors of a region's economy are more important than others. It is advisable, therefore, to weight each score in the matrix by some measure of the relative importance of each industry sector. For example, weights could be applied in proportion to the significance of each industry sector to gross regional product (GRP), or the proportion of regional employment by industry sector, or the proportion of regional exports generated by each industry sector. That would give a better indication of sectors in the economy which will be likely to be most affected by the impact of different types of risk. The weighted risk impact is calculated by multiplying the raw impact score by the weight. The advantage of this weighted scores adjustment approach is that it enables a better assessment to be made in regard to the risks which should receive precedence by reflecting the relative importance of the various industry sectors. The method of developing the index is the same as that described earlier, except the total maximum score for the risk column is multiplied by the average weights applied across all sectors; for example, if there are five sectors the maximum score would be multiplied by 1.2.

While the measurement of the potential risk *impact* and *likelihood* are important, they do not provide us with an indication of which risks should be given priority as far as risk management is concerned. Most regions and industry sectors will be concerned with managing risks that have the greatest potential to cause harm. If the likelihood of a high impact risk event occurring is very low, then it might be better to apply resources to the management of more frequent events that could have moderate impacts.

The intention of risk management is to apply resources to those risks that have the greatest potential impact and likelihood of occurring. To do that we need to develop another measurement of risk referred to as a*nticipated risk*. This is an assessment of how prepared a region or industry sector should be to manage specific types of risk. The technique involves multiplying the MSA weighted impact by the likelihood of the risk matrix to give the measure of *anticipated risk matrix*, with weights.

7.3.9 Identifying Potential Trade Development Opportunities

Typically a regional economy exhibits a degree of dependence on selected industry sectors, and the strategic issues are the diversification of industry sectors and their export performance. The MSA process can incorporate the development of a *relative market potential for export matrix*, which is derived by assigning a weighted score to indicate 1 = insignificant through to 5 = very high market potential. Each industry sectors is evaluated against key market destinations, for regional products and services. The matrix scores represent a market potential that needs to be derived through both a qualitative and a quantitative analysis of trade flows and market intelligence scoping.

From the matrix, two indices may be derived:
(a) an *index of potential industry development* derived by summing the matrix column scores; and
(b) an *index of export market potential* derived by summing the row scores.

7.3.10 Identifying Cross-Sector Industry Economic Development Opportunities

Input-output tables for a region (refer back to Chapter 4) can be used to progress to the next stage of regional analysis using the MSA process. This helps identify new opportunities for development and investment in a region through an analysis of business links between industry sectors. These may vary greatly between regions and between industries within a region. The mix and the magnitude of the inter-industry linkages provided by I/O analysis may be useful indicators of the relative strengths and diversity of a regional economy to support new export and economic development activities.

Here the MSA technique involves an assessment (which may be bi-polar or scaled, where, for example, 5 = significant and 1 = limited) of the *strength* (S) of the *potential* (P) for intra- and inter-industry sector linkages. This assessment will utilize I/O tables where available, and includes a consideration of the strength of commonly used regional infrastructure and the relative capabilities and strengths of industry sectors previously identified through other applications of MSA. However, a measurement of economic potential by itself is not particularly useful. Often the potential for collaboration between industries is viewed as high, but because of factors such as the size of industry, market potential and production costs, the economic value to a region may be very low.

To provide meaningful data for strategic planning purposes, it is necessary to gain some estimate or measurement of economic potential (P). That may be achieved by developing a second measurement related to the scale or economic value of opportunity (S), and then applying that as a weight to (P). Thus, (P) x (S) becomes a weighted measure of the magnitude of realizable economic potential. The scale factor (S) is usually a value which is equated to a scale of monetary measurement (for example, $0.5mX $10m=$5m). The matrix is filled in with the (P) x (S) score means (averages) for each industry sector linkage. The full matrix represents a two-way measurement of the magnitude of potential for collaboration within and between industries. For example, the magnitude of potential for collaboration between tourism/agriculture may equal 15[(P=5) x (S=3)], but the reverse between agriculture/tourism may equal 12[(P=4) x (S=3)]. Seldom will the potential for two-way collaboration between industries be equal. In developing the matrix, the average figures for two-way collaboration are used so that only half the matrix is filled in (see Fig. 7.5, in which hypothetical data is used). The diagonal in the matrix in the figure identifies opportunities for 'stretching'—that is, linking sectors within an industry or cluster. The other figures represent the potential opportunities for 'leveraging' between industry sectors to create new business activities, as was discussed in Sect. 5.4.3 in Chap. 5.

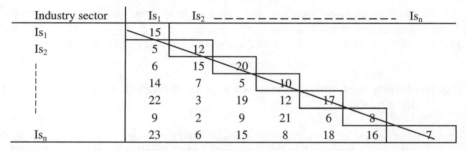

Fig. 7.5. Matrix of potential magnitude of inter-industry collaboration potential: hypothetical data

From that matrix, it is possible to derive for each sector an *index of potential linkage* with all other sectors of the regional economy. The value of this approach is both to identify linkages, which may be enhanced by improving innovation with industry sectors, and to identify those that are networked across one or more sectors.

7.3.11 Factor Analysis and Other Forms of Structural Analysis

The rich qualitative assessment data that may be collected through the MSA matrices discussed in the preceding sections may be analyzed in the simple forms discussed. However, more sophisticated multi-variate techniques may be used to interrogate the matrices. For example, *factor analytic* techniques may be used to identify underlying dimensions, which explain the pattern of correlations within a set of observed variables. Factor analysis is often used in the reduction of MSA-type data, by identifying a small number of factors, that explain most of the variance observed in a much larger number of variables. Factor analysis, when used in

MSA, identifies the deep structures that relate to the competitiveness of a region's economy. Deep structures are the factors that have a determining impact on the present and future structure of an economy.

7.4 The Process for Operationalizing the MSA Approach

Operationalizing the MSA technique through the derivation of the various indices discussed above in Sect. 7.3 involves a number of processes, including choosing data collection modes, the design of instruments and protocols, and the analysis of data and its interpretation. The process is complex, and it needs to be structured so as to ensure the generation of the various indices discussed above.

7.4.1 Data Collection

Data collection for undertaking a MSA of a regional economy may be gathered using three modes:

(a) survey questionnaires;
(b) expert panels; and
(c) a combination of survey questionnaires and expert panels.

Questionnaires are a preferred mode for collecting information in using the MSA methodology. They allow much more in-depth information to be collected, and tend to result in a broader perspective of views on an economy. However, questionnaires are time consuming, and it usually takes three to four months to collect, process and analyze the data.

Survey Questionnaire. MSA surveys comprise a questionnaire that is usually divided into several parts—for example, questions dealing with the core competencies, strategic infrastructure, regional risks, economic sector potential and marketing possibilities of the regional economy. It is important to ensure every question has two parts:

(a) The first part asks respondents to evaluate their perception of the *strength* (S) of factors related to their industry—for example, how strong is R&D in their industry sector?
(b) The second part to the question asks respondents to rank or score the *relative importance* (I) of the factor to the performance of their industry.

The latter part of the question is important for two reasons: first, it is used to develop the weights applied to different scores to develop the MSA indexes; and second, the relative importance score can be used to measure the extent to which the perceived *strength* (S) of factors match *importance* (I) to the performance of the industry. The difference between (I) and (S) provides an indicator of the relative competitiveness of economic development factors.

	Strength 1-5	Importance 1-5
E.1 Trade orientation		
How strongly <u>export/trade-orientated</u> is your industry (this includes all exports outside the region)?		
How important is <u>increasing exports</u> in the development of the industry?		
E.2 Strength of value adding industries		
How strongly orientated is your industry towards <u>increasing value adding</u> (i.e. adding to the value to products, goods and services produced in the region)?		
How important is <u>value adding to the development</u> of the industry?		
E.3 Industry diversification		
How <u>diversified</u> or broad based is your industry?		
How important is <u>diversification</u> to the development of the industry?		
E.4 Foreign investment		
How strong is <u>foreign investment</u> in your industry?		
How important is <u>foreign investment to the development</u> of the industry?		
E.5 Industry clustering or grouping		
How strong is the clustering or <u>grouping of similar types of industries</u> in the region?		
How important is the <u>clustering of industries</u> to the development and competitiveness of your industry?		

Fig. 7.6. Extract from a MSA questionnaire used in the FNQ region
Source: Roberts 1997

An example of a page from a MSA survey questionnaire used in the Far North Queensland region study in Australia is provided in Fig. 7.6.

The latter application [(b) above] of the MSA approach is particularly useful, as it is seldom possible to measure absolute competitiveness against known external benchmarks because of the way external benchmarks are developed, or because there are no external benchmarks. The assumption here is that most regional organizations and businesses understand—in some way—the level of importance of different factors that affect management, production, infrastructure and trade performance, and thus create their own benchmarks which they believe must be achieved to be competitive. A region that is highly competitive will attach great importance to those strengthening factors that are important in maintaining competitiveness.

The scoring system used to measure *perceived strength* and r*elative importance* factors are recorded using a Likert scale of 1 to 5, and a matrix of responses to the S and I components is derived. 'Not applicable (NA)' or 'don't know' are recorded as = 0 in the matrix. The points on the scale 5 = very strong through to 1 =

very weak. Weighted scores are calculated by multiplying the raw score (S) x (I_a). I_a is an average, calculated by taking the average of the sum of the I scores of factors recorded against a specific industry sector.

Expert Panels. The second mode of data collection using the MSA approach is to use a panel of experts representing a cross-section of specialists, government officials, business and community leaders, some of which trade or operate externally to the region. The experts are asked to fill in the matrix of strength and importance factors which will be compiled by a coordinator responsible for conducting an MSA evaluation of a regional economy. Expert panels may be organized in two ways:

(a) The first involves bringing a team of experts together to work in a group for one to two days.
(b) The second involves sending the questionnaires to the expert requesting completion in seven days.

The results of the expert panel may be further refined using a second-round assessment, similar to that used in Delphi analysis. The advantage of using an expert panel mode is that it consumes less of resources.

Combinations of Questionnaires and Panels Used for Quality Assurance. The combination of both questionnaire and expert panels leads to greater refinement of the MSA approach. Two such combination approaches may be used, and both involve external review panels in addition to panels from within the region:

(a) The first approach involves the questionnaire and asks the panels to make their own assessment. The results are used to measure if there are significant differences between the way respondents to the survey in a region perceive the relative importance and strength of different factors compared to external decision makers, investors and companies trading or doing other business in a region.

(b) The second approach involves providing a panel of experts with the results of the survey questionnaire and asks them to make their own assessment. A comparison can be drawn between the two assessments to check for congruency. This approach is akin to a modified Delphi.

The combination of internal and external assessment has many benefits:

(a) It acts as a check on regional bias and parochialism in answers to questions.
(b) It enables significant differences in competitiveness factors that may require attention to be identified.
(c) It enables latent strengths of some factors considered weak by those that live in a region to be realized.

There are no guidelines for the number of persons who should be involved in either survey questionnaire or evaluation panel modes for MSA. In panels, ideally at least one expert for each industry sector should be included to make an assessment, with some experts possessing knowledge and expertise over more than one industry. A panel should comprise no less than 20 experts, with panels of between

40 and 50 experts being desirable in order to achieve good results. Survey questionnaires may be sent to a small or large number of people judged to be appropriate for each industry sector. It would be possible to randomly select firms and individuals, but it is important to target key actors and stakeholders. It is not necessary for an expert panel to be assembled in one location, although there are advantages in group consultation and discussions. Large panels may be divided into small groups for working on specific sectors of a regional economy.

7.4.2 Stages in Conducting MSA

Undertaking a MSA of a regional economy takes significant time and resources. The steps involved in the process are discussed below:

Identifying Factors and Developing Questions. The first stage in conducting a multi-sector analysis exercise is to identify factors and questions to be included in the survey questionnaire or for consideration by the panel of experts. The factors listed above provide a useful basis for developing questions for the analysis. These factors are a guide, and organizations using MSA may wish to develop specific criteria for the purpose of conducting their own regional analysis. A sample size must then be developed for the survey work, or in the case of a panel, experts need to be identified and approached to seek their willingness to participate on the panel.

Developing a Sample Framework for Surveys or Panels. The number of questionnaires required for this type of analysis will depend upon the number of industry sectors to be investigated. For comparative analysis it is useful to adopt the same industry sector descriptions used for preparing regional input-output tables if these are available. Otherwise, the adoption of national industry standard classifications is suggested. In developing a sample framework for the questionnaire approach, a sample of 5 to 10 for each industry sector is probably sufficient, although more may be used. In some sectors it is desirable to conduct a larger number of surveys because of the diverse nature of the industry sector. When investigating industry sectors, achieving between 10 and 25 completed questionnaires for each one is desirable. For more detailed statistical analysis, more than 30 completed questionnaires per industry sector would be required. In large metropolitan regions, as many as 200 to 500 industries, organizations and associations should be surveyed. In such circumstances, it might be advisable to use a sampling design for each industry sector. Mail and electronic survey modes might be used to collect the MSA evaluations from firms in industry sectors.

Conducting the Survey or Panel Evaluation. After the design of the questionnaire has been completed, it is useful to run a small pilot survey of about ten organizations to remove bugs from the survey and data analysis processes. This will usually result in minor modification to questions, and the inclusion of new factors and questions. Questionnaires are normally distributed to leading business, government, professional and educational organizations and other to experts based on the level of sampling suggested above. The following steps are taken to conduct the survey once the questionnaire has been finalized:

(a) Advise business, professional and community organizations of the survey and involve them in its development. If organizations are aware of the existence of the survey and its purpose, the response rate normally will be high.
(b) Identify and select firms and organizations to be surveyed and contact names.
(c) Contact these to obtain their willingness to participate in the survey.
(d) Provide clear instructions, including examples in the questionnaire, of responses to questions and opportunities for organizations to add comment or clarification to answers.

In conducting this process it is important to sure to supply a 'hotline' number and completion date for the questionnaire.

Data Analysis. Data analysis is undertaken using a standard statistical package, such as SPSS, SASS or Access. Care should be taken to ensure checks on quality of the data input into the statistical package.

External Evaluation of Raw Data. Once the raw data results are processed, an external evaluation of data should be undertaken by an expert panel. The results of this evaluation may lead to adjustments in some of the data compiled and used for the MSA.

7.5 Examples Applying Multi-Sector Analysis

As a means of demonstrating the application of MSA, three case studies have been selected for different regions of the world. The first case study is of a small regional economy in Far North Queensland, Australia, which has experienced rapid growth in recent years through the internationalization of its economy. The second case study is of Northern Virginia, which is part of the National Capital region in the United States, a region which has experienced rapid growth in the development of producer services and high technology industries over the past 20 years. The third case study is of Ho Chi Minh City, Vietnam, which demonstrates the application of MSA in setting priorities for strategic infrastructure investment to improve competitiveness and support the development of a city's economy in a developing setting.

7.5.1 Analyzing Regional Risk in the Far North Queensland Region of Australia

Far North Queensland (FNQ) is one of Australia's fastest growing regions, with an internationalized regional economy having emerged since the mid 1980s through international tourism. The region has a population around 250,000 by the late 1990s, and an economy with a GRP of in excess of US$ 3bn. Cairns (population over 125,000) is the commercial hub for the region and the major city. The region has become a world class tourism destination because of its proximity to the World Heritage listed Great Barrier Reef and Tropical Rainforest, which are amongst the oldest in the world. The region has large mining, fishing and sugar industries. Because the economy had become so internationalized by tourism and national restructuring in the 1980s, it became necessary for FNQ to become skilled at maintaining its competitive advantage. A central question facing the region in 1993 as public agencies began to embark upon the preparation of an economic strategy was: What factors were important to the competitiveness of the economy, and what were the potential opportunities for economic development?

The pilot work to develop the MSA approach in the FNQ region of Australia (AHURI 1995) sought to understand what factors were responsible for regional competitiveness, and how to measure these. That research involved extensive consultation with expert panels to develop the series of MSA matrices to identify the competitive position of the region, risk and new economic development opportunities. The results of the early research used to develop the strategy are discussed in detail in Roberts and Stimson (1998).

Following that early work, a more detailed assessment regional risks facing the FNQ regional economy using the MSA approach was undertaken in 1998 by Roberts (2003). An analysis was undertaken of 25 risk attributes across 16 industries using a survey of 202 firms together with inputs from focus groups comprising managers of regional organisations. The results were used to develop regional risk management strategies for several export industry clusters in the region.

Weighted Risk Impact Assessment. The first stage of the analysis involved the derivation of a *weighted risk impact index* and its graphic representation. Figure 7.7 shows that a loss of telecommunication services, natural disaster, transport cuts, changes in consumer demand and government policy would have the highest impact upon the region's economy.

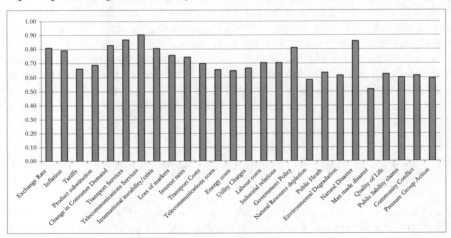

Fig. 7.7. Weighted risk impact index, FNQ region
Source: Roberts 2003

Other high levels of perceived risks to the FNQ regional economy are rise in production costs, exchange rate, inflation, and international crises. The results suggested that the highest risk factors tend to be concentrated in the export trade sectors. Anecdotal evidence uncovered during the study process suggested that many businesses in the region in these sectors operate on very low profit margins and do not manage regional risk well. Consultation with industry groups suggested there is a low level of risk awareness in many industries or knowledge on how to manage risk. Figure 7.8 shows the weighted industry risk impact index for 16 industry sectors of the FNQ economy. Food processing, mining, wholesaling and retailing are industry sectors perceived to have the highest risk impact attached to them. Other sectors of the economy with high levels of risk exposure are - tourism, business services, general manufacturing, land transport and fishing.

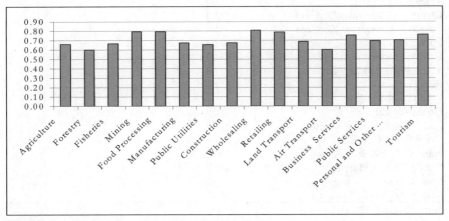

Fig. 7.8. Weighted industry risk impact, FNQ region
Source: Roberts 2003

Risk Possibilities Analysis. The second regional risks assessment conducted was about *risk possibility*. Table 7.3 shows the risk possibilities of the 25 risk attributes that have the potential to impact upon the FNQ economy. The table was derived from the average score of 202 responses for each risk category.

An ANOVA comparison of means test show a high level of significance between means for 16 of the 25 risk factors for the industry sectors evaluated. Exchange rate risk is the highest risk possibility factor, followed by natural disaster and changes to government policy affecting economic policy. These risk factors are perceived to have more than 80 percent possibility of occurring in the region over the next ten years. Other risk factors that have high possibilities of occurrence include: major international crises, inflation, rapidly rising interest rates, disruption to transport systems and loss of markets.

Table 7.3. Risk possibilities index, FNQ region

Risk Factor	Risk Possibility< 10 years
Exchange Rate	0.86
Natural Disaster	0.81
Tariffs	0.77
Government Policy	0.76
International instability/crisis	0.75
Inflation	0.74
Loss of markets	0.71
Interest rates	0.70
Transport Services	0.70
Labour costs	0.66
Industrial relations	0.66
Transport Costs	0.66
Change in Consumer Demand	0.64
Telecommunications Services	0.64
Public Heath	0.63
Utility Charges	0.63
Telecommunications costs	0.62
Energy costs	0.61
Product substitution	0.61
Quality of Life	0.59
Environmental Degradation	0.58
Community Conflict	0.58
Natural Resource depletion	0.58
Public liability claims	0.57
Pressure Group Action	0.56
Major Man made disaster	0.48

Source: Roberts (2003)

Risk factors perceived by respondents to have a low possibility of occurrence are wide spread actions by pressure groups, natural resource depletion, racial conflict, public liability claims, community unrest and manmade disaster. These findings suggest that the survey respondents perceived relatively high levels of social stability and sound management practice of the region's resources and infrastructure. This suggests that there is a very favourable climate for long-term investment in the region's economy.

The data in Table 7.3 provides an indicator of the likelihood of certain events occurring that will or might be likely to impact upon the FNQ regional economy. Some risks, such as natural disaster, can be measured more precisely. Historical analysis of the local Chamber of Commerce data indicated there are risk patterns for business related to cyclone activity in the region. For example, major destructive cyclones have hit FNQ at intervals of ten to twelve years. That suggests an 81 percent likelihood of a natural disaster occurring in the region in the next 10 years is reasonably reliable. Other risk factors - such as exchange-rate, changes in government policy, international crises etc. - are far less predictable, but developing some estimate of the possibilities of these occurring was important in formulating strategies for a regional economic risk management strategy.

Since the time when the survey was conducted in 1998, the exchange rate moved unfavourably for the tourism industry, government policy has introduced greater competition in the airline industry, and the region has lost some of the international airline carriers flying in and out of Cairns. International instability created by the war in Iraq and terrorism has affected Cairns as a destination and transport costs have increased.

Anticipated Risk Analysis. Table 7.4 and Fig.7.9 show the *index of anticipated risk factors* uncovered by the MSA process for the FNQ region economy. The table shows full anticipated risk. Exchange rate risk and natural disasters are the two highest anticipated risk factors affecting the region. Over 40 percent of the FNQ economy is linked to the use of its natural resources and capital assets. The impact of a natural disaster on the economy would, therefore, be very significant. The internationalization of the economy makes the region very susceptible to fluctuations in exchange rates, especially in the mining, agriculture and tourism sectors.

Table 7.4. Ranking of anticipated risk factors, FNQ region

Anticipated Risks	Anticipated Risk Index
Exchange Rate	0.70
Natural Disaster	0.69
Government Policy	0.61
Transport Services	0.61
International instability/crisis	0.60
Inflation	0.58
Telecommunications Services	0.58
Loss of markets	0.54
Change in Consumer Demand	0.53
Interest rates	0.52
Tariffs	0.51
Labour costs	0.46
Industrial relations	0.46
Transport Costs	0.46
Product substitution	0.42
Utility Charges	0.42
Telecommunications costs	0.40
Public Heath	0.40
Energy costs	0.39
Quality of Life	0.37
Environmental Degradation	0.35
Community Conflict	0.35
Public liability claims	0.34
Natural Resource depletion	0.34
Pressure Group Action	0.33
Man made disaster	0.25

Source: Roberts 2003

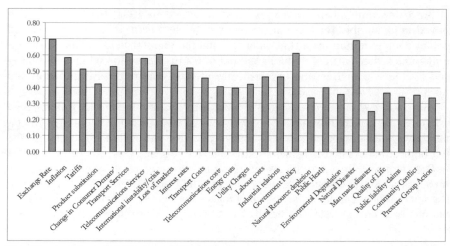

Fig. 7.9. Anticipated risk index, FNQ region
Source: Roberts 2003

The highly anticipated risks associated with disruption to transport systems, loss of markets, and telecommunication services are factors that are a product of location. FNQ is not densely populated and is geographically isolated; consequently, many consumer goods and services must be imported. Disruption to the communication and distribution networks would have a significant impact upon these and most other sectors of the economy. Other anticipated risk factors impacting upon the economy are inflation, international crisis, interest rates, labour costs and changes in consumer demand. The low level of anticipated social risks suggests the region, is relatively stable and free of factors that may cause concern about long term investment. Environmental risks, too, are perceived as low overall; however, some sectors of the economy have high anticipated environmental risk.

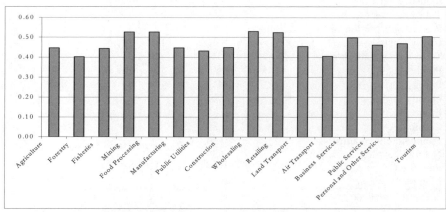

Fig. 7.10. Anticipated industry risk index, FNQ region
Source: Roberts 2003

Figure 7.10 shows the *anticipated industry risk index* for the FNQ economy. The wholesale trade industry has the highest anticipated risk exposure level in the region. Wholesaling has a high degree of interaction with other industry sectors, supplying a wide range of goods to the retail, tourism, construction, manufacturing and process industry sectors. The wholesale trade industry sector has both primary and secondary risk impacts. Rapid changes in exchange rates, changing markets in other industry sectors will trigger a major impact in the wholesale trade sector, so will secondary risk factors such as disruption to transportation. Retailing, manufacturing and business services will have similar effects. Food processing is the second highest anticipated risk industry, followed by general manufacturing, retailing and mining. The high level of risk exposure in general manufacturing is due to the greatly increased competitiveness of this sector resulting from tariff reduction, competitive labour costs and economies of scale and product change and substitution.

Most sectors of the FNQ regional economy have a low potential exposure to anticipated risk. They are predominantly endogenous industries including construction, community needs or small scale enterprises. These sectors are not highly exposed to competition and trade, except where major construction is involved.

The economy of the FNQ region has diversified rapidly over the past 20 years and this has had the effect of reducing the dependency upon one or two industries to support regional development. However, the high performance of the region's economy has been driven by the export sectors, and especially tourism and mining, which are high-risk industries subject to international market forces and competition. The region is thus much more vulnerable to risk, and this will likely increase in the future as the economy becomes more internationalized. That suggests that risk anticipation will increase in those sectors of the economy which are most exposed to production and international economic factors in future. These include exchange rate exposure, changing markets, product competition and substitution. This will call for a greater awareness of risk exposure to industry, and strategies to monitor and manage regional risk more effectively in future.

7.5.2 Identifying Industry Cluster Potential in Northern Virginia

Northern Virginia is a part of the Washington National Capital region of the United States. In this case study the application of a modified MSA as part of a strategic planning process was conducted in a highly time constrained, limited application context. The Northern Virginia economy is relatively young but has developed very rapidly in the high technology services since the early 1980s. Many of the new companies are small and medium sized, and therefore subject to forces of continuous innovation, making it difficult to involve senior company officials in the planning process. As a consequence, the MSA approach was modified so that a time constrained limited application could be employed and thereby tested.

Traditional quantitative methods commonly employed in regional analysis were used to identify the major industries in the Northern Virginia part of the National

Capital region. These included shift-share analysis and location quotients to identify the core industries in the region. Industry group leaders were organized to participate in a series of structured evaluation sessions with the objective of undertaking an analysis of the region and the performance of its industry sectors against a set of evaluation activities using the MSA methodology[1]. That included the use of a competitiveness survey instrument designed to elicit evaluation information on a variety of regional competitiveness factors, including the quality of the region's hard and soft infrastructure. Participants were selected from industrial directories and from economic development agency information bases to ensure that they represented senior officials from the region's major industries. The size of the expert groups ranged from five (aerospace and biotechnology) to eleven (tourism). The sampling methodology was a non-random expert sample. The survey was faxed and/or sent by email to the respondents, and data was collected through this mode (see Stough 1997; Stough 2001).

The survey asked respondents first to evaluate the region's competitiveness from the perspective of the person's company. Table 7.5 shows the results. The findings are organized by eleven economic sectors (column headings) and by 35 competitiveness and infrastructure factors (row headings). The data provide insight into the overall assessment by the 35 competitiveness factors and by major industry. They also may be used to assess the region's strengths and weaknesses by specific competitiveness factors of industries (or clusters) or both. The survey also asked respondents to evaluate the region's competitiveness from the perspective of a business resident of the region in general, in contrast to the perspective of their firm only. Again, this data provided information for a diversity of assessments at the sectoral and competitiveness factor levels.

The next step in the process was the presentation of the findings of the analysis of the survey results to the groups drawn from the survey respondents in meetings lasting about three hours. The presentation included data showing demographic and economic trends (including sector specific). With the modified approach used in this case, the results of the survey were also presented in the context of a modified Nominal Group Technique (NGT) (Delbecq et al. 1975). The NGT process generates expert data as derived from senior officials (experts) representing the region's core industry sectors. Each group was provided with the summary evaluation data from its sector (for example, aerospace, information technology, professional services, tourism). A major objective of the group meetings was for participants to verify and validate and, as necessary, modify and reinterpret the survey results. For example, local economic development efforts were viewed as rela-

---

[1] There are a variety of incentives for executives to participate in such as strategy development process. First, when an economy is very dynamic and uncertainty is high, participation can add more information of which some is inevitably useful. Second, peer pressure can be exerted through association memberships, e.g. Chambers of Commerce and in this case the local technology council. Third, some executives simply assume it is a duty and perform it as a community service. Finally, some feel that their company is facing considerable competitiveness pressure and hope that they can contribute to a process that will improve the general business climate and thereby the fortunes of their company.

tively weak by the information technology sector group, which indicated that the development community was making a great effort to reduce or eliminate the personal property tax levied on the industry (unfairly in their view); but these efforts have not been successful—that is, the tax remains in place.

Table 7.5. Responses to how factors affect the performance of your firm in the N.Va region

Competitiveness factors	Aerospace	Biotech	Information technology	System integrator	Telecomm	Transport	Association	Real Estate	Finance	Professional services	Tourism	Mean across factors
Infrastructure												
Adequate Highway System	3.429	4.000	2.143	3.542	3.625	3.091	3.556	4.500	3.444	3.750	3.667	**3.552**
Scheduled Air Service	4.000	4.375	2.571	3.292	3.500	3.182	4.000	3.250	2.889	4.250	4.667	**3.510**
Telecommunications	4.571	4.625	4.429	4.375	4.375	3.546	4.000	3.250	4.111	4.250	3.667	**4.095**
Environmental and Waste Mgmt.	2.570	3.000	1.857	2.250	2.625	1.818	2.667	3.167	2.000	2.000	2.333	**2.400**
Regional Quality of Life	4.143	4.125	3.571	3.958	4.500	3.273	3.444	4.417	3.889	3.500	3.833	**3.905**
Finance	**3.00**	**4.31**	**2.64**	**2.81**	**3.19**	**2.09**	**2.28**	**2.96**	**3.17**	**1.38**	**1.92**	**2.78**
Availability of Financing	3.143	4.375	3.000	3.125	3.625	2.455	2.556	3.417	3.667	1.750	2.000	**3.095**
Venture Capital	2.857	4.250	2.286	2.500	2.750	1.727	2.000	2.500	2.667	1.000	1.833	**2.457**
Human resource development	**4.00**	**3.45**	**3.49**	**3.94**	**4.18**	**2.45**	**2.98**	**3.25**	**3.62**	**2.80**	**3.87**	**3.50**
Higher Education /Training Svces	4.000	3.375	3.286	3.958	4.125	2.455	2.778	2.333	3.667	2.250	2.833	**3.286**
Availability of Skilled Labor	3.571	3.500	4.000	4.375	4.500	2.182	3.222	3.833	3.556	2.750	4.167	**3.705**
Availability of Prof. Employees	4.429	4.000	4.429	4.500	4.375	2.909	3.667	3.250	4.000	3.500	3.500	**3.924**
Flexible Labor-Mgmt. Relations	4.000	2.750	2.429	2.833	4.000	2.091	2.222	3.250	3.333	2.750	4.333	**3.010**
Competitive Wage/Salary Struct.	4.000	3.625	3.289	4.042	3.875	2.636	3.000	3.583	3.556	2.750	4.500	**3.591**
Technology and development	**2.14**	**3.15**	**2.11**	**2.06**	**2.63**	**2.60**	**1.69**	**1.70**	**1.60**	**0.95**	**1.83**	**2.06**
University Research Programs	2.000	2.875	2.429	2.125	2.625	2.273	2.000	1.917	2.000	1.750	1.667	**2.162**
University-Industry Partnerships	2.286	3.625	2.429	2.208	3.125	2.636	1.889	1.750	2.333	1.000	2.500	**2.352**
Federal Research Lab Programs	2.286	3.000	2.429	2.000	2.375	2.182	1.667	1.667	1.111	1.000	2.000	**1.971**
State Research Initiatives	1.857	3.125	1.571	1.833	2.250	3.364	1.333	1.583	1.111	0.500	1.667	**1.810**
Private Research Efforts	2.286	3.125	1.714	2.125	2.750	2.546	1.556	1.583	1.444	0.500	1.333	**2.000**
International trade orientation	**3.23**	**2.83**	**2.11**	**2.05**	**2.03**	**1.90**	**1.16**	**2.25**	**2.09**	**0.25**	**2.60**	**2.09**
Current Overseas Trade Activity	3.286	3.375	2.143	2.417	1.750	1.800	1.222	2.333	2.222	0.000	2.667	**2.212**
Foreign Investment into Region	2.429	2.125	1.286	1.625	1.500	1.400	1.222	2.333	2.444	0.500	3.500	**1.846**
Overseas Investmnt of Your Firm	2.571	2.250	1.143	0.917	1.750	1.300	0.889	1.833	1.556	0.000	0.500	**1.346**
Business Alliances (U.S. Firms)	3.857	3.250	3.714	3.250	3.250	3.000	1.333	2.500	2.444	0.500	3.667	**2.894**
Business Alliances (Foreign F.)	4.000	3.125	2.286	2.042	1.875	2.000	1.111	2.250	1.778	0.250	2.667	**2.144**
Government	**3.64**	**3.63**	**3.61**	**3.92**	**3.81**	**2.84**	**3.34**	**4.23**	**3.89**	**3.13**	**4.58**	**3.73**
Local Regulation of Business	3.429	3.375	3.571	3.917	4.000	3.000	3.444	4.167	3.556	4.000	3.833	**3.686**
General Business Climate	3.714	3.875	3.714	4.250	3.875	2.727	3.375	4.417	4.444	2.750	5.000	**3.914**
Local Econ. Development Efforts	3.429	3.625	3.143	3.250	3.500	2.636	3.000	4.250	3.889	1.750	4.667	**3.410**
Local Tax Structure	4.000	3.625	4.000	4.250	3.857	3.000	3.556	4.083	3.667	4.000	4.833	**3.904**
Regional economic strengths	**3.46**	**2.83**	**3.76**	**3.54**	**3.54**	**3.06**	**2.07**	**3.94**	**3.67**	**2.17**	**4.22**	**3.36**
Performance of your Ind. Sector	3.667	2.875	3.857	4.000	3.625	3.182	2.000	4.000	3.778	2.500	4.833	**3.567**
Strength of N.Va. Regional Econ.	3.143	2.875	3.857	3.542	3.714	3.091	2.444	4.500	4.222	2.500	4.833	**3.558**
Cross-Industry Information Flow	3.571	2.750	3.571	3.083	3.286	2.909	1.778	3.333	3.000	1.500	3.000	**2.962**
Your firm's mgmt. character.	**3.79**	**3.25**	**3.98**	**3.94**	**3.92**	**3.32**	**3.04**	**3.60**	**4.04**	**2.92**	**3.61**	**3.65**
Customer Service/Product Qual.	3.857	3.375	4.571	4.833	4.625	3.818	4.000	4.67	4.889	3.500	4.833	**4.381**
Inter-Business Networking	4.143	3.000	4.286	3.708	3.625	3.091	3.000	3.33	4.444	2.500	4.333	**3.600**
Available Mgmt. Consultants	2.714	2.625	2.857	2.609	2.250	2.818	2.222	2.417	3.000	2.000	1.500	**2.519**
Marketing Capabilities	4.000	3.250	4.000	4.000	4.250	3.455	2.778	4.167	3.889	2.250	4.333	**3.762**
Entrepreneurship	3.429	3.500	3.857	4.125	4.250	3.182	2.667	4.000	3.889	3.000	3.500	**3.686**
Info/Telecommunication Systems	4.571	3.750	4.286	4.375	4.500	3.546	3.556	3.000	4.111	4.250	3.167	**3.933**
Mean through firm sectors	**3.393**	**3.377**	**3.088**	**3.226**	**3.375**	**2.681**	**2.517**	**3.161**	**3.145**	**2.157**	**3.305**	

Values are average scores of responses to 5 point importance ratings (5 most important)
Source: Stough 2001

Following the verification and reinterpretation of the survey findings, the groups then identified new business opportunities for the future of their sectors, and assessed the risk associated with developing these options. Out of this exercise it was possible to create alternative proposals for deepening, and stretching and leveraging the sectors. An illustration of the deepening opportunities for the information technology and telecommunications (IT&T) sector is given in Fig. 7.11. Stretching and leveraging possibilities are illustrated for the information technology and telecom industry sector in Fig. 7.12, and for the tourism industry in Fig. 7.13. A fully integrated vision of future industry clusters for Northern Virginia is summarized in Fig. 7.14.

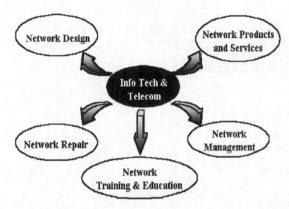

Fig. 7.11. Sectoral deepening (specialization) development for the Northern Virginia region
Source: Stough 2001

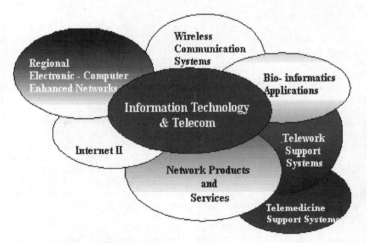

Fig. 7.12. Information technology and telecom cluster in the Northern Virginia region
Source: Stough 2001

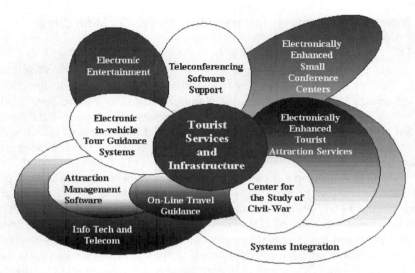

Fig. 7.13. Tourism cluster in the Northern Virginia region
Source: Stough 2001

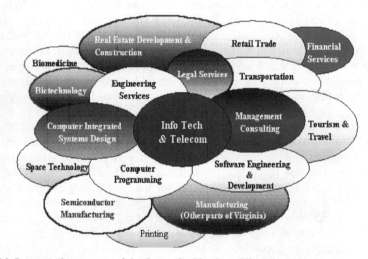

Fig. 7.14. Integrated economy of the future for Northern Virginia region
Source: Stough 2001

7.5.3 Identifying Strategic Infrastructure for Ho Chi Minh City, Vietnam

A final example of the application of the MSA technique is a study to identify industry clusters in Ho Chi Minh City (Saigon) in Vietnam. Vietnam had reached a critical phase in the transition from a centrally planned to a more open market economy. The rapid period of economic development leading up to the Asian financial crisis in 1998 resulted in great prosperity and improvements to the quality of life of people living in the city. However, as Ho Chi Minh City then faced a period of uncertainty over its development, there is need to reflect and carefully consider how best to develop a sustainable economic development path for the future. Of critical importance to any future economic strategy for the city was the need to improve industry competitiveness and develop value adding industries to boost the economy. There was a danger if the city did not take steps to do this, then it risked remaining as a low-cost manufacturing centre for predominantly foreign based and relatively footloose enterprises, which would move on if the operating costs were to rise in the future. This would be a great tragedy for a nation and city that had significant entrepreneurial spirit and the capability to become a modern advanced economy in the 21st century.

In 1996, the United Nations provided assistance to the Ho Chi Minh City's People Committee to improve the Planning and Management of the city and boost economic performance. The Institute for Housing Studies (IHS) in the Netherlands and the Australian Housing and Urban Research Institute (AHURI) were commissioned to undertake research and training to improve strategic planning in all development agencies of the city. A major issue addressed by the planning process was how to set priorities for infrastructure investment to improve economic performance.

It was decided to use the MSA technique to determine priorities for strategic infrastructure investment in 21 industry clusters that had been identified through a shift-share analysis and the calculation of location quotients, and using GIS analysis of industry data sets (Lindfield 1998; Roberts and Lindfield 2000). The priorities and analysis for the development of 15 elements of strategic infrastructure to support the 21 industry clusters followed a procedure similar to that used in the Northern Virginia case study discussed in Sect. 7.5.2. Table 7.6 reproduces the MSA matrix which shows the scores derived from the assessment process whereby an expert group evaluated the strategic infrastructure priorities for the industry sectors of Ho Chi Minh City, plus the two indices derived from it, namely the 'industry sector priorities index' and the 'strategic infrastructure priorities index.' Figure 7.15 plots the scores for the *industry sector priorities index*, and Fig. 7.16 plots the scores for the *strategic infrastructure priority index*. Business and marketing development, financial development and technology upgrading are the highest priority strategic infrastructure elements requiring upgrading. The industry thus identified sectors requiring the greatest need of support for strategic infrastructure were: food processing; wood processed products; textiles; and plastics. These also happened to be the fastest growing sectors of the city's economy.

Table 7.6. MSA of strategic infrastructure priorities in Ho Chi Minh City

Industry cluster	Transport infrastructure	Water supply, sanitation	Electricity supply	Telecommunications	Piped gas and fuels	Other infrastructure	Estate development	Technology upgrading	Research & development	Storage and transport	Business and marketing	Education and training	Finance development	Regulation reform	Environmental management	Total	Index score	Rank
Agriculture	5	3	3	3		4		4	2	5	4	4	4	3	2	46	.6	10
Food Processing	5	5	5	3	4	3	3	4	4	5	5	3	4		4	57	.8	1
Chemicals	2	5	5	3	5	4	3	4	2	4	4	2	4		5	52	.7	4
Garment	4	3	4	3	1	3	4		4	5	2	4			3	40	.5	13
Textiles & Plastics	4	5	5	3	4	4	3	4	3	3	4	2	4		5	53	.7	3
Tourism	5	5		5		4	4		4		5	4	4	4	5	49	.7	8
Leather Products	3	4	4	3		3	3	2		4	4	3	3		5	41	.5	12
Construction	5					4	5	4	3	4	3	4	4	5	4	45	.6	11
Education				4		5	3		5			5				22	.3	21
Transportation	5					5	3	4		5		2	4			28	.4	18
Metal Products		4	5	3	4	4	3	4	4	4	4	5	4		4	52	.7	4
Rubber and Plastics		5	5	3	5	4	3	4	2	4	4	2	4		5	50	.7	6
Wood Processed Products	3	4	5	3	4	4	3	4	4	4	4	5	4		4	55	.7	2
Non-Mineral Products						3	4		4	2	2	3			5	23	.3	20
Furniture	4		4	5			3	4		4	4	4				32	.4	15
Paper Products		5	5	4	4	4	3	3		4	4	3	4		5	48	.6	9
Printing & Publications				4		4	3	4		4	4	2	2		4	31	.4	16
Machinery Equipment		4	5	3	4	4	3	4	4	4	4	5	4		2	50	.7	6
Electronic Appliances		5	4				3	5	4	4	4	3	3		2	37	.5	14
Medical Instruments						4	3	5	5		4	4	4			29	.4	17
Cycles and Motor Cycles Vehicles	5					4	3	1	1		3	3	2		3	25	.3	19
Total	50	52	60	56	34	69	63	72	47	66	75	69	73	12	67			
Index Score	.2	.3	.3	.3	.2	.3	.3	.4	.2	.3	.4	.3	.4	.1	.3			
Rank	12	11	9	10	14	4	8	3	13	7	1	4	2	15	6			

5 = Highest Priority
Source: Lindfield 1998

Further analysis of the matrix enabled priorities to be set for each industry cluster. This analysis was linked to a Goals and Achievement Matrix Assessment (GAMA) of priority development projects for Ho Chi Minh City to determine which strategic infrastructure projects and programs should feed into the overall development plan for the city.

Given the paucity of data for conducting economic analysis in Vietnam, the qualitative evaluation that was undertaken using the MSA technique provided a feasible, useful and relatively quick methodology for helping inform the development priorities and to help support economic development in a developing country.

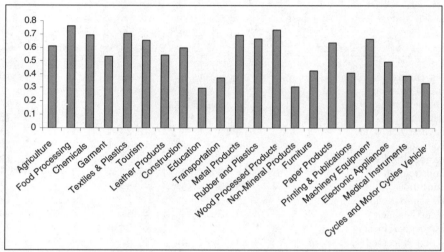

Fig. 7.15. Industry sector priorities index in Ho Chi Minh City
Source: Lindfield 1998

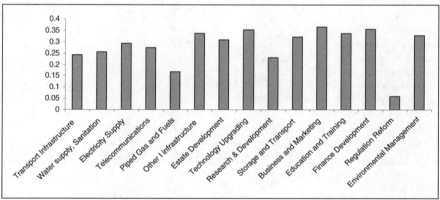

Fig. 7.16. Strategic infrastructure. Priorities index for Ho Chi Minh City
Source: Lindfield 1998

7.6 Future Development of Multi-Sector Analysis

Of importance for the future development and application of the MSA methodology will be the use of a multitude of major decision maker groups, drawn from the industry sectors both internal and external to a region. This will permit comparison between industry sector groups within a region, as well as between groups internal and external to a region, in order to measure the degree of congruence that exists in the qualitative assessments of industry sector potentials and for the evaluation of the performance on competency criteria. In particular, it is important to determine the degree of congruence between the assessment and evaluation of groups both internal and external to a region. For example, if the assessment of industry sector potential or of regional competence by an external group diverges from that by an internal group, then there could be a problem for the region in that incongruence exists in their perceptions about the region's performance and/or its potential. That could mean that investment flows from outside the region are not likely to be forthcoming at levels sufficient to develop an industry sector or to encourage the development of industry clusters to the extent that locals officials would like to see happen.

Formulating strategy for regional economic development needs to embrace multiple methodologies, including both quantitative and qualitative tools of analysis to assess the current performance of a region, to identify its core competencies, and to evaluate its economic possibilities in order to develop a strategic intent for a region's future development. That might involve evaluating alternative futures utilizing the MSA approach. But to realize strategic intent it is necessary to build on the core competencies of a region, and to develop regional strategic leadership, infrastructure, resources and governance, while providing marketing information systems enabling the development of industry clusters to happen. Roberts (1997) proposes the way these strategic elements come together to drive the economic development process is through building the strategic architecture of the region, as discussed in Chap. 5. The 'game plan' for realizing strategic intent is to:

(a) define core competencies to build or maintain on a continuing basis;
(b) identify what sectors to develop or maintain;
(c) identify what strategic needs should be provided;
(d) describe those competitive resources that need to be protected and managed more effectively and efficiently;
(e) address what approach to marketing intelligence should be developed;
(f) identify deficiencies in governance; and
(g) identify the risk to be managed.

The strategic infrastructure (architecture) needs identified through the processes discussed in Chap. 5 and through the MSA methodology outlined in this chapter will supply the tools and support systems necessary to facilitate and support the sustainable economic development of a region, focusing on the development of industry clusters (as discussed in Chap. 6) with good potential for success and to enhance a region's competitive position. The whole process is dependent on the participation of key decision makers in those industry sectors that are important to the region, with participants being drawn from organizations both internal and external to the region.

Of course completing the regional economic development process requires that strategy be implemented through plans and appropriate mechanisms involving the development of regional information systems, partnerships, networks and alliances. That is likely to require substantial institutional reform. Implementation will be highly dependent on the support of key political and business interests in a region, as well as investors and decision makers external to a region. Their involvement in strategy formulation should enhance commitment to implementation as well as development of leadership in a region. These issues and others related to capacity building and leadership in regional economic development and are taken up in Chap. 8.

8 Capacity Building, Institutions and Leadership for Regional Economic Development

8.1 An Emerging New Perspective on Regional Capacity Building

Developments in the new economic growth theory (Romer 1986; Nelson and Winters 1982; and Krugman 1990), and subsequent research, have made it increasingly clear that local development effort is very important part of regional economic development (Stough 1998). Unlike in earlier times, when comparative advantage and factor cost differentials were viewed as the determinants of regional economic development, today a diversity of *institutional* factors, including *leadership*, are viewed as important intervening variables and catalytic agents in the mix of endogenous processes shaping and building capacity and capability to enhance regional growth and development.

This chapter examines some recent advances in regional development theory which are viewed as providing the springboard for a new approach to regional economic development planning and practice. Assessing local capacity and capability traditionally have been aspects of conducting a regional economic audit as discussed in Chap. 2. But in the new approach, local effort and the capacity of *institutions* are viewed as absolutely crucial components in regional economic development. Further, contemporary thinking views social capital as one of the major factors that affects regional leadership and institutional capacity. Thus, after a summary assessment of the new economic growth theory, the chapter examines the construct of *social capital* and its relevance for regional economic development. This is followed by an examination of the role of institutions and leadership and their relationship to capacity building for regional economic development.

In Chap. 5 reference was made to K-cycles of innovation in technology and economic development, noting how the contemporary era is one of transition from the fourth cycle to an emerging fifth cycle, involving a shift to artificial intelligence and bio-engineering as the driving new technologies for economic development following on from the electronics technology drivers of the 4th cycle. The rise of information technology, the Internet and e-commerce have had a huge impact on how modern economies and societies operate. The consequential restructuring of society that is now well underway may be compared in magnitude to changes that occurred during the industrial revolution. However, the contemporary pace of transformation has been much swifter thus pushing institutions, values and culture to change equally rapidly. Regional economic development, like many other processes, has been forced into a rapid response mode to adjust to such changes. Yet, like most other processes, regional economic development historically has been conducted in a time context that allowed for relatively slow and, therefore, incremental change. This is no longer the case due to the rapid rate at which technology is being substituted for other factors and in particular labour. Hence, regional economic development process and strategy planning approaches need to enable and promote rapid, flexible responses to changing conditions. To be competitive in the contemporary information age, regions must be ready to re-engineer rapidly and continuously. Thus, a significant part of this chapter deals with the re-engineering problem and rapid response issues. Central to the successful rapid response capability of a region is the strength and coordination of social and business networks and strategic alliances, and networked resources to support re-engineering. Part of the chapter examines the role of these networks and alliances and their linkage to a region's social capital. The chapter concludes with a discussion of infrastructure with an emphasis on the concepts of smart and virtual infrastructure.

8.2 Endogenous Processes

As we outlined in Chap. 1, regional economic development involves a wide array of factors the importance of which varies across regions and over time. Thus, a regional economy that is resource dependent will have different defining characteristics than one that is service based. Likewise, a regional economy embedded in a transition to knowledge-based, information-intensive industries will exhibit different attributes than at an earlier time when it was, for example, a labour and energy intensive manufacturing economy. This understanding is not new. However, as intimated several times in previous chapters, two types of factors are essential to the understanding of regional economic systems: *endogenous* and *exogenous*. In all regional economic systems, these factors are entwined and intermingled yet functionally inseparable.

We saw how recent work in the area of *endogenous growth theory* (Romer 1986, 1990; Johansson et al. 2001b) separates endogenous and exogenous factors for analytical purposes. Such an exercise leads to a conclusion that it is possible for even closed economic systems–for example, hypothetical regional and national economies– to grow and develop. In short, in a fundamental sense, regional growth and development are not ultimately dependent on external factors and higher order conditions; however, they may be conditioned by them. Thus, as the knowledge base of a region is enhanced from within—for example, through learning– (Arrow 1962) it becomes a continuous and internally created source of competitive advantage and monopoly power (Romer 1986, 1990; Lucas 1988). Such internal learning may, at the same time, lead to a decision to build new infrastructure—for example, a road or improved wastewater treatment—to enhance development. Nonetheless, *knowledge creation through learning* appears to be the central proposition of the endogenous growth formulation. Through learning it is possible to envision how a closed regional economic system could survive, develop and sustain itself.

It is important to note that *exogenous* factors—such as trade, labour mobility and migration, knowledge and innovation diffusion, foreign exchange, business cycles, and capital mobility—are important to a region's economic performance. Yet these factors are not substantially under the control of local development efforts. What is significant about *endogenous growth theory* is that it emphasizes the importance of *local* factors in creating and maintaining sustained development as opposed to ones external to the region. Such a cohesive argument for the importance of local factors has been lacking in much of the theory of regional economic development. Perhaps that is why the fields of community development on the one hand, and regional science on the other, have minimally overlapped in the past. Endogenous growth theory provides a way to see a broad array of community, institutional and non-traditional economic variables—including learning, leadership and social capital—as major inputs to a successful regional economic development process.

Endogenous growth theory embodies the notion that it is possible for regional economic growth and development to be sustained by local internal forces. While these forces include a wide array of factors, some of the more important are learning, leadership and institutions, physical infrastructure and human capital. Through these it is possible for even totally endogenous or closed economic systems with feedback to become self-sustaining and to experience the phenomenon of dynamically increasing returns (Arrow 1962; Arthur 1994; Kilpatrick 1998). This in turn fuels and sustains regional growth and development.

8.3 Social Capital

Social capital has emerged as a non-economic concept that is gaining increasing interest as an important way to think about the role of more intangible development factors such as institutions, networks and trust in regional economic development. This concept emerged out of the thinking of such luminaries as Jane Jacobs, and was formally defined by Coleman (1988, p. S98) as that which "inheres in the structure of relations between actors and among actors." Malecki (1998) explores a variety of definitional extensions, noting that social capital as a concept has been difficult to measure and, therefore, to apply. This continues to be a problem despite a number of efforts to operationalize the concept and to link it to economic development (see, for example, Abel and Stough 1999; Fukuyama 1995; Irwin et al. 1998; Putnam 1993).

The underlying importance of social capital—the so-called '20 percent' solution as Fukuyama (1995) calls it—is that it is seen to reduce considerable friction in market transactions in regional systems—that is, it reduces transaction costs. Malecki (1998, p. 11) notes such friction is reduced in three ways:

(a)   the creation of a system of generalized reciprocity;

(b)   the establishment of information channels, providing sorted and evaluated information and knowledge; and
(c)   the simplification of market transactions by instituting norms and sanctions by which economic exchanges can occur, bypassing costly and legalistic institutional arrangements associated with market transactions.

The vision of the role of social capital in economic development is that it contributes to increasing returns to the regional economy across the board by reducing transactions costs. Abel and Stough (1999) have attempted with some success to test its effect on the economic development of metropolitan regions.

Fukuyama (1995) and others (see, for example, Granovetter 1985) view social capital as evolving from and as a function of trust. Like social capital, trust is difficult to measure, but it might be viewed as being the component making it possible for non-routine transactions to take place with a minimum of friction. Some regions, such as Emilia Romagna (the Third Italy), have high levels of trust among firms and the institutions upon which competitively productive activities rest—for example, finance. Yet such regions are capable of maintaining global competitiveness through networks of small and medium firms, rather than being anchored in one or more large-scale corporations. This is because of their level of trust and, therefore, social capital. While the evidence of the important role social capital plays has been acquired largely through deductive reasoning and case study analyses, increasingly its importance is becoming viewed as a fundamental factor of regional competitiveness. Regions with high levels of social capital and, therefore, trust appear to be more competitive because they can better adjust to the rapidly changing conditions prevalent in the contemporary high flux era of technology-led economic development.

A usual result of high levels of trust and of social capital is *effective leadership* for regional economic development. Where trust levels are low it is difficult for leadership and supporting associations and organizations to develop. Yet from an endogenous perspective, local leadership is central to the successful economic development of a region.

8.4 Leadership and Institutional Capacity Building

Research on the role of leadership in regional economic development has been sparse, with relatively few studies published (however, see Stough 1990; DeSantis and Stough 1999; Stough 2001; and Rees 2001). Drawing on these studies and more recent work by Stimson et al. (2005) a picture of a leadership/institutions oriented approach to modelling regional economic development is emerging. Below we present our thinking at this time on the role of leadership. It is important to note that little empirical research has been conducted to date other then the original work by Desantis and Stough (1999). Here we present the conceptual model.

A leadership model by DeSantis and Stough (1999) for regional economic development has been tested across a sample of data from 33 metropolitan areas in the United States. That research along with earlier work by Stough (1990) identified a number of attributes of successful development initiatives:

(a) local initiative is critical for initiating and sustaining regional economic development;
(b) local initiative is consistently undertaken by non-government or intermediate regional/community organizations;
(c) such intermediate organizations are effective economic development planning organizations;
(d) economic development plans are a basis for cross-sector collaboration; and
(e) successful regions have access or create access to a broad range of local and extra-local national resources.

Work by DeSantis (1993) and DeSantis and Stough (1999) note additionally that local economic development effectiveness is also related to:

(a) the degree of political jurisdiction fragmentation;
(b) the degree of cooperation among local stakeholders (public, private, intermediate and individuals);
(c) the tendency for a region to participate in local problem-solving (a distillation of Stough's five factors); and
(d) the availability of resources locally for economic development.

This leads to a conclusion that economic development effectiveness is influenced by resource endowments and leadership, where the basic elements of the leadership construct consist of Stough and DeSantis' factors except the last one, which defines the resource endowment construct. Research conducted by DeSantis and Stough as referenced above found support for a model that argued *ceteris paribus* that resource endowments (traditional but also contemporary, e.g., labour quality, entrepreneurship, etc.) was the strongest explanatory variable accounting for some 50 percent of the variance in performance in a 33 metropolitan region sample and that the explained variance increased to about 70 percent when the leadership variable was added as a third variable using both a static and a dynamic model where the resource and leadership variables were lagged five years. In short, there is some empirical evidence that leadership amplifies economic performance. These results were encouraging and thus led to new work that aimed to incorporate leadership into a more inclusive conceptual model for regional economic development. We now present that model and its background.

8.5 Leadership and Regional Economic Development: A Conceptual Model

A long term objective of regional economic development is to internalize a process that ensures a *competitive* and *entrepreneurial* region and one that achieves *sustainable development*. Such a process is likely enhanced through a regional economic development strategy that enables a *proactive* approach to development, as against a reactive approach, to plan for and manage development and to managing risk in adjusting to changing circumstances. In this paper we propose that *leadership* and *institutions*, and how they interact to facilitate *entrepreneurship*, are crucial in achieving sustainable development. Of course a city or region's *resource endowments* and its 'fit' vis-à-vis *market conditions* are also important factors affecting regional economic development, growth and performance, but leadership and institutional factors may serve to enhance or detract from the effectiveness and efficiency with which those resources are used and markets are captured.

8.5.1 Model Framework

The idea of sustainability - paying explicit attention in regional development to what is being called the "triple bottom line" - may be conceptualized as a *virtuous circle*. The circle (as presented in Stimson et al. 2005) is maintained by *effective leadership* as it is used to *change and adjust institutions* in order to adapt the *structure, processes* and *infrastructure* of a regional economy to meet and anticipate changing circumstances and to facilitate the optimal use of its resource endowments and to assist industries to tap their full market potential (Fig. 8.1). In this context endogenous growth theory is viewed as providing the context and intellectual platform for the propositions underpinning this proposal.

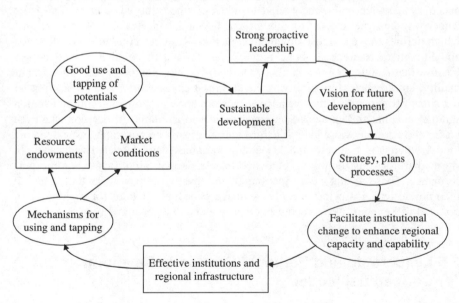

Fig. 8.1. The virtuous circle for sustainable regional development
Source: Stimson et al. 2005

In this view *strong leadership* means a region will be *proactive* in *initiating regional economic development strategy* to *monitor* regional performance, set a *vision* for the future development of the region, and *implement plans* and *processes* which *facilitate institutional change* and *encourage and facilitate entrepreneurship*. This, in turn, will enhance the *capacity* and *capability* of the region to *positively adjust* to changing circumstances, to attain a good and/or improved fit with market conditions, and to harness its *resource endowments* in order to maintain and improve its performance and to achieve *sustainable development* as a *learning region* and to be one that is *competitive* and *entrepreneurial*.

While the argument here is normative, i.e., we are advocating for this process while at the same time we recognize that while it often is used in whole or part for strategic development initiatives, all too often it is used in a less than thoughtful, coherent and pre-planned way. The argument here is derived from the notion that the *presence of leadership* in regions that are performing well, or have been re-engineered and turned around from performing poorly to perform better, has been crucial in providing the appropriate policies and creating and facilitating the right environment. For example, the Silicon Valley region may be viewed in this perspective as having channelled resource endowments into efficient allocations (Leipzieger 1997). In such places, leaders have initiated crucial institutional reforms, policies, projects and facilitated the creation of an environment that benefited business and citizens in general (Rowen 1998).

The sections that follow provide a discussion and description of the nature of the components of the model framework set out in Fig. 8.1.

8.5.2 Institutions

Institutions are crucial in providing the 'rule structures' and the 'organisations' within which a society operates. 'Government' is the system by which a nation state or city or region is governed, while "governance" is the act or manner or process of governing and the office or function of governing.

North (1990) argues that the institutional framework determines the incentive structure of a society:

> ...Institutions, together with the constraints of economic theory, determine the opportunities of a society" (p. 4).

The economic performance of a city or region over time is fundamentally influenced by the way institutions evolve, how they decrease uncertainty, how they allow individuals to have access to information, and how they decrease market imperfections that increase transaction costs. "They can provide the stability in collective choices that otherwise would be chaotic" (Clingermayer and Feiock 2001, p. 3).

The choices that political and economic actors make are shaped by the rules, conventions such as values and beliefs embodied in things such as constitutions, property rights and informal constraints that, in turn, shape economic performance. The nature of those institutional factors and the degree to which they impose constraints or help facilitate action in the pursuit of opportunities are seen as conditioning the capital accumulation process and as a result the economic development of cities and regions (Vazquez-Barquero, 2002; 12). That is because their behaviour can:

(a) reduce transformational and production costs;
(b) increase trust among economic and social actors;
(c) improve entrepreneurial capacity;
(d) increase learning and relational mechanisms; and
(e) reinforce networks and cooperation among actors.

While institutions may be viewed as contributing positively to all of these outcomes it is important to recognize that there are threshold issues relating to the role of institutions. One may envision a loose collectivity with no or few institutions on the one hand and on the other one that is so bound up with institutions that it becomes difficult to act in any way without violating or challenging an institution. Between these extremes lies a U-shaped relationship whereby at first increasingly strong institutions reduces transaction costs, builds trust, etc. but at some point diminishing returns set in and thereafter the cost of institutional constraints increase and begin to detract from or reduce the benefits. The ideal situation is to have a sufficiently strong system of institutions to gain the benefits but not so strong as to constrain or reduce them.

Blakely (1994) refers to the necessity of having appropriate institutional arrangements to manage and fund the regional or local development strategy process and to ensure the implementation of plans and actions. Thus, the capacity and the capability of local institutions to initiate, undertake and carry through plans and decisions are fundamental to that process. Institutional capacity-building is now seen as a fundamental factor in regional economic development. That is now being discussed as well in the context of the creation of 'learning infrastructure' and the *learning region* (Simmie 1997; Jin and Stough 1998; and OECD 2000).

One issue with our formulation and the role of leadership as a mediating variable is that a number of institutional scholars such as Durlauf and Peyton (1945), Granovetter (1985), North (1990), Storper and Scott (1992), Streeck (1992) and social capital scholars such as Putnam et al. (1993) and Fafchamps (2004) do not make a distinction between leadership and institutions. Rather, leadership is viewed as a basic characteristic provided by institutions and not as a separate variable or category. In short, the performance of a region (good or poor) would provide the conditions, incentives and leadership for the development of economic activity. While this may be true when viewing performance over the long run (although it is not clear even then due to perpetuity or accumulated effects of past behaviour patterns) it is not necessarily so over the short run where equilibrium like conditions are punctuated as has occurred over the past 30 years or so with the rise of the information or knowledge age, the growth of a generic technology like ICT and/or the period of de-industrialization that accompanied it. In such situations responsiveness to changed conditions must be driven by some vector or agent. Herein, we argue that the agent is leadership and affiliated entrepreneurship because these are inherently faster changing variables than most of the others that make up an institutional system. Yet, it is important to note that places that have strong institutions and social capital, *ceteris paribus,* are more likely to create the leadership and entrepreneurship needed to address the changed conditions. But at the same time, while it may be argued that high quality institutional and social capital type places may have made the best adjustment of their institutions to equilibrium like (or slow changing) conditions thus have created a powerful ability to operate in that situation, they may perhaps be less fit to respond to rapidly unfolding or punctuated changes such as de-industrialization, technological change (for example, ICTs), and wars and pestilence! In these situations, in particular, we argue that leadership and entrepreneurship are the agents from which direction and guidance for the ways in which other institutions such as values, cultural traits, constitutions, laws, regulations and informal practices change or are changed to make a successful adaptation. In short, we may think of leadership and entrepreneurship as those parts of the stock of institutions that are potentially fast changing and thus provide the dynamics for adjustment in the face of changed conditions.

Stimson et al. (2005) further develop this line of thinking by examining how institutions relate to issues such as city or local governance, social capital, strategic alliances and community collaboration, and central local relations. The interested reader is referred to this article for more elaboration on these important issues. Here we move on directly to other parts of the core of the model. Suffice it to say here that institutions underpin business attractiveness of a region, facilitate network alliance formation to support collaborative advantage and underlie learning infrastructure that is needed to support strategy development and implement.

8.5.3 Leadership

Leadership is in most analysts' view one of several if not many institutions in the development frame or process. However, in our view leadership is a special institutional variable in that it is relatively potentially powerful and is capable of driving change in the corpus of the institutional system that supports and maintains a functional regional economic system.

The institutional system is an informal and formal hierarchically organized network of rules that provide prescriptions for acceptable behaviour and actions in a societal context (Williamson 1994). Elements of this institutional system can be durable and hard to change (for example, values and cultural traits, constitutions, executive orders) or relatively non-durable and easy to change (for example, traffic regulations, land use requirements) as well as a vast array of various combinations in between. Major adaptive change in the functional regional system however usually requires change in one or more relatively durable institutions, that is, in the ones that are difficult to change. This involves "bucking the system" and rarely can be achieved by incremental change of the less durable institutions. Although, in periods of relative stasis or equilibrium change occurs incrementally and can occur via change in the less durable institutions, that is, it involves mostly fine tuning a generally acceptably working regional system. However, major concern in regional economic development is how to achieve re-generation and sustainability in the face of major and systemic change in a regions environment or context driven by, for example, war, recession or depression, new technology, globalization, systemic structural change, natural disasters and man made disasters (for example, terrorism). While this sort of events or processes serves as catalysts for change, we argue that leadership is the catalytic agent (hopefully autocatalytic) that drives adaptive change and thus sustainability in the face of such stressors. In this sense the kind of leadership of interest here has a contingent character and thus is consistent with contingent leadership theory (see Hughes et. al. 1998). Next, we explore the leadership construct for regional economic development in a more broad based examination of the literature.

Parkinson (1990) views as

...the capacity to create stable and durable mechanisms and alliances that promote economic regeneration and identifies a range of micro-level skills and macro-level resources that can generate that capacity (p. 241).

While it is common for leadership to be viewed in terms of a 'great person,' it might be more appropriately seen as an expression or result of collective action. Thus, in regional economic development, leadership usually occurs not as a 'starring role' but as a 'collaborative' action (Fairholm 1994; Heenan and Bennis 1999). Leadership may thus be defined as "the tendency of the community to collaborate across sectors to enhance the economic performance or economic environment of its region" (De Santis and Stough 1999).

Heenan and Bennis (1999) point out that, in the new economy of increasing interdependence and technological change, collaboration is not just desirable; it is crucial. At an earlier time in history influence, power and decision-making often depended on single individuals, and leadership was based on a traditional hierarchical authority relationship between leader and follower. But today, power, influence and decision-making are more dispersed among power stakeholders working together towards a common goal (De Santis and Stough 1999; Heenan and Bennis 1999; Judd and Parkinson 1990). It is through collaboration and collective processes that cities and regions will have sufficient flexibility and knowledge to adjust to shocks and continuous changing conditions (Saxenian 1994; Stough et al. 2001). In this sense, we suggest the following statement regarding leadership:

...leadership for regional economic development will not be based on traditional hierarchy relationships; rather, it will be a collaborative relationship between institutional actors encompassing the public, private and community sectors - and it will be based on mutual trust and cooperation.

It will be about shared power, flexibility and entrepreneurialism to 'energise' a city or region to meet its competitive challenges and adapt its environment to the needed challenges (Porter 1990). All of this involves the capacity to engage in risky (with respect to the *status quo*) or stretching behaviour (Doig and Hargrove 1987; Hofstede 1997).

Because leadership plays such a prominent role in our formulation it is important to specifically note the attributes of good leadership for regional economic development in our view. These attributes are:

(a) recognizes and anticipates problems, especially large scale 'equilibrium' threatening ones;
(b) induces collaboration and consensus building patterns among diverse stakeholders;
(c) guides strategy development;
(d) elicits participation in strategy implementation;
(e) elicits commitment of "slack" institutional resources to strategy goals; and
(f) requires monitoring of implementation to assess progress.

8.5.4 Resource Endowments (Economic and Social Resources) and Market Conditions

It is widely recognized that economic growth and performance is related or tied to resources. "The more endowed a region is in terms of resources the better it should perform *ceteris paribus*" (Stough et al. 2001). Thus, the capacity of local leaders to act will be considerably dependent on the resources available to them. Such resource endowments are diverse and differ from place to place and include capital, natural resources such as materials and regional locational and environmental assets, historical economic base, competitive position, human capital, technological infrastructure, etc. (Fainstein 1983, p. 32; Judd and Parkinson 1990).

Traditionally resource endowments of a city or region were seen to bestow either a comparative advantage or disadvantage on a place. However, a well endowed city or region might succeed even if it has few or relatively poor resource endowments or if there are few opportunities for economic expansion (Jessop 1998, p. 96), and this may be achieved through the interaction of strong leadership and effective institutions acting as the catalyst and facilitating entrepreneurial activity to stretch and leverage those resource endowments that exist, to source others from outside the region and to enhance market capture. Conversely, poor leadership and inadequate or inappropriate or ineffective institutions often mean those resource endowments are not being used effectively and that market opportunities are not effectively pursued and tapped. In this way a city or region might experience a competitive advantage or disadvantage.

Of course scale factors relating to the size, wealth, population and diversity of a place and the market opportunities they represent as well as the external markets (and their size and scope) that a place potentially and feasibly might tap will be of considerable importance for the nature and rate of economic development and growth in a city or region. Thus, we are not underestimating the effects of scale and agglomeration in our model. Indeed, as seen in much of the recent work in theories of endogenous growth, local externalities (Scott 1988; Feser 2001) are key factors in the regional economic development process.

Special importance is now being placed in those resources that the public and private sectors and NGOs can direct towards community economic development or community problem-solving (Stough et al. 2001). The degree to which such actors and decision-makers commit resources into the community and as well as the availability of resources for economic development will determine the scope and scale of local action, thus potentially enhancing the resource endowments of a city or region.

Global and national processes of economic and political restructuring increasingly are imposing new challenges and opportunities to cities and regions. For example, deep-seated sectoral shifts have redefined the economic base of advanced capitalist economies. In places such as North America, Western Europe and Australia, these shifts have manifest themselves in the stagnation and decline of many mass production labor-intensive activities such as textiles and heavy manufactures. As a result, many cities and regions have experienced unfamiliar uncertainty as they could no longer rely on past practices but had to search for new economic activities and development strategies. For example, Pittsburgh's steel jobs practically disappeared as firms closed and residents left before its reemergence as a centre for information technology based activities and   producer services (Sheppard and Leitner 1998, pp. 286-287). The revolution in information and communication technologies (ICTs) and the accelerating pace of technological change, along with the mobility of capital, exacerbate that uncertainty and the rate and scope of the transformation that may occur in a city or region (Sheppard and Leitner 1998, p.287).

These new challenges mean that cities and regions - or even locations within them - need to offer a favourable set of conditions among the intervening variables in our model. Those regions that do offer a favourable set of conditions that result in strong leadership and effective institutions and which encourage and facilitate entrepreneurship are more likely to become places with a competitive advantage (McGuirk, et al.1998, p.110).

8.5.5 Entrepreneurship

Leadership in a regional economic development context needs to show entrepreneurial characteristics. Derived from the ideas of Schumpeter (1934) and Kirzner (1973) idea of entrepreneurialism, a city or region might be considered as being entrepreneurial if community leadership shows the following characteristics:

(a) believing in change and initiative to "energize" it to meet competitive challenges and   to keep progressing; and
(b) possessing insights to enable it to identify opportunities and pursue innovative ideas to improve or adapt the region's environment to meet the needed challenges facing it through "new combinations" or innovation in institutional arrangements (Jessop 1998, pp. 84-85; Jessop and Sum 2000, p.2290; McGuirk et al.1998).

These entrepreneurial characteristics can be achieved if attention is focused on the following (Jessop 1998, p.85):

(a) Using new methods to create location-specific advantages for producing goods/services or other urban activities to shift in the economic base of the city. Examples include technopoles, agglomeration economies, etc.

(b) Introducing of new types of urban place or space for producing, servicing, working, consuming, living, etc. Examples can include gateways, intelligent cities, multicultural cities, creative cities, etc.

(c) Refiguring or redefining the urban hierarchy and/or altering the position of a given city within it. Examples include the development of a regional gateway, hubs, etc.

(d) Finding new sources of supply to enhance competitive advantage. Examples include attracting inward investment or reskilling the work force.

Therefore, the focus on this factor will be on the tendency shown by:

(a) the community to undertake entrepreneurial local initiatives;
(b) opening new markets, whether by place marketing specific cities in new areas and /or modifying the spatial division of consumption through enhancing the quality of life for residents, commutes or visitors; and
(c) finding new sources of supply to enhance competitive advantages.

Examples include changing the cultural mix of the cities, finding new sources of funding, or reskilling the workforce.

In each case, entrepreneurialism in the context of the region contains the element of uncertainty that many see as the very essence of entrepreneurial activity. In this sense, as suggested by Jessop (1998):

...it is speculative in design and therefore dogged by all the difficulties and dangers which attach to speculative as opposed to rationally planned and coordinated development (pp. 84-85).

8.5.6 Outcomes

Taking into account the proposition that cities and regions inevitably are influenced by their political institutions, leadership, social composition, economic structure, and the degree of entrepreneurial activity, all of which interact and evolve in a unique manner over time and display a unique set of circumstances and a particular outcome state at any point in time, the conceptual model framework depicted in Fig. 8.2 stresses the dynamic uncertainty of the reality that confronts cities and regions in the contemporary world. Regional economic development (RED) over time and the outcome state of those factors and processes that affect RED may be measured and evaluated through performance indicators relating to:

(a) the competitive performance of a city or region *vis-à-vis* other places;
(b) the degree of entrepreneurial activity occurring; and
(c) the degree to which it has attained sustainable development vis-à-vis 'triple-bottom-line' economic growth and performance, social equity and environmental quality indicators.

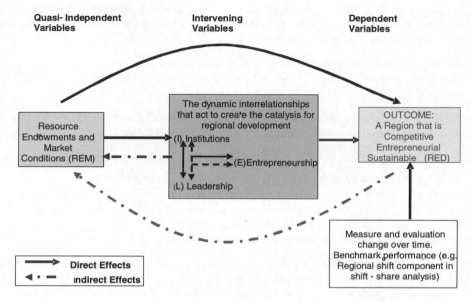

Fig. 8.2. A new model framework for regional economic development
Source: Stimson et al. 2005

A way to conceptualise that outcome for a city or region at any point in time
and its progress in economic development and its performance through time is to
envisage its path through the regional competitiveness cube (RCPC) (see Fig. 8.3).
Using the three dimensions defining the axes of the RCPC, and in addition giving
explicit consideration to the importance of *entrepreneurship* (E), we propose a
new model framework depicted in Fig. 8. 3. This can be represented as:

$$RED = f [REM \text{ mediated by } (L, I, E)] \tag{8.1}$$

Where:
L = Leadership;
I = Institutions; and
E = Entrepreneurship.

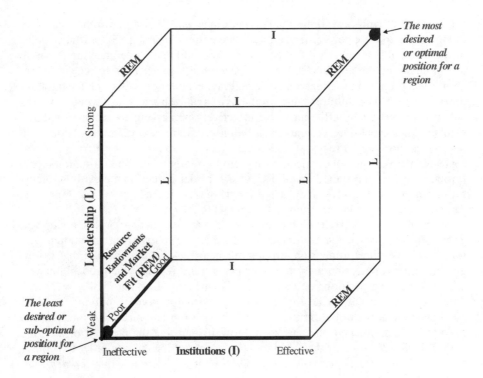

Fig. 8.3. The Regional Competitiveness Performance Cube (RCPC)
Source: Stimson et al. 2005

In this model the outcome of the regional economic development process (RED) is the degree to which a city or region has achieved competitive performance, displays entrepreneurship, and has achieved sustainable development. Those *outcome states* are defined as the *dependent variables* in the model. That outcome state is conceptualized as dependent on a set of *quasi- independent variables* relating to a city or region's resource endowments and its 'fit' with market conditions (the REM axis in Fig. 8.3), that being mediated through the interaction between sets of *intervening variables* that encompass factors defined as leadership and institutions (the L and I axes in Fig. 8.3) which may interact to facilitate, encourage or suppress entrepreneurship (E). Importantly, the model framework represented in Fig. 8.3 incorporates both *direct* and *indirect* effects in the interactions between REM (the quasi-independent variable) and L, I and E (the intervening variables).Also, the interactions between the intervening variables L, I and E may be both direct and indirect.

At any given time a city or region's economy will fall somewhere within the sphere of the RCPC. Places will vary greatly in the REM dimension, particularly concerning the magnitude, quality and mix of a location's resource endowments, and as well with respect to the prevailing market circumstances and to competitiveness of their industries, and effectiveness of their institutions in seeking to achieve a "fit" with prevailing market conditions, and, therefore, to their capacity to tap into market opportunities and facilitate entrepreneurship. Few, if any, cities or regions will have a perfect fit because markets and market demand are dynamic due to changing circumstances, both *endogenous* and *exogenous,* to the city or region. Our proposition is that a city or regional economy needs to be trying at all times to adjust its institutions and productive organizations so as to maintain and enhance market fit by efficiently and effectively harnessing its resource endowments to be competitive, and thus to sustain itself. Some regions do this better than others; and how well a regional does it can change dramatically over time, for better or for worse. Thus, the trajectory of a city or region over time through the performance space represented by the RCPC will be dependent on the evolving interactions between the efficiency and effectiveness with which L and I provide catalytic processes and situations to harness its REM.

Stimson et al. (2005) develop a rationale that argued that strong leadership and good performance on the L dimension and the way it impacts institutional performance - i.e. the I dimension - represent major *endogenous* factors that distinguishes a good performing from a poor performing region. Our proposition was that how a region performs on these three dimensions - L, I, and REM - will condition its position within the RCPC. We argued that regional economic development strategy needs to be formulated, and that appropriate plans and mechanisms need to be implemented, that are geared towards shifting the position of a region within the RCPC towards the top-right hand corner of the cube in order to achieve a position that reflects *performance optimality* for a *sustainable development* outcome.

8.6 Re-Engineering Regions and Rapid Response

The agility or response capability of a region might be viewed as dependent upon its *learning infrastructure*. As such, it is assumed that the competitiveness of a region's firms depends on their learning capability, which in turn rests upon the learning infrastructure of the region. Thus, *learning process and learning capability*, are seen not only as core concepts underlying regional rapid response capability but also more broadly as the core concepts of an emerging 'new organizational paradigm' for a knowledge based world economy (Jin and Stough 1996). Thus it is useful to develop the concept of an *agile region*—that is, a region with rapid response capability. This requires identification of a set of defining characteristics of the agile region and a discussion of an agile or rapid response transport infrastructure as a component that supports learning. The discussion that follows draws on the work of Jin and Stough (1996 and 1998).

8.6.1 The Concept of Agility and Rapid Response in Regions

Several times throughout this book it has been noted how information and telecommunications technology are profoundly transforming the industrial society into a knowledge society, and how the driving force of regional economic development increasingly is the endogenous capability of a region to learn and innovate (Jin and Stough 1996; Saxenian 1994). The learning capability of a region (the so-called *learning region*) is heavily dependent on its *learning infrastructure* (Simmie 1997; OECD 2000). This includes:

(a) technology and knowledge infrastructure;
(b) information and telecommunications infrastructure;
(c) transportation infrastructure; and
(d) institutional infrastructure (including universities, business networks, and various inter-linkages among business sectors, governments and universities).

A superior learning infrastructure enables effective knowledge creation, transformation, synthesis and diffusion. This effectiveness in knowledge creation, transformation, synthesis and diffusion makes urban regions and their sub-regional jurisdictions *agile* in making business and economic decisions and agile in organizational and technological innovations which provide new competitive edges for a region's business in a globally competitive world. A superior learning infrastructure requires not only superior hard systems (for example, transportation and telecommunications and a system of technological sectors with mutually reinforcing core and complementary competencies and capabilities) but also soft systems of institutional linkages and trustful environments that make possible effective cooperation in knowledge creation, transformation, synthesis, and diffusion. A superior learning infrastructure not only means effectiveness in learning and innovation, but also it means effectiveness in reducing the transaction costs of doing business and in knowledge creation, transformation, synthesis and diffusion.

An agile region is composed of learning agents—for example, individuals, firms, government agencies, voluntary organizations and universities—that are rich in their stock of knowledge and superior in their learning capabilities. Moreover, these agents are interconnected with one another such that the transaction costs involved in knowledge creation, transformation, synthesis and diffusion are low, while the knowledge base and learning capabilities of these agents are mutually reinforcing. Any individual agent's knowledge-base and capacity to learn contributes to the learning capability of other agents in the region because of the high level of effectiveness and low levels of cost in the exchange and synthesis of knowledge.

An agile region requires not only rich *cross-sectoral linkages* (among business sectors) and *cross-functional linkages* (among business, government agents and universities) but also cross-spatial linkages within and across the region. The linkages that are essential to the function of agile regions are physical and technical, plus institutional and psychological. They include trustful relationships, inter-linked financial stakes, effective information and knowledge flows, and cooperative arrangements including consortia. This web of linked agents and institutions lowers transaction costs and increases knowledge exchange and diffusion, and thereby increases the core competencies and learning capabilities of individual agents in the region.

The evolving agile region engages not only in *flexible specialization* (Sable 1989) and *lean production* (Womack et al. 1990) but also in *agile manufacturing* (Kidd 1994) and agile service production by both the private sector and government agencies. The competitive advantage of an agile region lies in its learning infrastructures and in the learning capabilities of its agents. The metropolitan expression of the concept of an agile region on the one hand reflects the fact that localities below the metropolitan level (for example, local political jurisdictions, neighbourhoods) do not have a sufficient concentration of self-organizing and mutually reinforcing competencies and capabilities. On the other hand a metropolitan region expresses and achieves its *agility* in its self-organizing, autocatalytic and mutually reinforcing nature of the spatial (geographic) agglomeration of institutionally linked competencies and capabilities.

8.6.2 The Elements of Regional Learning Infrastructure

Regional learning infrastructure includes (but not exclusively) the following elements:

(a) Transportation infrastructure, including intelligent transportation systems (ITS), which increases the accessibility and mobility of people and goods in a region.

(b) Communications and information infrastructure, which provides technical channels for the quick and effective storage, movement and exchange of information and knowledge.

(c) Technological and knowledge infrastructure, including a complete set of agglomerated competencies and capabilities, which provide complementary and core competencies and capabilities for the effective co-evolution of technological innovations.

(d) Dense business networks and a high trust business environment, which make possible organizationally, effective and quick knowledge creation, transformation, synthesis and diffusion.

(e) Institutional infrastructure—including business networks, public/private partnerships, business-university-government linkage—that provide effective organization linkage and institutional trust among business sectors, among government, business and research universities and between business and employees. These linkages, coupled with a trustful environment institutionally, reduce the transaction costs of doing business and enhancing the learning capability of government, business sectors and individuals.

(f) Effective information infrastructure for the exchange of information and knowledge and for monitoring the changes in regional industries and employment structure, in regional competencies and capabilities, and in regional economic development.

(g) Existence of agile regional governments, so that essential decisions and coordinative actions can be made to improve and strengthen the learning infrastructure of the region.

(h) Existence of agile communities and associations—including effective cooperation among stakeholders in knowledge creation, transformation, synthesis and diffusion—and the cooperative provision of training and education to enhance the knowledge and skill stocks in a region.

8.6.3 Smart Infrastructure for Smart Regions

The concept of agility and rapid response in regions and their need for regional learning infrastructure has been approached in a somewhat different way through the concept of smart infrastructure for most regions. Smart regions monitor best practice in other regions to become more competitive. Competitiveness does not mean that regions act in isolation; rather they learn to stretch and leverage resources in order to develop a competitive niche within the distribution and supply network of both manufacturing and service industries. For a region to be successful, adopting principles of best practice and adapting these two local conditions is critical in laying down a path to the future.

Smart infrastructure is a term that has emerged in the context of globalization. It refers to a *milieu* of hard and soft infrastructure and services needed to support value-adding industry and advanced producer services. As we have noted on several occasions throughout this book, originally 'infrastructure' only referred to 'hard' or physical support systems, such as roads, walls, canals that were readily identifiable in structure and purpose. Research by the World Bank suggests a one percent increase in asset value of basic infrastructure contributes 0.5 to 1.5 percent to gross product, depending upon the development status of a country (Kessiadies 1993). Other research using a production function approach to measuring the effect of infrastructure on development finds similar results (Aschauer 1989; Munnell 1991, Cook et al. 1991).However, the term 'soft' or 'social' infrastructure came into common use as it became apparent that facilities and services such as schools and health facilities perform the same task (reducing input and transaction costs per unit output) for enhancing human capital. But these two elements do not fully explain the success of some regions in fostering what have been called technopole activities that have emerged in the contemporary 5th K cycle era of technology and economic development (referred to in Chap. 5). Today new types of infrastructure which underpin such technology led regions are less visible and have taken on a role much harder to define than traditional physical and social infrastructure. Infrastructure supporting competitive enterprise now needs to be faster and smarter in terms of its ability to be agile in adapting to changing circumstances in order to enhance productivity in urban centres. It needs to facilitate the incorporation of, and adaptation to technological change. It also needs to foster economically, socially and environmentally sustainable development that can support the requirements of population change in the context of sustainable development thresholds. This means that, in addition to cities and regions needing to promote themselves in terms of their traditional comparative advantages of transport links, communications and other physical and social infrastructure, there is also a need for them to provide an *innovative milieu* and *quality lifestyles*, including good housing, cultural attributes and tourism opportunities.

Thus, this new *smart infrastructure* becomes crucial to economic development (Smilor and Wakelin 1990). Typically in today's urban environment, key smart infrastructure includes international airports and seaports, telecommunications, education and research and development (R&D) facilities, and cultural facilities and services. This infrastructure is perhaps better described as strategic as this it embodies less the technology aspect and more the importance of its role in the economy. No matter how defined, it involves: integrated road, rail and air network linkages with other 'world cities' and with similar regions within the nation; minimization of energy and utility service costs; access to capital market institutions; and flexible and responsive institutional arrangements to enhance a competitive business culture and innovation.

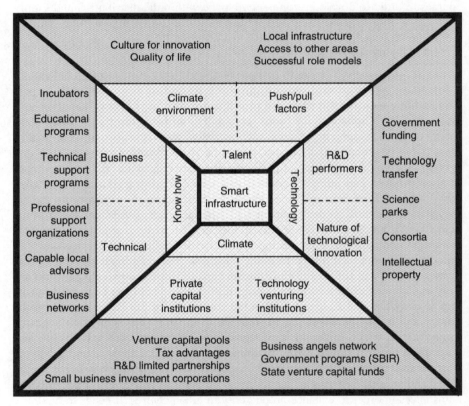

Fig. 8.4. Policy implications for smart infrastructure
Source: Smilor and Wakelin 1990

Human resource development is a fundamental component of urban infrastructure. Smilor and Wakelin (1990) refer to *smart infrastructure* as comprising talent, technology, climate and know-how, with public policy being influential in determining how these key factors are harnessed in promoting technological and design innovation, and value added production (see Fig. 8.4). Institutions, public, private and intermediate, are crucial smart infrastructure for facilitating an *innovative milieu*. The institutions involved include business angels' networks, venture capital pools, consortia, and professional support organizations. Regulatory and government support environments need to facilitate:

(a) enlightened but quality-oriented planning and development controls;
(b) business incubators;
(c) technical support programs;
(d) competitive taxation arrangements;
(e) intellectual property protection;
(f) maintenance of quality indicators; and
(g) advanced transportation and telecommunications systems access.

All of these measures either increase productivity and throughput or reduce transactions costs.

The key characteristics of smart infrastructure that make it different from other infrastructure are linked closely to the concept of *linkage* and *synergy* in production. Smart infrastructure has physical components – such as computers or fibre optic cables – but the essential characteristic is that it provides new and more efficient ways to combine inputs/resources for businesses and other organizations and more generally for all industrial sectors so as to reduce transactions costs and increase productivity.

Thus, business and other organizational networks are needed which:

(a) help spread productivity-enhancing technology and lobby for improved physical infrastructure;
(b) provide venture capital to lower the finance costs;
(c) enhance the improvement of local quality of life which increases the productivity of human resources; and
(d) provide R&D networks which produce productivity enhancing technology, all constitute smart infrastructure

8.6.4 An Example: The Role of Intelligent Transportation Systems Assisting the Formation of Agile Regions

Intelligent transportation systems (ITS) refer to the application of information technology to the physical infrastructure of transport systems and to the vehicles that use this infrastructure. The system level impact of ITS on the formation of agile or rapid response regions can be understood in terms of several dimensions:

(a) Reducing transaction costs in the private provision of transportation infrastructure and in the collection and dissemination of information about the movement of people and goods in the region.
(b) Providing knowledge bases about the movement of people and goods in the metropolitan region for adaptive policy and individual learning.

(c) Providing certainty for just-in-time delivery.
(d) Increasing mobility for people to interact with one another face to face and therefore enabling the establishment of trust and networks.
(e) Reducing delay and safety costs and therefore increasing productivity and increasing attractiveness of the region for highly skilled workers and human capital.
(f) ITS itself is a potentially competitive sector in some regions.

One of the greatest bottlenecks for metropolitan economic development is the problem of congestion and environmental pollution caused by the lack of agility and flexibility in urban transport systems and policy making. While the limitation in urban land use drives-up exponentially the cost of building and maintaining new transportation infrastructure, the limitation of public funds for the expansion of urban transport systems has further complicated solving congestion and pollution problems in most major metropolitan regions in the United States (Downs 1992). The economic development and competitiveness consequences of metropolitan congestion and pollution are many (Stough et. al. 1994 and Stough 2001). In its first order effects, metropolitan congestion incurs loss of valuable time, causes uncertainties in travel time and in just-in-time delivery, increases safety costs, and gives rise to more air pollution. In its second order effect, severe and persistent metropolitan congestion decreases the attraction of the region for talented people who value a high quality of life and for private investors who value highly just-in-time movement people. The results have been the distraction and relocation of investors and talented people to some other regions with better transportation infrastructure. These in turn damage the potential of regional economic development and negatively impact regional real estate markets.

In essence, the deployment of intelligent transportation systems adds agility and flexibility to the conventional inflexible physical transportation infrastructure. This occurs through the impact of on-time provision of traffic information, the deployment of various intelligent transportation systems (for example, advanced traffic control systems, on-board real time traffic information systems) upon the existing metropolitan physical transportation infrastructure, which enables agile adaptation of traffic control, agile accident response, and agile change in the choice of travel routes by drivers in real time. The result is a reduction in traffic delay and an increase in traffic throughput (Stough et. al. 1995 and Stough 2001). The latter in turn increases the attractiveness of the region to talented people and agile investors.

A more intangible benefit in the deployment of ITS comes from its provision of a knowledge base for agile response and effective adaptive learning. In the case of the traffic control system in Montgomery County in Maryland (United States), Klein (2001) finds that the system has enabled a traffic controller to change traffic control lights agilely. In yet another case study, Stough et al. (1994 and 2001) find that the deployment of an electronic toll collection system provides on-time traffic database and effective policy control tools for multiple purposes of agile response. For example, an electronic toll collection system can be used efficiently and agilely for the purpose of congestion and peak-time pricing. With additional sensing equipment, it can also be used to track cars with high pollution levels, thereby in-

creasing the enforceability of emission standards. When properly used, the traffic
data collected by electronic toll collection systems can also be used for agile business decisions.

Further, Jin and Stough (1996) suggest that electronic toll collection technology
can reduce the transaction costs involved in the public and private provision of
tolled transportation infrastructure. Electronic toll collection systems also reduce
the number of required toll gates, thereby reducing the use of labour and limited
urban land. The reduction in transaction costs, labour costs, and land costs in the
provision and maintenance of tolled transportation infrastructure helps to overcome both the constraints of limited land and limited public funding for the provision of new transportation infrastructure in metropolitan regions.

An even greater intangible impact of ITS is that the increase in mobility of people, through ITS deployment, will increase the incentive for residents in the metropolitan region to become engaged in more face-to-face interaction and thereby
enable the establishment of trustful relationships and networks. This is especially
the case when aided by advanced communication systems.

8.7 Networks, Strategic Alliances and Network Resources

Increasingly networks and strategic alliances are part of the strategic infrastructure
not only for business but also for regions enabling them to operate successfully in
the new economy. *Networks* are human chains of communication that pulsate to
pass on, receive, direct and disperse information and knowledge between people.
Networks use hard (telecommunications) and soft infrastructure (institutions and
associations) as a means of transmitting information. *Alliances* are closely related
to networks. However, alliances are generally a more tactical aspect of networking
and involve (usually commercial) relationships entered into for mutual benefit between two or more entities having compatible business roles (Segil 1996). Alliances are strategic when planning processes between alliance parties are manoeu-
vred to ensure the most advantageous position. Different types of alliances and
network structures play key roles in the regional economic development process.

8.7.1 Networks

There has developed a considerable literature on the role of networks in supporting regional economic development (see, for example Silicon Valley Joint Venture
Network (SVJVN) 1995; Humphrey and Schmitiz 1995; Sohal et al. 1998). Cooke
and Morgan (1993) tell how the development of networks in the United Kingdom
and Europe are important instruments of economic infrastructure for regional and
local economic development. Networks are not fixed assets; they are constantly
changing, rebuilding and dissolving in response to changing conditions and the
needs of people and organizations.

Several different forms of networks support regional economic development.
The most basic form is the *community* (Laumann and Pappi 1976), which creates

the network for common knowledge and sharing information (Borman et al.1994). Common knowledge also enables people to be better informed, be better educated, and to become wiser in making decisions related to public and private sector investment, production and trading. Common knowledge leads to the development of a second form of networking that is important to economic development in the form of networks of association involving learning (Cooke and Morgan 1998) and capacity building to improve knowledge, production and industry development.

Historically, *networks of association* relating to regional economic development have been formal structures, where membership was required to participate in the network. Chambers of Commerce, industry associations, professional organizations, and service clubs (such as Apex, Kiwanis, and Rotary) are examples of what might be called the 'old' economy networks. These play an important role in supporting the local development processes. However, in the past couple of decades, economic networks of association have become increasingly specialized. The old network structures primarily have been used for information sharing, for professional development, and in some cases as devices to protect industries from change and competition. Most networks of association are localized chapters or branches, which have been organized and linked to other chapters by hierarchical structures. Information flows within these old network structures are often slow as well as being controlled.

Globalization and the information technology revolution have fundamentally changed the meaning and role of networks. The new economy networks have become more open, aspatial and informal. Information transactions within networks have become more rapid and channelled through hub and spoke structures. Key people and organizations have become the catalysts for collecting and distributing information. There is emerging a new form of virtual network brokering that has replaced space based functions of old network structures. This has led to the emergence of totally different kinds of network structures supporting economic development.

The new networks of association have both local and global characteristics, and may be both spatial and a-spatial in nature. The generators and transponders of new networks tend to revolve around people or organizations that have a key role in the development, receiving and transferring of information. There is emerging a new form of catalytic structure that links people and organizations locally, nationally and internationally between suppliers and users of information. Normally a 'hub' organization establishes and maintains the network. Information technology industries provide the infrastructure hub for servicing both global and national network structures. Local industry clusters and network organizations (Abramson 1998; Audretsch and Feldman 1996; Anderson 1994; Humphrey and Schmitiz 1995) are emerging as key catalysts in cultivating economic development opportunities.

Networks are now important elements of *strategic infrastructure* for development in the new economy. Networks with suppliers and customers provide the most important sources of innovation. They also play an important role in value adding to an economy (Borman et al. 1994). Networks are in between markets and producers, involving a complex array of relationships between firms. Competing is achieved by positioning oneself in the network, rather than by attacking the environment. Strategic networks are those firms or regions a region can use to position itself for a stronger competitive position. A network allows a firm to specialize in those areas in the value chain that are essential to its competitive advantage (Jarillo 1988).

Network associations are emerging as a power means of overcoming resource, market scale and capacity shortfall problems. The Blue Berry Network (Nova Scotia Department of Agriculture not dated) in Nova Scotia is an example of regional business establishing a local network to facilitate marketing and sales of local agriculture product. Another example is the Silicon Valley Network Partnership in California, which began in 1995, and has developed several successful enterprises through local partnerships.

In Australia, *collaborative research* programs are playing a significant impact on the structure of national innovation systems by creating and strengthening networks that are essential for breeding innovation clusters. Collaborative research programs involve linking universities from around the nation with concentrations of expertise with industry partners to function as a network in addressing both applied and generic research issues central to business and product development and innovation. These networks involve both technology and market stakeholders and are extended to include industry, research and technology producers. Network activities have resulted in setting priority in research and linking research fields that have high potential to coalesce into distinct technological clusters. Liyanage (1995) examines the work programs of over 50 Australian Co-operative Research Centers and concludes these clusters and networks of research enable public policy makers to identify complementarities between generation, acquisition and diffusion of knowledge across a range of innovations rather than a single innovation.

8.7.2 Strategic Alliances

In the business world, *strategic alliances* are agreements between organizations that aim to increase market share of sales or business between parties that form the alliance. Strategic alliances are a response to competition to gain access to global markets and protect local market position. Parties forming an alliance benefit from each other by offering to transfer goods, services and customers onto other parties in the alliance.

The airline industry is a good example of strategic alliances in practice, which have expanded in recent times to form several large global partnership groups, for example, Star Alliance and Global Alliance. Here one airline co-shares routes or transfers passengers to another alliance partners' routes to expand the customer market of partners forming the alliance.

Strategic alliances take different forms, ranging from a *memorandum of understanding* to cooperate on specified matters, to *contractual obligations* between alliance parties. Strategic alliances are an essential marketing tool to gain access to networks in the global economy.

Strategic alliances are phenomena that have also been emerging in the context of regional economic development. Strategic alliances require well-established networks and regional organization structures to be firmly embedded in a regional economy. The network acts as a catalyst that grows in importance in future as regions seek to be the highest priority initiative to facilitate business development. The most common strategic alliances have been *sister city* relationships. These normally provide for softer areas of exchange in areas of culture, sport, information sharing and technical cooperation. The Silicon Valley Joint Venture Network, referred to above is one of the more advanced strategic alliance network structures to have emerged to date.

Several new strategically aligned organizations have emerged from the network. In Australia, in 1999 the Cairns Regional Development Corporation successfully spawned several network companies that are now involved in developing alliances for global marketing of education, information technology and fish products.

Strategic alliances may assist in the economic development of regions by developing a virtual critical mass of core business services in advance of providers being located on site by:

(a) building immediate access to networks, markets, information, services for investors and developers;
(b) enabling the development of incentive packages to potential customers and investors; and
(c) giving regions the key element of business architecture that will create competitive advantage over competition form regions elsewhere.

A key element of building strategic alliances is the alliance catalysts or hubs. Catalysts are the persons or organizations that act as brokers and transponders of information that package information, products, resources etc for alliances. These catalysts are the 'brains' of the alliance. They may operate internally or externally to the alliance. A key role for regional economic development organizations in future will be their role as a regional catalyst, as opposed to their traditional role as a regional marketing and promotion organization. Some of the key strategic alliances that a regional economic development organization might need to develop in the future to facilitate and encourage are listed in Table 8.1.

Table 8.1. Some key strategic alliances for regional economic development

Technology alliances
Information technology firms
Telecommunications firms
Environmental sciences firms and organizations
Bio-technology firms and organizations
Engineering sciences firms and organizations
Education alliances
Universities
International education, training and R&D institutions
Private training providers

Table 8.1. (cont.)

Franchise training providers
Schools
Financial services alliances
Merchant banks
Venture capitalist
Transportation services alliances
Freight industry
Long-haul transport operators
Airlines
Shipping lines
Public transport operators
Government services alliances
Government development agencies
Education
Health
Community services
Conveyance
Labour and professional organizations alliances
Labour organizations
Professional institutes
Business services alliances
Building design professions
Marketing
Business and accountancy professions
Insurance
Employment and recruitment firms
Legal services
Community organizations alliances
Religious organizations
Clubs and societies
Media alliances
Local media

Source: The authors, based on experience consulting with local communities in regional economic strategy planning

8.7.3 Developing Networks and Strategic Alliances

While there is a considerable literature on the principles and practices of developing networks, clusters and alliances (see, for example, Humphrey and Schmitiz 1995), there are, however, no universal principles or means of ensuring the success of these key elements of infrastructure needed to support economic development in the new economy. However, several important elements are needed in the development of these structures.

Common Knowledge. As discussed above, the first element is common knowledge which provides the basis for open exchange. If knowledge is not common, networks do flourish and soon fall apart because the basis of trust and equity dissipate. Common knowledge forms the basis for the second important element of network building: the development of inter-organizational networks. Inter-organizational networks provide the basis for collaboration and formal exchange. However, Borman et al. (1994) warns of two concerns about the development of inter-organizational networks of which firms and regions need to be aware. First, there is the implicit assumption about the development of common knowledge. The emergence of common knowledge is not necessary secured simply by the flow of information between firms. Second, common knowledge does have costs involving access to information. The central argument here in encouraging inter-organizational networking for economic development is that unless the structures for the development of common knowledge are created—and exit and entry costs are minimized—inter-organizational networks will reinforce existing hierarchical or market based networks. The networks created will only service a local or organizational benefit, rather than serving to benefit the whole industry or region.

Catalysts and Hubs. The second infrastructure element in network building is the development of catalysts and hubs. There are no precedents for developing network hubs. The two key elements of hub construction are the physical infrastructure the network facilitator or alliance carrier. Fibre optics is becoming increasingly important in building network communities that increase common knowledge (Bianchi 1996). The importance of there being places for people to meet for conducting social and business transactions are also important elements of network infrastructure. The development of network facilitators or carrier alliances is much more difficult. This requires the building of special competencies related to entrepreneurship, leaderships, innovation, style, flare, risk averseness and many other less tangible factors. Of course cultural, societal and political factors will interplay with these processes.

Networking may be hard to reconcile with traditional theories of strategy based on competition. Networks and alliances introduce the theory of *collaborative competition* that is perverse to the concept of *competitive advantage*. However it is highly likely that networks and strategic alliances will provide the energy required to run the economies of the 21st century (Jarillo 1988). How to develop the networks and the strategic alliances infrastructure needed to support regional economic development will become a central focus of much of the strategic planning for regional economic development organizations.

8.8 Conclusions

New theories such as endogenous growth (Romer 1986) and evolutionary economics (Nelson and Winters 1982) have been made it abundantly clear that factors such as local effort, leadership and institutions, and the networks that support them are important in addition to factor cost or price differentials in local regional economic development. In this chapter we have attempted to examine the thesis that endogenous factors shape local economic development outcomes. We have done this by examining the concepts of social capital, leadership, smart infrastructure (focusing on non physical infrastructure) and individual and organizational networks.

Social capital, despite some definitional issues that are yet to be worked out, has emerged as one of the more critical variables in contemporary economic development policy making and programming. Social capital relates to the trust that does or does not build up within organizations and individuals that makes it possible for them to reduce transaction costs involved in the achievement of community development objectives. Regions with high trust (or high social capital) appear to initiate and execute economic development strategies more easily and more effectively than low trust regions. Difficulty in measuring the trust variable however means that more empirical research is needed to better understand how to design effective social capital or trust intensive regional economic development policies.

Leadership, like social capital, has a definitional history that continues to raise questions of clarification. To move this discussion along a conceptual model of leadership and institutional driven regional economic development was provided after reviewing the earlier empirical work by DeSantis and Stough (1999) on the role of leadership in a regional economic development context. While this model helps to pull many factors together that drive regional economic development and adjustment to change, it remains at this time a conceptual model and waits empirical testing.

Providing physical infrastructure was a major economic development policy applied in the industrial period of the twentieth Century. It was believed that such infrastructure promoted and supported economic development and there is considerable evidence supporting that this supply side approach to economic development produces net benefits to society. At the same time, in the knowledge and services oriented and more globalised economy of the late twentieth and twenty first Centuries, the importance of physical infrastructure appears to be decreasing while the importance of soft infrastructure (human capital and institutional investment) is increasing rapidly relative to the more traditional hard infrastructure. The smart infrastructure case study about Regional Intelligent Transportation Systems (ITS) was presented to provide more contextual insight into this apparent shift in priority. The literature suggests that the shift will become more pronounced and will continue.

Finally, at the heart and core of the knowledge or information age is the merging of computer and communications technologies (ICTs). This convergence has resulted in the creation of rapidly forming and highly flexible communication networks making information and knowledge available to an ever wider group of individuals and organizations at a cost that is almost insignificant. Given that the effort and cost of building and maintaining such networks has decreased considerably, it is no wonder that there has been a decentralization of organizational form within large organizations and in the form of large integrated supply chains of companies (or as some have observed supply chain communities). Thus, the latter part of this chapter examined the role of computer and information technology generally in society and more specifically through a case study that examined the role of ITS. Concepts such as networks, strategic alliances and network benefits are important elements of this discussion.

9 Decision Support Tools to Inform Regional Economic Development Analysis and Strategy

9.1 Using Spatial Information Technology to Build Decision Support Systems

Decision making in regional economic development necessarily is multi-faceted, involving economic, social, environmental and political considerations. And the implementation of regional economic development strategy plans typically will involve an interface with regional plans concerned with the provision of infrastructure and land to support economic activities. Thus the development of decision support systems (DSS) and associated analytical tools is important to enhance the regional economic development strategy planning and implementation processes. Fortunately there are several tools that have been developed to help decision makers, including those methods of regional economic analysis we have discussed in previous chapters, and these are dependent on having access to spatial information at various levels of scale.

The development of decision support tools has been revolutionized by the innovations in spatial information technologies, commonly referred to as *geographic information systems* (GIS), and include major advances in our ability to integrate complex and very different types of spatial information at various levels of aggregation and disaggregation, including the building of GIS-based systems that integrate regional economic and social data sets with land use information, transportation networks, flows of people, goods and information, and environmental information. These GIS-based data systems are also able to be integrated with information derived from satellite imagery now widely available for regions. When interfaced with spatial modelling algorithms—including, for example, spatial optimization approaches to allocation-location models and network analysis models—we have the capacity to undertake sophisticated analyses of spatial interrelationships between phenomena, to model actual (existing) conditions, to simulate potential future conditions, and to evaluate these against planning criteria related— for example, to maximizing some form of spatial efficiency or maximizing a social welfare requirement. In addition, when these approaches are linked to new developments in visualization techniques, it becomes possible to simulate in real time the potential future spatial outcomes of trends the impacts of a specific change on a region, including aspects of its economic development.

Thus, what we are talking about is how advances in spatial information technology are making it possible to build *spatial decision support systems* (SDSS) which enhance our capacity both to improve the scope and detail of analysis of spatial and other data and to enrich the way we might inform the regional economic development strategy planning process.

Non-GIS based integrated decision support systems also have been advancing enabling us to integrate traditional and new regional analysis models – such as shift share analysis, Data Environment Analysis, spatial econometrics, and input-output analysis, as discussed in Chaps. 3 and 4 – through analytical hierarchical process models linking econometric analysis with social and political factors, thus producing more policy relevant decision frameworks.

But it is when these types of approaches are linked in new SDSS that we can achieve incredibly powerful representation of model out puts integrating different layers of data with visualization of output.

It is necessary to say something briefly about DSS in general. The goal of DSS is to focus ill-structured problems. As Densham (1994, p. 207) notes, DSSs are important when decision-makers "find it hard to define...problems and to articulate characteristics that they would like to see in a solution". He argues that attributes of DSS include:

(a) support for the identification and use of data;
(b) the ability to represent complex relations among data which are needed for search, modelling and visualization;
(c) an architecture that is flexible enough to enable the user to combine modelling and data in a variety of ways;
(d) an ability to generate a variety of types of output;
(e) a single, integrated, user interface which supports a variety of decision making styles; and
(f) an architecture that supports the addition of new capabilities as user needs evolve.

A system with these attributes can be used to support decision research and decision-making—that is, to define the problem, generate and evaluate decision alternatives, and select an alternative. The emphasis on DSS is on iteration of problem definition and redefinition, and search and evaluation of alternatives to find an acceptable solution. As such, DSS are "systems to think with" (Webber 1984), and thereby to aid the investigation of problems and decision-making. Thus, a successful DSS is one that enables its users to redefine and resolve ill-structured problems through alternative representations, manipulations and operations and aids to memory (Densham 1994).

This chapter first refers to recent developments in GIS technology. This is deemed important as it is the technology that best supports DSS, and in particular SDSS, for regional economic development. A case study analysis of a jurisdiction level based GIS used to support economic development planning and policy development is presented. The chapter then describes a regional economic modelling system that enables planners, policy makers and industry groups to conduct virtual experiments on the consequences of alternative simulated policy and investment interventions. Finally, the chapter looks at how a regional SDSS integrating socio-economic data and satellite imagery data can be modelled to simulate the allocation of land for future growth in a region, and how this might be presented using a visualization technique. The chapter should not be regarded as anything more than serving the purpose of illustrating the potential power of these new decision support tools, and the reader should peruse in detail the references cited to gain a more in-depth outline of the approaches discussed.

9.2 Geographic Information Systems

Because regional economic development problems have a spatial geographic dimension, development of a SDSS for the regional context can help inform the processes of regional economic analysis and development strategy planning. GIS technologies are used to achieve this. Nijkamp and Scholten (1993, p. 87) describe a GIS as:

> a coherent representation of a set of geographic units or objects which – besides their location – can be characterized by one or more attributes. Such information requires a consistent treatment of basic data, from their collection and storage to their manipulation and presentation.

More specifically, Church and Murray (1999) define GIS as:

> ...a particular form of information system which combines geographically (spatially) referenced data as well as non-spatial attribute information. Traditional information systems, such as Access, FoxPro, Paradox, dBase, etc., process and manage only attribute information that is not explicitly represented and located in space. As an example, an aspatial database may contain information on company characteristics, like address, net worth and number of employees, for business engaged in engineering consulting. However, it is not possible within a traditional information system to determine one of the most obvious relationships existing in the database like distance between companies. This is the major contrasting feature of GIS because geographic extent is an explicit and important component of all information being managed and processed. Thus, GIS may be considered to be a hybrid information system which structures data and summarizes features based upon an inherent characteristic of the data being managed—geographic extent (p. 8).

Thus, GIS is a system composed of computer hardware and software along with procedures that support geographical or spatial decision making. A GIS does this through several processes, including capture, management, manipulation, analysis, modelling and display of geographically referenced information. The reader is referred to work by Laurini and Thompson (1992), Worbys (1995), Chrisman (1997), Chou (1997), and Burrough and McDonnell (1998), for a technical discussion of GIS and related processes as it is not possible to into these issues in detail here.

Given this background on the nature of GIS, it is a straight forward extension to view it as a SDSS for regional economic development decision making. The great benefit of GIS lies in the "ability to handle large, multi-layered, heterogeneous data bases and to query about the existence, location and properties of a wide range of spatial objects" (Fischer and Nijkamp 1992). However, until quite recently the lack of more powerful analytical and modelling capabilities integrated with GIS has been an impediment to the widespread use of GIS in DSSs to aid decision-makers in thinking and deciding about complex problems. Clarke (1990), Oppenshaw (1990), Anselin and Getis (1992), Fischer and Nijkamp (1992), Nijkamp and Scholten (1993), and Densham (1994) provide a detailed description of the bottlenecks that have been and continue to be faced in the full utilization of GIS for decision support purposes.

Until recently, spatial analysis and GIS have evolved separately as their focus and intents were quite different. In particular, GIS focuses on data management, creation and manipulation whereas spatial analysis focuses on particular management and analysis problems, and mathematical modelling. While there has always been recognition of the need to combine both activities, more recent commercial GIS packages (such as ArcView) are making such integration more of a reality. In fact integrated tools to support and model spatial decision-making are beginning to emerge. SpaceStat (Anselin and Bera 1997) and EMSPlan (Enache 1994) are examples of analytical routines linked with ArcView which considerably enhance faster than real time spatial analysis, simulation, modelling, and forecasting. Despite these recent developments and the fact that the dynamics of the future development of GIS will be in this integration of tools, nonetheless technical, operational and conceptual issues will continue to limit its use for the near future.

The integration of GIS into DSS to provide a SDSS to assist the regional economic analysis and strategy planning processes will likely evolve rapidly over the coming years. As the integration of analytical tools into GIS occurs, more robust capabilities will enhance and improve the ability to provide decision-makers with high quality decision support services across a broad range of areas.

9.3 Case Study of a GIS to Support Economic Development Decision Making

9.3.1 ESMPlan

An interactive GIS software system called ESMPlan, developed by Enache (1994), makes it possible to both describe and analyze spatial data arrayed at the political jurisdiction level. EMSPlan has since gone through a number of redesigns, resulting in a system that now uses ArcView as a primary platform and permits a diverse array of descriptive and analytical operations for data arrayed by jurisdiction, such as U.S. Census of the Population data. EMSPlan has been calibrated for a number of countries through projects supported by the World Bank—including the Philippines, Costa Rica, El Salvador, The Dominican Republic, Romania, Mali, Angola and Egypt—as well as for the Washington National Capital region in the United States. This system makes possible the near instantaneous display of pattern data—for example, population, economic, social and political in cloropleth mapping formats using colour and texture gradations for presentation clarity. Importantly, it also provides modelling and hypothesis testing routines enabling the examination of associated patterns. The capacity of the model is quite large as it accommodates relatively large scale databases—that is, several hundred variables for several hundred jurisdictions. The rapid solution time and the high quality graphics routines enable the evaluation of alternative scenarios in a real time format. In short, as quickly as decision-makers can create scenarios, EMSPlan can model them. Likewise, as hypotheses are formulated they can be tested with regression techniques and the results plotted and presented to decision-makers. The system is capable of providing comparison data across a set of modelled alternative scenarios, thereby enabling decision-makers also to conduct sensitivity analyses.

9.3.2 Application to the Washington National Capital Region

A version of EMSPlan developed for the Consolidated Washington-Baltimore Metropolitan Statistical Area (CMSA) in the United States, referred to as the Washington National Capital Region is now discussed. It comprises 38 jurisdictions and forms the fourth largest geographic market in the United States with a population of more than six million. Until the 1990 decennial Census, the Washington, D.C. and Baltimore MSAs were considered separate markets. Their designation as a Consolidated Metropolitan Statistical Area (CMSA) increased the need to describe and model conditions across the former two regions. Some EMSPlan analyses conducted for the Washington National Capital Region over the period 1960 to 2000 are presented below to illustrate this SDSS. These illustrate how GIS can be used in the regional audit component of the process of regional economic development strategy planning, including the testing of hypotheses. We have not run the maps based on the additional 2000 year data as it would have provided little additional value to the reader. The focus here is on demonstrating the application.

9.3.3 Examples of GIS Data Display

Like traditional GIS, EMSPlan makes possible the description of all of the information in the database. It also enables the presentation both electronically (visually) and in print. Fig. 9.1 shows the EMSPlan generated population distribution for the Washington National Capital Region in 2000. Figures 9.2., 9.3, 9.4, and 9.5 show EMSPlan generated population change over the period 1960–2000 at ten-year intervals. The results of continued regional out-migration over the full period are apparent with the region's core jurisdictions (Washington D.C. and Baltimore, Maryland) population losses prominently illustrated. Population growth in the outer jurisdictions is also displayed prominently. A comparison across the four time periods shows the location of new residents was most intense in the inner suburbs in the 1960s and 1970s, with higher growth in population occurring in the outer suburbs since then. In short, geographic decentralization of population has continued to occur throughout the period with the spatial locus of this process spreading increasingly outward over time.

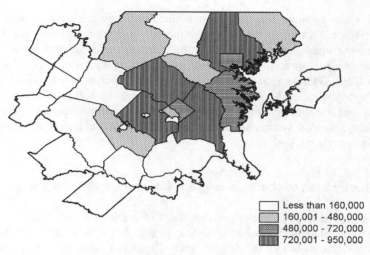

Less than 160,000
160,001 - 480,000
480,000 - 720,000
720,001 - 950,000

Fig. 9.1. National capital region population levels in 2000

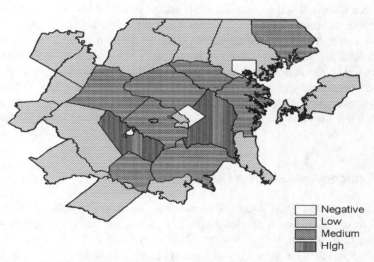

Negative
Low
Medium
High

Fig. 9.2. National capital region change in population: 1960-1970

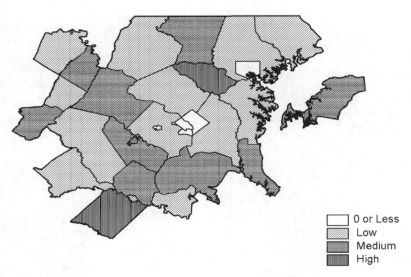

Fig. 9.3. National capital region change in population: 1970–1980

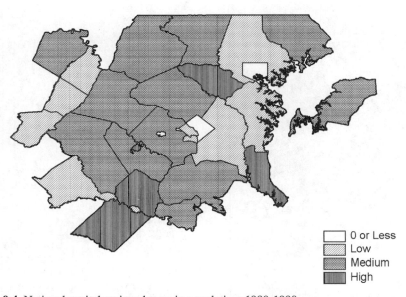

Fig. 9.4. National capital region change in population: 1980-1990

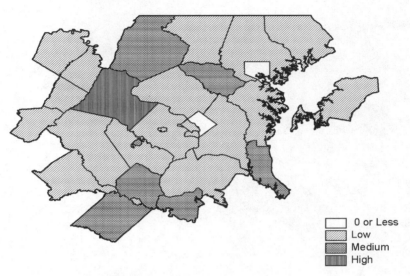

Fig. 9.5. National capital region change in population: 1990-2000

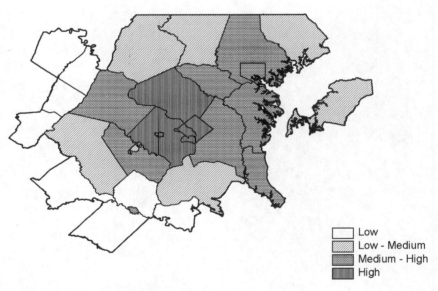

Fig. 9.6. National capital region local taxes

9.3.4 Hypothesis Testing and Analysis

Figure 9.6 shows the distribution of tax burden throughout the region where the highest burden levels are found to be in central and suburban jurisdictions. This suggests a hypothesis that tax burdens are highest in those areas where population is growing most rapidly (suburbs) with the exception of the two geographic core cities. EMSPlan enables the investigator not only test this hypothesis, but also to formulate it in a more sophisticated manner.

It assumed that population growth is greatest where land prices are lowest, *ceteris paribus* (a classic tenet of urban land rent theory), and that this will be in non-core or more peripheral parts of the region (again, from classical theory and supported by empirical research) yet not on the far periphery. Those facing the largest income constraints are apt, again *ceteris paribus*, to be recent migrants who in turn are more likely than existing residents to be in the family formation stage. EMSPlan enables the testing of this hypothesis through its regression routine. The hypothesis is tested by regressing population growth rates in the most recent time period (the dependent variable) on housing value, family age and size variables. The results produce a multiple R square of 0.77 and the beta coefficients are all significant and in the expected direction, although there is considerable intercorrelation between the family size and age variable. Re-running the analysis using only the housing value and family size variables, similar results are obtained. Most positive residuals are for the peripheral and faster growing counties and negative residuals for the slower growing inner jurisdictions. This means that the relationship is non-linear. Using a sine transform on the dependent variable and re-running the analysis produces a higher R square of 0.91 and generally uncorrelated residuals.

A fairly obvious observation from this analysis would be to suggest that the attraction to peripheral jurisdictions of disproportionately large numbers of families in the formation stage is probably increasing significantly the demand for services and thus also the tax burden. A simple regression of tax burden on family size shows an R square of 0.87, with a positive and significant coefficient. In short, the data and analysis suggests that relatively low land costs are attracting those in early family formation stages and with lower incomes. Further, these groups are demanding relatively more services (for example, more children means more public schools and teachers) and, as a consequence, the need for increased public investment relative to other areas despite lower housing and land costs.

So what are the implications? The peripheral jurisdictions could continue to support growth policies but suffer the probable outcome of providing inadequate services. They could also adopt an aggressive industrial growth strategy in hopes of increasing the commercial tax base faster than the growing demand for family services. The short run effects of this would likely be to create further stress in the quality of and general provision of services. The jurisdictions could also adopt policies to raise the cost of residential development—for example, increase the cost of building permits or reduce the number of permits granted for low cost dwelling construction—in an effort to reduce the cost differential with the less peripheral parts of the region. Jurisdictions might also adopt a consolidated strategy emphasizing the expansion of the commercial tax base and at the same time restricting residential development. The ESMPlan SDSS could be used to evaluate the impact of such policies. However, more micro-level analysis would be required at the individual jurisdictional level to fully evaluate the effects of such alternative policy stances.

9.4 A Spatial Decision Support System

In Chap. 4 the methodology for regional input-output analysis was discussed, including how it may be used in regional economic development forecasting. Recent extensions to the methodology have enhanced computational ease, and linked it with GIS to produce a DSS. Beginning in the early 1990s, in the United States, Hewings et al. (work reported in 1998) developed an econometric input-output model of the Chicago economy (CREIM) that could be run in a Windows environment. This feature, one of several new capabilities, makes it possible for the first time to rapidly simulate the effects of alternative policies and or investment strategies on the economy and its various sectors. Later Campbell and Stough (1994), with the assistance of Hewings, constructed a similar model for the Northern Virginia part of the US National Capital Region. More recently models of the whole region have been constructed that have near real time simulation capabilities (NCREIM). By the late 1990s, commercial regional model building systems— for example, REMI—have adopted this policy simulation feature. Here we use as a case study the Northern Virginia model (NVREIM) following the work of Campbell and Stough (1994).

9.4.1 The North Virginia Case Study Model: NVREIM

Over the last twenty years the Northern Virginia region in the United States has experienced substantial growth by virtually any measure. Growth in gross regional product, investment, income and employment has made Northern Virginia one of the fastest growing areas in the country. Additionally, this growth has not taken place in low wage, low skill sectors of the economy, but rather in high-end portions of the service sector where skilled labour and keen entrepreneurial spirits are essential. Northern Virginia now enjoys one of the highest per capita incomes and levels of educational attainment in the nation. While most local governments only dream of the growth experienced, for example, by Fairfax County, it has become a reality in Northern Virginia. While the benefits of rapid growth are clear, there are associated costs which if left unattended, could curtail the region's rate of future growth. Thus it is important to have a tool permitting the performance of the region to be monitored and evaluated using a DSS which has the capability to measure potential future impacts.

Though only recently adapted for Northern Virginia, NVREIM has its roots in a modelling methodology that has been successfully used in the state of Washington since 1977 (see Conway 1990). Since that time the model has been calibrated for several mid-Western states as well as the regional economy of Chicago. The modelling methodology fuses a regional input-output model commonly used for economic impact assessment with an econometric model used for forecasting and projection. By integrating the two methodologies attractive features of both are retained while overcoming their respective constraints.

NVREIM is structured as a set of double-entry industry accounts. Thus, each sector of the regional economy is viewed as both producer of outputs and consumer of inputs. Households are treated similarly to consumers of final goods and services and producers of labour services through their participation in the labour market. One major advantage of input-output models in general is the detail with which industries can be depicted. The NVREIM contains 27 industry sectors and seven sources of final demand. The current input-output model for the United States contains over 500 industrial sectors, thereby making it the most detailed picture of the American economy possible. While there are trade-offs in choosing the number of industry sectors to portray, the detail of a regional input-output model is limited by the availability of data, the number of industries actually present in the region (not all industries are represented in each region) and the cost of producing such models.

While input-output models have many attractive features, they also suffer from some restrictive assumptions. Set out in matrix form, the double-entry accounting system of input-output models implies linearity in production and consumption functions as well as constant production technology. Further, because there is no explicit temporal dimension to these models, they are considered static. In other words, while we can use input-output models to estimate the magnitude of anticipated economic change, we do not know the time frame over which economic impacts will unfold. For these reasons, we rely on statistical procedures to trace over time changes in income, employment, output, productivity and technology. Time series analysis of those factors allows separate equations, one for each industry, to be estimated and used to forecast probable levels of economic activity. The same econometric approach, however, also allows the estimation of various labour force and demographic features of the regional economy, thereby rendering a more complete economic and demographic picture of the Northern Virginia region. The end result of linking input-output and econometric methodologies is a dynamic capacity for economic forecasting and impact analysis.

The structure of NVREIM is presented in Fig. 9.7. Broadly, the model consists of five blocks of equations: Industrial Output, Employment, Income, Population and Labour Force, and Final Demand. Although during simulation each of the blocks interacts with one another, we may conceptualize the workings of the model by referring to the blocks in that figure.

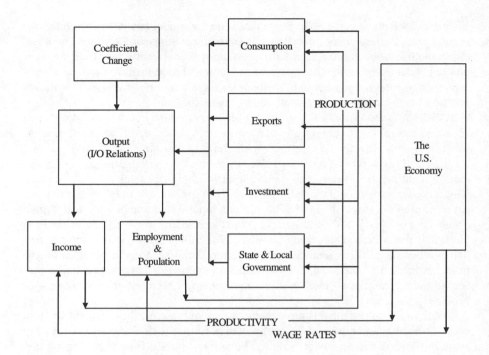

Fig. 9.7. The structure of NVREIM
Source: Center for Regional Analysis, George Mason University

Step 1. During baseline simulation, exogenous forecasts of economic activity at the national level are used to provide the initial stimulus to the Northern Virginia economy. The initial stimulus takes the form of extra-regional demand for Northern Virginia products and services. In this sense, the model is export, or demand-driven. Demand for Northern Virginia exports set in motion a number of effects.

Step 2. With new demands for Northern Virginia product, changes in industrial output take place. In a standard input-output framework, the total demand associated with the initial stimulus is a linear function of the initial change. This approach assumes that production and consumption functions are linear and that technology is unchanged from one year to the next. Because these factors are not linear or static, NVREIM simulates the process of technical change through a series of equations that relate actual output to the output that would exist if technology were static. The difference between these equations reflects processes of technical change that are then simulated into the future. This step removes a major criticism of input-output models—their static and linear assumptions. In Fig. 9.7, this is represented by the box labelled 'coefficient change'. These coefficients are the regional direct requirements coefficients from the standard input-output model, whose change is simulated through the equations describing the relationship between actual and expected output. From this step, we move away from the static input-output analysis and toward a dynamic model of the region.

Step 3. Changes in output give rise to changes in the demand for labour. There is not a one-to-one correspondence in these demands as increased labour productivity means that less labour may be required to produce a unit of output. Thus, estimates of labour productivity are used to dampen the demand for labour associated with new levels of output. Ultimately, employment and labour force will determine new levels of population. However, at this stage only the initial effects are working through the model.

Step 4. Just as employment by industry is determined by changes in sector-specific output, income by industry is related to industry employment given prevailing wage rates. Income by industry, property income, transfer payments, contributions to social insurance and residence adjustments combine to determine total personal income in Northern Virginia. Step 4 completes the first set of internal demands, which originated from the external demand outside the region.

Step 5. With the first set of internal demands completed, a second set of internal demands are generated. Changes in employment have given rise to new income levels and population responses through net migration. As wages and salaries are paid by regional employers, they are spent by regional households in the form of personal consumption expenditures. Further, some regional businesses will invest part of their receipts in new plants and equipment while changes in population will spur investment in residential structures and stimulate state a local government spending. All the induced effects are captured in a second 'round' of demands internal to the region. These demands then cause industrial output to grow and steps two to five are repeated until new changes in economic activity have been exhausted.

This completes the model sequence for a single year! The same process is then completed for each forecast year for a 15 year time horizon.

With an understanding of the basic model structure and a historical perspective on the region, the following sections turn attention to the baseline forecasts of the model and to estimate what lies ahead for the region.

9.4.2 Baseline Forecasts Using NVREIM

Many factors can influence the future growth of the region. Availability of capital for investment, federal government down-sizing/relocation, compliance with federal air quality standards, disasters and availability of appropriately trained labour are a few examples. However, without certain knowledge of these factors, reliance is on past historical trends in productivity, wage rates, and the region's relationship to the United States. Given these factors, should the region be expected to continue growing at its historic rate? The answer depends on how we measure growth. In terms of population, the baseline forecasts suggest the region will grow at a rate similar to recent history. In terms of gross regional product, employment and income our forecasts indicate a continuation of strong growth, but at a somewhat lower rate. In the discussion below, several indicators of the region's growth are presented. For each indicator or variable two sets of forecasts are given: the 'baseline' forecast which provide the standard estimates produced with NVREIM, and an 'alternative' forecast which assumes that the rate of growth is only half that predicted by the model. These estimates are also referred to as 'high' and 'low' estimates in Tables 9.2 and 9.3 to follow and illustrate two different scenarios produced and evaluated by the model. These tables also include the actual figures for 2000 and 2005 showing that the forecasts were quite good when compared to the high forecast figures with the exception of unemployment for 2000.

Table 9.1. Northern Virginia population and employment - actual and forecasted values 1995, 2000, and 2005

	High			Low		
	1995	2000	2005	1995	2000	2005
Population	1,750	922	2,167	1,622	1,700	1,806
Actual		1,870	2,100		1,870	2,100
Labour Force	1,055	1,216	1,401	973	1,045	1,123
Actual		1,068	1,185		1,068	1,185
Employment	1,034	1,192	1,373	954	1,025	1,100
Actual		1,051	1,160		1,051	1,160
Unemployment (%)	2.0	2.0	2.0	2.0	2.0	2.0
Actual		3.1	2.4		3.1	2.4

Source: Center for Regional Analysis, George Mason University: NVREIM; unpublished analyses; actual figures from U.S. Bureau of the Census & U.S. Bureau of Labor Statistics 2000 and 2005

Projections for Northern Virginia's population and labour force are presented in Table 9.1. Under the standard (high) model run, we see that population is projected to exceed 2 million by the year 2005 (also see Fig. 9.8). Labour force and employment (by place of residence) are expected to grow at a similar rate. The growth in population from its 1990 level of 1,466,000 to 1,922,000 represents a 31 percent increase to the then next census period in 2000. Population growth to the year 2005 is expected to break the two million mark, implying an average annual growth rate of 2.7 percent. That contrasts slightly with its 1969–1990 average growth rate of 2.3 percent. The alternative (low) estimates, assuming the region grows at only half its estimated rate suggests that by 2000 the region would contain 1.7 million and just over 1.8 million by 2005. How close are these estimates to other published forecasts? Estimates by the Virginia Employment Commission of regional population in the year 2000 are 1,746,462. Those projections fall between the ones presented in Table 9.1. Further, actual figures are presented in Table 9.1 and are with one minor exception all within the range of the model forecasts.

Labour force and employment (by place of residence) are expected to grow at an average annual rate of about 3 percent. As reflected in the unemployment rate, the labour market is expected to remain tight over the forecast period.

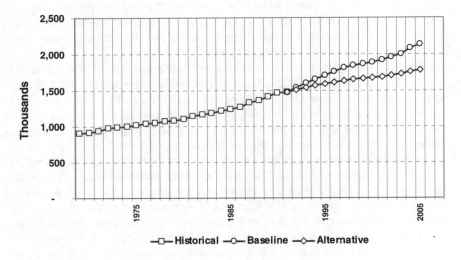

Fig. 9.8. NOVA population: baseline and alternative forecast
Source: Center for Regional Analysis, George Mason University: NVREIM; unpublished analyses

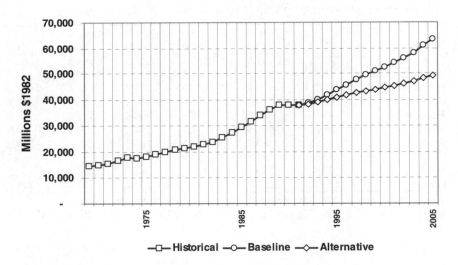

Fig. 9.9. NOVA gross regional product: baseline and alternative forecast
Source: Center for Regional Analysis, George Mason University: NVREIM; unpublished analyses

Driving this growth are projections of Gross Regional Product (see Fig. 9.9). Over the historical period it is estimated that Northern Virginia's GRP grew almost twice as fast as the Nation's GNP. From 1969 to 1990, Northern Virginia's

GRP grew at an average annual rate of 4.7 percent. The baseline forecast suggests growth of 3.8 percent per year while the alternative forecast implies 1.9 percent per year. Coincidentally, the growth rate implied by the alternative forecast is similar to projections of United States GNP.

Given these overall rates of growth, it is instructive to examine how that affects employment and its industrial distribution. The data are presented in Table 9.2. By either of the forecasts, total regional employment (by place of work) is expected to surpass the one million mark by 2005 (by 2000 with the 'high' scenario forecast - see Fig 9.10). This will be achieved as employment is projected to grow by 2.9 percent per year. That contrasts with its historical growth rate of 4.4 percent per year. Growth in the government sector is expected to be modest. Most employment in government will come from state and local governments, while military employment is forecast to decline throughout the forecast period. By the end of the forecast period, government is expected to slip in rank from the second largest source of employment to the third as the trade sector becomes the second largest. This is not the case under the 'low' scenario as the rate of growth in trade is curtailed relative to government. In both scenarios, more jobs will be found in services than in any other sector. The actual employment figures are consistent with the 'high' scenario forecasts (Table 9.2).

Table 9.2. Total Northern Virginia employment - forecasted and actual values, and Employment projections by sector 1995, 2000 and 2005

Sector	High			Low		
	1995	2000	2005	1995	2000	2005
Total	991.5	1,137.5	1,155.7	929.1	994.5	1066.0
Actual	940.9	1,050.5	1,160.0	940.9	1050.5	1,160.0
Extraction	7.2	8.8	10.3	6.7	7.4	8.0
Construction	65.5	70.9	76.4	59.1	6.16	63.9
Manufacturing	48.0	54.2	68.3	44.4	47.2	52.8
TPU	58.7	68.5	71.1	54.7	59.3	60.5
Trade	196.8	230.7	284.1	186.3	201.8	224.2
FIRE	83.3	102.7	118.7	67.4	74.9	80.7
Services	302.8	350.3	407.3	288.7	310.5	334.5
Government	229.2	251.5	274.5	221.8	231.8	241.4

Source: The Center for Regional Analysis, George Mason University: NVREIM; unpublished analyses; actual figures from U.S. Bureau of Labor Statistics 1995, 2000 and 2005

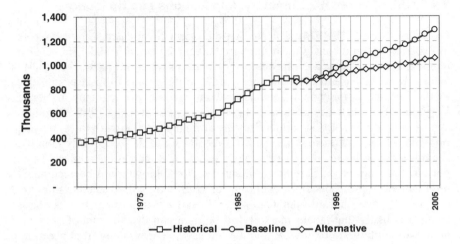

Fig. 9.10. NOVA employment: baseline and alternative forecast
Source: The Center for Regional Analysis, George Mason University: NVREIM; unpublished analyses.

Similarly, growth in income by sector is presented in Table 9.3. As in Table 9.2, services, government and trade top the list as sources of income. Income in the trade sector is projected to grow as fast as the government sector, leaving the relative rankings unchanged. Total personal income is expected to grow by 3.8 percent per year compared to its historical rate of 5 percent per year. According to NPA Data Services (1993), total personal income is expected to grow by 2.3 percent per year. Thus projected growth rates are in line with other projections for the region, both being below the historical average.

Table 9.3. Total Northern Virginia income - forecasted and actual values, and Income projections by sector 1995, 2000 and 2005

Sector	High			Low		
	1995	2000	2005	1995	2000	2005
Total	39,934	47,320	55,088	35,887	39,097	42,211
Actual	23,010	43,528	47,000	23,010	43,528	47,000
Extraction	159	194	228	144	160	173
Construction	1,665	1,798	1,938	1,512	1,572	1,633
Manufacturing	1,847	2,493	3,580	1,631	1,893	2,266
TPU	1,851	2,406	2,671	1,646	1,881	1,991
Trade	3,554	4,422	5,805	3,305	3,691	4,235
FIRE	1,923	2,521	3,176	1,564	1,793	2,013
Services	8,265	10,103	12,455	7,714	8,524	9,449
Government	6,675	7,138	7,774	6,480	6,680	6,943

Source: The center for Regional Analysis, George Mason University: NVREIM; unpublished analyses; actual figures from U.S. Bureau of the Census 1995, 2000 and 2005

9.4.3 NREIM Forecast: Summary, Implications and Epilogue

Projections using NVREIM indicate that the North Virginia region was expected to experience significant growth through the year 2005 and it did. However, the regional economy was not expected to grow as fast as it did from 1969 to 1990. While all of the baseline forecasts are similar in magnitude to forecasts produced by the Virginia Employment Commission, for example, the economy of Northern Virginia was expected to grow at a slightly faster rate over the 1995-2005 decade and again, it did. Table 9.4 summarizes the average annual growth rates for the period 1969–1990 and those produced by NVREIM for 1991–2005. Only population is expected to grow at an above average rate, and that increase is modest. The implications of the forecast are that continued regional growth will be accompanied by relatively strong incomes while income growth will exceed employment growth. However, because population is not expected to grow as fast as employment, it is likely that labour markets will be tight and that the rate of unemployment will continue to be low. While the forecast is encouraging, it is based upon the assumption that factors which generated growth in the past will continue in the future. Without perfect foresight, it is therefore recommended that such forecasts be viewed as approximate, or 'best guesses' given the historical growth path of the region.

The employment forecasts that 137,000 to 164,000 jobs would be added to the region between 1995 and 2005. In other words, the region would be adding between 13,700 and 16,400 jobs per year. Assuming 2.5 persons per household, this means that there will need to be between 55,000 and 66,000 new dwellings to support these households. That also means considerable increases in vehicles with each household now owning nearly two vehicles we would expect 100,000 to 130,000 new vehicles in the region and a considerable expansion in trips, and, therefore, emissions unless technology outpaces the effect this increase in trips will create. To support the new jobs that will be created, it will be necessary to expand other features of the physical infrastructure of the region including commercial office space, roads, sewerage, water supply, parks and recreational facilities and softer infrastructure items such as education, health services, library services and recreational services.

There are a number of barriers to accommodating the new jobs that are forecast for the region. These range from infrastructure financing to air and water quality regulations to competition from other areas to intergovernmental relations (for example, the every increasing constraining impact state and federal mandates are having on local government capacity) to labour supply and training. It will take much more cohesive regional leadership to manage (for example, how to find a way to satisfy the Clean Air Act Amendments and yet accommodate growth) the constraining effect these barriers will have on development. Without cohesive leadership, the region will likely be constrained from attaining the expected development path.

The forecasts using NVREIM were made in 1994. Given that more than ten years have passed since then, it was possible to compare the forecast values with actual outcomes. The actual data presented in Tables 9.1, 9.2 and 9.3 demonstrate that the model performed quite well.

Table 9.4. Summary of average annual growth rates (%), historical and predicted

	1969–1990 (historical)	1991–2005 (predicted)
Total Population	2.3	2.7
Gross Regional Product	4.7	3.8
Total Employment*	4.4	2.9
Total Personal Income	5.0	3.8
Labor Force	4.1	3.2

Source: Center for Regional Analysis, George Mason University: NVREIM; unpublished analyses

9.5 Integrating Quantitative and Qualitative Analysis

A major challenge in the development of DSS is how to incorporate political and other qualitative factors with quantitative assessment. Dinc et al.(2001) have used the Saaty (1980) Analytical Hierarchy Process (AHP), a popular tool for multi-objective decision making and planning introduced in the mid 1970s by Saaty (1980), to integrate outputs of these quantitative models—shift-share (S-S) analysis, data envelopment analysis (DEA) and input-output (I/O) analysis—into a DSS to assist policy choices. AHP incorporates judgments and personal values in a logical way using a multi-level hierarchy. It requires the decision maker to provide rate scale comparisons between various objectives, and calculates the vector of weights implied by those comparisons. The decision maker provides pair-wise comparisons between alternatives with respect to each objective, and the implied rankings are calculated. This produces a ranking of alternatives, a combination of the rankings under the individual objectives and the relative priorities of the objectives. The final rankings enable the decision maker to choose the best alternative.

9.5.1 The DSS Framework

Dinc et al. (2001) have set out to develop a DSS to provide policy guidance for prioritizing economic sectors by local or state government actors according to the specific utility of the decision maker. Their first step in using the AHP was to establish the decision hierarchy used as the basis of prioritization as shown in Fig. 9.11. Two objectives were considered at the first level; economic and socio-political. The decision maker seeks to maximize the utility of each component. At the second level, these objectives are refined by specific criteria, which represent the aggregation of factors that are compared only to each other when evaluating the socio-political objectives.

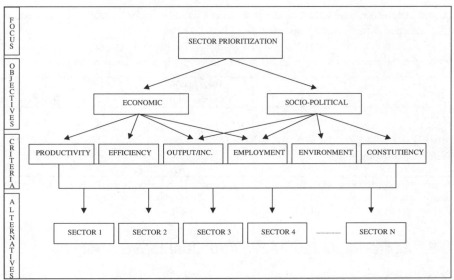

Fig. 9.11. The decision hierarchy for sector prioritization in an SHP study of Virginia
Source: Dinc et al. 2001

The economic criteria are defined as productivity of the sector, income generated and employed due to the sector, and the relative efficiency of it. These are
quantitative assessments obtained from S-S, DEA and I/O model outputs. The
socio-political criteria share employment, but also consider environmental issues
and constituency related factors, covering both quantitative aspects of the decision
model (such as productivity and efficiency), and qualitative aspects (such as constituency). At the base of the hierarchy are the sectors considered for prioritization, and these sectors are evaluated through the criteria defined in the upper level
of the hierarchy. The evaluation of the decision maker is made through the results
of the first phase and the decision maker's judgments. As Dinc et al. (2001) say,

> ...decomposing the problem into more definable components thus enables the decision
> maker(s) to view it within a structured framework. Once the hierarchy is established, the
> decision maker(s) judgments are required to measure the importance of any one criterion in
> relation to all other criteria. The process results in a set of pair-wise comparison matrices
> which is used as input to produce relative criteria weights at each level of the hierarchy. In
> the final steps of the AHP, these weights are aggregated to produce a ranking of the deci-
> sion alternative, namely the sectors for this problem. Since AHP works with ratio scale
> comparisons, the different matrices that result from S-S, DEA and I/O analysis do not cre-
> ate a problem (p. 12).

Applying this DSS to the State of Virginia in the United States, Dinc et al.
(2000) used a two-stage process:

(a) first, the process uses data from a S-S model, DEA and ten analysis conducted
by Dinc and Haynes (1999);

(b) second, the outcomes of the first stage—the sectoral employment change resulting from output and productivity variations, relative efficiency scores, employment, income and output multipliers—are incorporated in the AHP framework.

9.5.2 The Data

The S-S model used US Bureau of Economic Analysis CA25+CA27 files for the labour input (measuring the total number of workers employed in a sector during a year). Sectoral value added was used as the output measure, and productivity was defined as output for work based on the share of labour in a given sectors total value added. Output data for the manufacturing sectors were taken from the Census of Manufacturers and the Annual Survey of Manufacturers. Dinc and Haynes (1999) estimated non-manufacturing sectors output via a simple ratio relating the output of earnings of national industries to earnings of state industries. Wages (earnings) data were obtained from BEA CA07 and CA05 files. Value added by industry data were deflated using the BLS industry price deflator.

The DEA model identified two outputs and five inputs, these being derived from data files of the IMPLAN (1993). The output measures were:

(a) Total sector outputs: the dollar value of the total product produced by a sector during a year.
(b) Exports: The total dollar value of the product sold outside the region, giving an indicator of the sector's competitiveness in the State.

The five input measures were:

(a) *Labour costs:* total employee compensation, including wages, salaries, and benefits paid by state industries.
(b) *Proprietary income:* The sum of income from self-employment, corporate income, rental income, interest and corporate transfer payment.
(c) *Employment:* The number of annual average jobs in a given industry in the state, including self-employed.
(d) *Materials:* The dollar value of purchases of commodities by a given industry as raw materials and intermediate input.
(e) *Imports:* The dollar value of all foreign and domestic imports of commodities by industry. For the State, this variable may be seen as an indicator of the lack of competitiveness of a regional industry.

The analysis used two labour related inputs—total labour costs and number of employees—to capture the impact of labour on efficiency.

The Dinc and Haynes (1999) study showed the top five growth sectors in the Virginia economy for the period 1969 to 1992 to have been business engineering and management services, health services, retail trade, eating and drinking establishments, and the construction sectors—all except the latter being nationally growing sectors. The bottom five sectors were livestock and agricultural products, membership of organizations and household services, chemicals and allied products, apparel and textile products, and the leather and leather products—declining both regionally and nationally. The extended S-S model (as discussed in Chap. 3) showed how in both declining and growing sectors, output growth and productivity gains played an important role in overall employment growth, although non-labour factors such as capital investment or disinvestments considerably impact on employment change in these sectors. The I/O analysis enabled the efficiency of Virginia's industrial sector to be determined, with employment, output and income multipliers representing direct, indirect and induced change resulting from change in final demand.

9.5.3 AHP Findings

The hierarchy in Fig. 9.12 was used by Dinc et al. (2001) to determine priorities, with the weight assessments of the first two levels being all decision maker specific judgments, and the sector evaluations being done using the ratings model of Expert Choice Software, which uses AHP for its methodology.

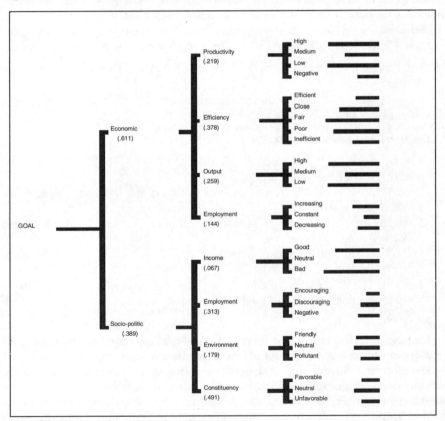

Fig. 9.12. The hierarchy, with the intensities and weights, for an AHP study of Virginia
Source: Dinc et al. 2001

They explain how the ratings model allowed them to elicit ratings for all the industry sectors without comparing them to each other, combining the power of the hierarchy with the capability of rating a large number of alternatives. At the lowest level of the hierarchy, on top of the alternatives (industry sectors), intensity nodes were inserted. In the model the scale of intensity for each criterion appears as a group of nodes under the criterion. The priorities of the intensity nodes are determined by pair-wise comparison, with the intensities first being constructed for each criterion, and then used to rate the alternatives. The researchers point out that as the results of S-S, DEA and I/O analysis can be in both absolute and relative figures; this method serves the purpose well, with the Expert Choice software allowing the entry of all the values direct from the previous Dinc and Haynes (1999) study. The upper levels of the hierarchy are prioritized in a ratio scale. The rating assignments of the sectors with respect to economic criteria are done using the results of the S-S, DEA and I/O analysis. In their Virginia case study, Dinc et al. (2001) did this by taking the outputs from the Dinc and Haynes (1999) analysis to numerate the hierarchy while assigning hypothetical weights to socio-economic facts. Figure 9.12 shows the hierarchy with its weights and sector prioritization allocations.

Economic objectives were assigned a higher weight (0.611) then socio-political objectives (0.389). Weights were assigned for criteria using pair-wise comparison matrices, comparing each set criterion (such as constituency vs. environment) with respect to the objective (such as economic) on a scale of 1 to 9. When all the pair-wise comparisons were made, AHP produced the weights for each criterion, with this procedure being repeated for all the other objectives and the criterion. At the lowest level of the hierarchy, in this case study the authors grouped the numbers generated by the model into a small number of original intervals, "giving a more realistic reflection of a thought process for the decision maker in evaluating this many sectors especially in the prioritization of qualitative factors such as constituency and environments" (Dinc et al. 2001, p. 19).

Table 9.5 lists the final ratings for industry sectors on the economic and socio-political criteria for Virginia. Health services, retail trade, eating and drinking places, educational services, business, engineering and management, hotels and other lodging services, and insurance agents lead the list. Thus, the services sectors with high-income prospects dominate, with the highly prioritized sectors in the AHP model having positive or at worst neutral feedback from socio-political criteria. Dinc et al. (2001, p. 20–21) say that because it is easy to follow the local and global weights of each criterion, the justification of each rating can be traced throughout the hierarchy.

This traceability of characteristics in the AHP is one aspect that makes a great deal of sense in public sector decision-making. We all recognize that most decisions in the public sector contain elements of subjectivity for one or more reasons, but this process makes such decision elements explicit and traceable.

Table 9.5. The final ratings of the sectors in an AHP study of Virginia

SECTORS	TOTAL	ECONOMIC				SOCIO-POLITIC			
		PRODUCTIVITY	EFFICIENCY	OUTPUT	EMPLOYMENT	INCOME	EMPLOYMENT	ENVIRONMENT	CONSTITUENCY
		Weights of the Criteria .							
		0.134	0.2309	0.1583	0.0878	0.0259	0.1217	0.0503	0.1912
Health Services	0.866	NEGATIVE	EFFCIENT	HIGH	INCRSNG	GOOD	ENCOURAG	FRIENDLY	FAVORBLE
Retail Trade	0.841	HIGH	CLOSE	HIGH	INCRSNG	GOOD	ENCOURAG	NEUTRAL	FAVORBLE
Wholesale Trade	0.759	LOW	EFFCIENT	MEDIUM	INCRSNG	GOOD	ENCOURAG	NEUTRAL	FAVORBLE
Eating & Drinking Places	0.748	NEGATIVE	EFFCIENT	MEDIUM	INCRSNG	GOOD	ENCOURAG	NEUTRAL	FAVORBLE
Educational Services	0.738	LOW	EFFCIENT	LOW	INCRSNG	GOOD	ENCOURAG	FRIENDLY	FAVORBLE
Business, Engineering & Mgmt Services	0.731	MEDIUM	FAIR	HIGH	INCRSNG	GOOD	ENCOURAG	FRIENDLY	FAVORBLE
Hotels & Other Lodging Places	0.712	NEGATIVE	EFFCIENT	LOW	INCRSNG	GOOD	ENCOURAG	NEUTRAL	FAVORBLE
Social Services 7/	0.712	LOW	EFFCIENT	LOW	INCRSNG	BAD	ENCOURAG	FRIENDLY	FAVORBLE
Insurance Agents, Brokers, & Services	0.712	LOW	EFFCIENT	LOW	INCRSNG	BAD	ENCOURAG	FRIENDLY	FAVORBLE
Motion Pictures	0.712	LOW	EFFCIENT	LOW	INCRSNG	BAD	ENCOURAG	FRIENDLY	FAVORBLE
Livestock, Livestock Prod. & Agrc. Prod.	0.623	HIGH	EFFCIENT	HIGH	DECRSNG	NEUTRAL	DISCOURA	POLLUTNT	NEUTRAL
Water Transportation	0.612	LOW	EFFCIENT	LOW	INCRSNG	NEUTRAL	DISCOURA	NEUTRAL	FAVORBLE
Membership Orgs & Household Services	0.581	HIGH	FAIR	MEDIUM	DECRSNG	BAD	ENCOURAG	FRIENDLY	FAVORBLE
Tobacco Products	0.568	NEGATIVE	EFFCIENT	LOW	DECRSNG	NEUTRAL	ENCOURAG	POLLUTNT	FAVORBLE
Transportation Equipment	0.56	LOW	EFFCIENT	LOW	INCRSNG	GOOD	ENCOURAG	POLLUTNT	NEUTRAL
Rubber & Misc. Plastics Products	0.56	LOW	EFFCIENT	LOW	INCRSNG	GOOD	ENCOURAG	POLLUTNT	NEUTRAL
Communications	0.557	LOW	FAIR	MEDIUM	INCRSNG	GOOD	ENCOURAG	NEUTRAL	FAVORBLE
Construction	0.557	LOW	FAIR	MEDIUM	INCRSNG	GOOD	ENCOURAG	NEUTRAL	FAVORBLE
Paper & Allied Products	0.543	LOW	CLOSE	LOW	INCRSNG	GOOD	ENCOURAG	POLLUTNT	FAVORBLE
Food & Kindred Products	0.532	LOW	EFFCIENT	LOW	INCRSNG	GOOD	ENCOURAG	NEUTRAL	UNFAVRBL
Elect. Equiprnt, Excl. Comp. Equipmt	0.522	LOW	FAIR	LOW	INCRSNG	GOOD	ENCOURAG	NEUTRAL	UNFAVRBL
Instruments & Related Products	0.522	LOW	FAIR	LOW	INCRSNG	GOOD	ENCOURAG	NEUTRAL	FAVORBLE
Printing & Publishing	0.522	LOW	FAIR	LOW	INCRSNG	GOOD	ENCOURAG	NEUTRAL	FAVORBLE
Personal Services	0.522	LOW	FAIR	LOW	INCRSNG	GOOD	ENCOURAG	NEUTRAL	FAVORBLE
Machinery & Computer Equipment	0.522	LOW	FAIR	LOW	INCRSNG	GOOD	ENCOURAG	NEUTRAL	FAVORBLE
Legal Services	0.521	NEGATIVE	EFFCIENT	LOW	INCRSNG	GOOD	ENCOURAG	NEUTRAL	UNFAVRBL
Insurance Carriers	0.51	LOW	FAIR	LOW	INCRSNG	BAD	ENCOURAG	FRIENDLY	FAVORBLE
Security & Commodity Brokers & Serv	0.51	LOW	FAIR	LOW	INCRSNG	BAD	ENCOURAG	FRIENDLY	FAVORBLE
Amusement & Recreation Services	0.499	NEGATIVE	FAIR	LOW	INCRSNG	BAD	ENCOURAG	FRIENDLY	FAVORBLE
Depository & Nondepos Credit Inst.	0.496	LOW	FAIR	LOW	INCRSNG	BAD	ENCOURAG	NEUTRAL	FAVORBLE
Transportation By Air	0.486	LOW	FAIR	LOW	INCRSNG	GOOD	ENCOURAG	POLLUTNT	FAVORBLE
Real Estate	0.485	NEGATIVE	FAIR	LOW	INCRSNG	BAD	ENCOURAG	NEUTRAL	FAVORBLE
Railroad Transportation & Services	0.466	LOW	CLOSE	LOW	DECRSNG	BAD	ENCOURAG	NEUTRAL	FAVORBLE
Electric, Gas, & Sanitary Services	0.449	NEGATIVE	FAIR	LOW	INCRSNG	BAD	ENCOURAG	POLLUTNT	FAVORBLE
Misc. Manufacturing Industries	0.419	LOW	EFFCIENT	LOW	DECRSNG	BAD	ENCOURAG	NEUTRAL	UNFAVRBL
Agricultural Services, Forestry & ...	0.413	NEGATIVE	FAIR	LOW	INCRSNG	NEUTRAL	DISCOURA	FRIENDLY	FAVORBLE
Petroleum & Coal Products	0.39	LOW	EFFCIENT	LOW	CONSTANT	NEUTRAL	DISCOURA	POLLUTNT	NEUTRAL
Pipelines, Except Natural Gas	0.39	LOW	EFFCIENT	LOW	CONSTANT	NEUTRAL	DISCOURA	POLLUTNT	NEUTRAL
Primary Metal Industries	0.384	LOW	EFFCIENT	LOW	INCRSNG	NEUTRAL	DISCOURA	POLLUTNT	UNFAVRBL
Trucking & Warehousing	0.358	LOW	FAIR	LOW	INCRSNG	GOOD	ENCOURAG	POLLUTNT	NEUTRAL
Fabricated Metal Products	0.334	MEDIUM	FAIR	MEDIUM	DECRSNG	GOOD	ENCOURAG	NEUTRAL	UNFAVRBL
Other Repair Services	0.332	LOW	FAIR	LOW	INCRSNG	BAD	ENCOURAG	POLLUTNT	NEUTRAL
Auto Repair, Services, & Parking	0.321	NEGATIVE	FAIR	LOW	INCRSNG	BAD	ENCOURAG	POLLUTNT	NEUTRAL
Mining	0.293	NEGATIVE	CLOSE	LOW	INCRSNG	NEUTRAL	DISCOURA	POLLUTNT	NEUTRAL
Local & Interurban Passenger Transit	0.275	NEGATIVE	FAIR	LOW	DECRSNG	NEUTRAL	DISCOURA	POLLUTNT	FAVORBLE
Stone, Clay, & Glass Products	0.218	LOW	FAIR	LOW	INCRSNG	NEUTRAL	DISCOURA	NEUTRAL	UNFAVRBL
Lumber & Wood Products	0.182	LOW	FAIR	LOW	CONSTANT	BAD	DISCOURA	POLLUTNT	NEUTRAL
Apparel & Textile Products	0.181	MEDIUM	FAIR	MEDIUM	DECRSNG	BAD	DISCOURA	POLLUTNT	UNFAVRBL
Chemicals & Allied Products	0.158	LOW	FAIR	LOW	DECRSNG	NEUTRAL	DISCOURA	POLLUTNT	NEUTRAL
Furniture & Fixtures	0.125	LOW	FAIR	LOW	DECRSNG	BAD	DISCOURA	NEUTRAL	UNFAVRBL
Leather & Leather Products	0.095	LOW	FAIR	LOW	DECRSNG	NEUTRAL	DISCOURA	POLLUTNT	UNFAVRBL

Source: Dinc et al. 2001

9.5.4 The Importance of Linking Quantitative and Qualitative Assessment in Decision Support Systems

The integration of traditional regional economic analysis tools such as S-S, DEA and I/O into the AHP provides a useful DSS for policy makers at the state or local level as it incorporates quantitative with qualitative factors in the assessment process. "Comparison of the results of individual models with the results of the complete model showed that qualitative factors such as environment and constituency can change the importance of rankings of sectors. This, in time, could affect regional development policies. Therefore, the complete model can provide guidelines for regional in industrial policy makers in terms of which sectors should be promoted, or recruited" (Dinc et al. 2001, p. 22). A complete AHP model forces the decision maker to explain and justify their choice through the weights of their objectives, criteria and sub-criteria, making the decision process more transparent, and bringing more accountability to the final decision. Thus, the AHP model represents a DSS with considerable potential linking quantitative analysis with qualitative assessment of political factors to inform the decision process.

A challenge is how to link the type of approach used by Dinc et al. (2001) to a more disaggregated level incorporating sub-regional models within a GIS based SDSS which can permit decision makers to make industry development policy decisions that take account of spatial impacts. The earlier discussion of the NREIM model for the North Virginia component of the National Capital Region represents a promising approach incorporating GIS based analysis of growth and investment impacts.

9.6 A Spatial Optimization Modelling and Visualization Capacity

Much of the work on spatially disaggregated DSSs to date has focused on using GIS technology to help inform the planning processes such as allocating land to meet the needs of regional growth, including residential land to accommodate projected population growth, and allocating land for industrial and commercial purposes. This may be informed by the type of econometric modelling discussed in Chaps. 3 and 4 and in the previous section of this chapter, whereby future directions and magnitudes of industry sector development and employment might be modelled and their land requirements assessed.

Approaches to using GIS to monitor and plan for land supply are discussed in volume edited by Vernz-Moudon and Hubner (2000). Many of those contributions make mention of the highly politicized nature of local decision making regarding land allocation through planning and zoning and development approach processes as well as identifying the importance of regional development and planning strategies needing to be informed by the implications of structural shifts in regional economies and the implications for the types of built environment spaces, and as a result demand for land and its relationships with infrastructure provision, that those shifts might generate. In addition, considerable attention is given simply to the issue of planning to manage regional growth, and dealing with the attendant urban sprawl it generates. When planning for growth management incorporates sustainable development principles, even bigger challenges emerge for designing SDSS to help inform the development and planning process, including the building of SDSSs to simulate future urban growth scenarios. The case studies presented by Vernez-Moudon and Hubner (2000) include planning studies incorporating GIS based SDSS for Oregon in Portland, the Puget Sound Region in Washington State, and Montgomery County in Maryland.

The following example examines an Australian case study demonstrating how a SDSS may be linked to spatial optimization modelling. The modelling exercise also incorporates visualization techniques to simulate future impacts of growth on land requirements. This is an important issue both for public agencies and for developers in a region that is experiencing rapid population growth. The methodologies outlined below may be used to estimate future land requirements for various economic functions or land uses, the nature of which may be informed through the type of econometric input-output forecasting analyses discussed above.

9.6.1 A Modelling Framework

The modelling approach discussed here is a prototype SDSS based on an approach by Densham (1994) that could be applied to address many issues in regional development and planning. It provides a way of developing spatially detailed land use plans associated with population projections. Economic, social and environmental constraints associated with sustainable urban growth can be implemented to meet activity target thresholds. The approach offers a means of evaluating the multiple concerns and criteria associated with the often conflicting and competing objectives of a planning process. 'What if' type questions associated with development and planning alternatives can be evaluated geographically and interactively using a highly visual interface.

The modelling framework was developed by researchers at the University of Queensland (Ward et al. 2000). It uses an *optimization model* that addresses issues of sustainable urban development at a regional level. This is integrated with a *cellular automata* (CA) model (Batty and Xie 1994) which simulates local decision making processes associated with regional growth. The modelling approach involved a two-stage process:

(a) In the first stage, the optimization model produces regional land use allocations at an aggregate planning unit level. Planning units can be any aggregate spatial units—such as Statistical Local Areas (SLAs) or suburbs, which may comprise the planning units for a region.
(b) In the second stage, the CA model is applied at a local planning unit level to produce spatial realizations of growth scenarios.

Numerous issues associated with regional urban growth are addressed in a two stage modelling approach. The first stage optimization approach is applied to address regional economic, social and environmental issues associated with population growth. Any number of regional economic, social and environmental criteria may be implemented in the model to meet activity target thresholds such as regional open space criteria. In the second stage of the modelling approach, sustainability issues, associated with the local influences of accessibility and cost of infrastructure provision, are addressed: for example, accessibility to services and transport have a range of sustainability issues associated with travel distance, cost of infrastructure provision; similarly, equity issues are addressed, such as those associated with access to services and transport.

The CA model addresses those issues based on local friction-of-distance functions in the allocation of urban development units. For example, the localized influence of distance from service centres and distance from transport networks can be reflected in the spatial realizations of planning scenarios.

9.6.2 The Regional Scale Optimization Model

Population projections data typically form the basis upon which planning for urban growth is considered. As well, for factors such as infrastructure provision, a major consideration in the planning process is the land required to meet population projections and the associated needs for housing and other urban functions and land uses. Considerations in identifying areas for future potential development might include issues of land suitability (for example, flooding) and serviceability (for example, cost of water supply infrastructure). The amount of time required to accommodate new increasing population and new economic activity is typically related to the planning process by considering things such as housing density, including both for new broad hectare developments and infill/redevelopment in existing urbanized areas.

One objective of a strategy planning process might be to spatially allocate land for development to meet a specified population target in such a way that economic, social and environmental target thresholds are met. Given population projections over a set of planning time periods for a particular region or a component of it—for example, a Local Government Area—an optimization approach may be used to simulate the allocation of land uses. The development of a planning based optimization model involves deciding on one or more objective functions, such as minimizing the total deviation from projected regional population targets while, at the same time, minimizing the deviation from specified economic, social or environmental targets. Linear programming (an operations research technique) is applied to develop feasible and optimal solutions.

By specifying population targets, environmental zoning targets, initial density and a value for the density change parameter, the model optimally allocates the fraction of each land use zoning to each planning unit. The result is the determination of an optimal fraction of each zoning (residential, industrial and commercial) for each planning unit required to meet the overall projected population target while maintaining the environmental target for each planning unit.

9.6.3 Local Scale Cellular Automata Model

Local scale factors associated with the suitability of the land for urban development—such as proximity to services and transport networks, as well as factors such as land slope—play an integral role in the morphology of urban systems (Clarke et al. 1997). Ward et al. (2000) developed a stochastically constrained CA model that provides a means for simulating a local scale decision making process associated with urban growth. In this CA model, local neighbourhood transition rules were developed to produce realistic urban morphologies associated with transport networks. The local neighbourhood rules are applied within a broader region (or LGA) in which friction-of-distance limitations and constraints associated with distance from transport networks and service centres, as well as land slope, are stochastically realized.

Given an area target for development associated with a set of population projections, the model produces spatial realizations of possible future urban form for a projected planning period. The model functions simply at a local level in that the likelihood of a cell to be developed decreases as the distance from major and minor roads, the distance from service centres and land slope increase. The model is not conceived of as a predictive tool. However, for a test region, Ward et al. (2000) show that the model can produce realistic growth patterns with correlation's of 60 percent to 70 percent between actual and simulated growth patterns over a seven year time period.

9.6.4 Integration of Regional and Local Scale Models

The optimization modelling approach described above operates at a regional level to meet regional population projections and associated economic, social and environmental thresholds. The model provides as output the optimal fraction of area of each zoning allocated to each planning unit. This information is still in aggregate form, as no knowledge of the spatial distribution of a zoning within a planning unit exists. The CA model discussed above operates at a local level and incorporates friction-of-distance factors such as distance from service centres, transport networks and land slope.

By treating the planning process as a two-stage process, the two models can be integrated to develop spatially detailed land use plans. For example, by running the optimization model for a region (say a LGA comprising many planning units) and producing the optimal allocation of zonings to meet some population targets and environmental thresholds, a regional scale land use plan can be developed. Given that the optimization model provides output by planning unit, it is possible to then run the CA model within a planning unit to locally allocate land use based on local constraints associated with friction-of-distance factors, topography and institutional controls. The result is a spatial realization of an optimal planning scenario associated with a population projection and specified population densities.

9.6.5 An Example: Applying the Model to the Gold Coast in South East Queensland

To demonstrate the integration of the optimization and cellular automata models and their potential to inform the development and planning process, Ward et al. (2000) apply this modelling process to Gold Coast City, which is one of the most rapidly urbanizing regions of coastal eastern Australia, and is located in the southern component of the Brisbane-South East Queensland metropolis region (discussed in Chap. 4).

To apply the model, population projection data were used to estimate future growth in demand for housing and how that projected need might be optimally allocated across planning units (SLAs). 'Low', 'medium' and 'high' population projection series through to the year 2026 were taken, using destinations produced by the Queensland Government. The medium series was used, projecting growth in the population from 356,440 in 1996 to 563,600 in 2016. These population projection data were simply translated into future potential demand for new house unit construction by taking the existing (1996) pattern of average household size, and projecting that as a constant factor over the successive five year (to 2016) planning periods. These data are then adjusted according to results derived from modelling household structure and size changes over the projection period using an age cohort approach. The base data (1996) for these parameters are available in a GIS framework.

To demonstrate the modelling approach, two growth scenarios were considered by Ward et al. (2001):

(a) Scenario A: low density diffuse growth (one which basically perpetuates past urban growth processes)
(b) Scenario B: high density compact growth (one which seeks to both limit the diffusion pattern of future new urban growth and also seeks to accommodate growth through redevelopment of the existing urban area).

For growth scenario A, current population densities for each planning unit (SLA) are maintained. The growth compactness parameter is set such that optimal allocation allows large development projects to take place. This is the current form of growth in Gold Coast City. For growth scenario B, the density parameter is set such that overall density increases by 28 percent. To demonstrate a more compact form of growth, the compactness parameter is set such that, where land is available in a planning unit, the area of urban extent will increase by ten percent in each time period.

To demonstrate the incorporation of environmental thresholds in the optimization model, an open space provision target of ten percent was specified in each planning unit. So as to provide services and employment to the projected increase in population in Gold Coast City, a ratio of residential area to commercial and industrial area was set as 8:1. This ratio could have been varied, but is based on 1995 area ratios of residential to commercial and industrial land uses for the Gold Coast. The model also assumes progressive growth with no 'deaths'; that is, once a planning unit's area is urbanized it remains urbanized. For the purposes of this application only four broad land use zonings are used: residential, commercial, industrial and open space.

To develop planning units, advantage was taken of existing studies in the Gold Coast that identify residential land suitability, availability and serviceability. Planning units were designated by combining land identified by the state government planning agency's BroadHecatre Study as being suitable, available and serviceable for development. The residential, commercial and industrial zonings in the Gold Coast City Town Plan were used, all of which are available on a GIS. This combination represented all the land potentially available for development. SLAs were then used to assign this land to planning units; that is, all the land that is potentially available for development within each SLA was assigned a unique ID, and thus was referred to as 'planning units'. Planning units form the base units for allocation of land use in the optimization model.

The optimization process produces an area allocation for each zoning type by planning unit for each of the three five year time periods from 2001 to 2016. The results of the optimization process are shown in Figs. 9.13a and 9.13b, where the bars represent the total area of change in each planning unit for each time period:

(a) Figure 9.13a shows the results of the analysis for growth scenario A, the low density diffuse growth scenario.
(b) Figure 9.13b shows the results of the analysis for growth scenario B, the high density compact growth scenario.

As seen in Fig. 9.13a, most of the total area of change required to meet the population target and its associated demand for new housing construction is allocated to a small number of planning units. This type of low density diffuse growth under Scenario A has been the most common form of development in Gold Coast City with land developers requiring large areas of land for planning for large-scale developments. For the compact higher density growth scenario B shown in Fig. 9.13b, growth is distributed across a larger number of planning units. A consequence of this is that growth would be more compact because it would be distributed across a larger proportion of the urban matrix.

Figures 9.14a and 9.14b are maps showing the optimization modelling results for residential usage for the spatial pattern of new residential land required to accommodate the population projected for Gold Coast City and the associated potential demand for new housing unit construction. These are maps showing the optimization results for which the local influences of distance from service centres, distance from major and minor roads, and land slope have been stochastically realized using a cellular automata model. Figs. 9.14a and 9.14b show growth scenarios A and B respectively for 2016. While the results depicted in these maps are *stochastic realizations*—and as such will change slightly each time the CA model is run—they can represent growth realistically (Ward et al. 2001), and are an effective way to present the results of the optimization process shown in Fig. 9.13. The maps shown in Figs. 9.14 show potential growth as it might actually occur. The potential growth pattern can be mapped alongside the existing urbanized areas as shown, in Figs. 9.14a and 9.14b.

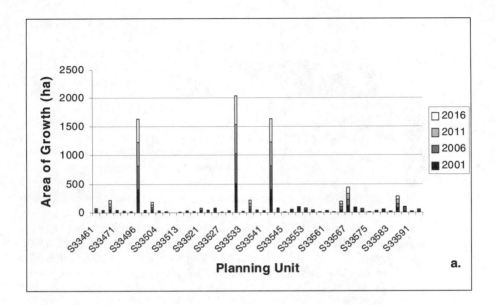

Fig. 9.13a. Optimization results showing area of growth by year by planning unit - growth scenario A low density diffuse growth
Source: Ward et al. 2000

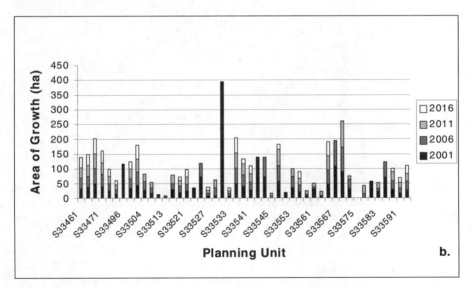

Fig. 9.13b. Optimization results showing area of growth by year by planning unit - growth scenario B high density compact growth
Source: Ward et al. 2000

Fig. 9.14a. CA realizations of optimized growth scenarios to 2016 - growth scenario A: low density diffuse growth
Source: Ward et al. 2000

Fig. 9.14b. CA realizations of optimized growth scenarios to 2016 - growth scenario B: incorporating a high density component growth
Source: Ward et al. 2000

9.7 Spatial Analysis of Geographic Clustering of Industry Clusters

In Chap. 6 industrial clusters were identified and measured in terms of economic structure and performance variables. There is, however, another body of literature (not entirely separated from the functional or structurally oriented concept of clusters) that examines the hypothesis that the sectors that compose industrial clusters should also be spatially clustered (see, for example, Steiner 1998).

Stough et al. (2001) have examined high technology clusters in three metropolitan areas in the United States: Austin, Texas; Boston, Massachusetts; and Washington, D.C. Some 33 technology sectors at the 3 and 4-digit SIC level were identified as important dimensions of technology led development in these regions. The location of each establishment for all 33 sectors was plotted using a GIS package, ArcView 3.2a (see Fig. 9.15 showing the centres of gravity for the Washington region).

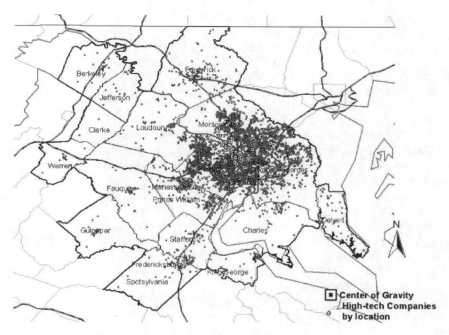

Fig. 9.15. Distribution of technology sector companies in the Washington metropolitan region and center of gravity of technology sectors
Source: Mason Enterprise Center, School of Public Policy, George Mason University, Fairfax, VA; unpublished data and analysis results

Geographic centres of gravity were computed for each of the 33 sectors which allowed the estimation of the geographic distance between these points. These distances were regressed (OLS) on the input-output coefficients for the different pairs of sectors. The results showed sizeable R square values for each of the analyses with the largest for the Boston region and smaller values for both the Austin and Washington regions presumably because these latter two are more recently formed technology regions. Thus, this implies that there has not been sufficient time for stronger proximal relations to develop in these two regions. This set of analyses supports a hypothesis that the stronger functional interaction levels among industry sectors the more proximal the location of the sectors.

This research by Stough et al. (2001) also examined the shape of the distributions of the technology cluster distributions using an algorithm based on the analytical geometry of gravity centres. With this method testing was conducted to determine if the shape was quasi-circular or quasi-ellipsoidal. The Washington case showed a maximum quasi-ellipsoidal shape compared to the other two with Boston having a near quasi-circular shape. The analysis concluded that the shape of these distributions is closely related to topographic features and consequently transportation infrastructure that has been influenced by the topography.

9.8 Conclusions

This chapter has referred to only a limited range of approaches being used to integrate tools of regional economic analysis into a DSS, which when linked with GIS technologies, produce a SDSS. This is indeed an exciting field for research and application. Development of the type of methods illustrated by the case studies we have provided in this chapter have the potential to help decision makers make better informed decisions regarding regional economic development and regional growth policies, and to test strategies against planning outcome criteria. It is a fast developing technological, methodological field that will increasingly enable the integration of econometric tools of regional analysis and forecasting, the measurement of economic performance and impact with both quantitative, qualitative socio-political assessment criteria, occurring at various levels spatial aggregation/disaggregation enabling those types of data to be linked with land use data and the land requirements for economic activity and housing and with transportation and other spatial infrastructure and services provision. This permits the development of SDSSs which can be used by both researchers and policy analysts and by policy makers and planning decision makers to assist the process of regional economic analysis and regional development and planning strategy formulation.

10 Emerging Issues for Regional Economic Development

The final chapter of this book canvasses a number of issues that are likely to be crucial for the future direction and development of analytical tools and strategies for regional economic development. One of the most challenging issues regions face in their future economic development is how to ensure more sustainable and equitable outcomes and how this will affect the way we think about competitiveness. Another is how to develop and implement strategies for regions to be fast and flexible in creating knowledge and become learning regions by developing competitive industry clusters. Future directions of regional economic strategy will require the development of additional tools to support both analysis of regional economic competitiveness and the management of economic development, including how to manage regional risk.

The chapter refers to the emerging 'virtual world' of economic development as we move into a new age where virtual trade, organizational structures and infrastructure will grow in importance. In this context, it will be important that the human dimension of economic development is not lost in the headlong rush to an embrace the virtual world. This raises the question of the *digital divide* in society, whereby some people and some places are being 'left out' of the new information economy, resulting both in increasing gaps between the *haves* (or wealthy) and the *have-nots* (or poor) and in an increasing differentiation between those regions that are linked to the global economy and those that are not. The chapter also suggests that the increasingly pervasive paradigm of sustainable development is presenting significant challenges for regions and the actors in the regional development process to address in incorporating sustainability principles within the strategic intent of regional development and planning strategies.

10.1 A Rapidly Changing World: The Need for a New Paradigm

The impact of globalization upon regions in the early part of the 21st century will continue to be profound. World trade forecasts indicate that by the year 2010 nearly 60 percent of international trade is likely to be in services. This compares to just 22 percent back in 1970. Thus, the focus of most regional economic activity in the future will be in the services sectors. The new rounds of trade negotiations are expected to increase the openness of world markets and the flow of trade. Improvements in telecommunications, involving virtual and e-commerce, will accelerate the development of the service economy. Those factors, together with increased travel and mobility–particularly of skilled labour but also of unskilled labour (including transnational mobility)–will place growing pressure upon regions to improve their industry and organizational competitiveness.

The quest for improved competitiveness seems to be moving towards greater specialization in regions based on natural and other assets—and in particular on human capital, and on industry created competitive advantage. This is a paradox in the information age when IT was supposed to reduce spatial barriers of distance and enable the wide diffusion of economic activities. The race for competitiveness is bringing about unparalleled rates of change in the way we plan for the development and management of our cities and regions, as well as how we will work, travel and live in future. In the early 1990s, Reich (1992) forecast that, by the end of the first decade of the 21st century, as many as one-half of the new jobs that will be created in developed economies many involve employment activities that did not exist in the last decade of the 20th century. Reich said that many of these jobs would rely on technologies and infrastructure that had still to be invented, and would be undertaken by a workforce that would need to be trained and retrained in new skills and to be more flexible.

At the same time, as societies have been facing greater pressure to become more competitive, populations in some parts of the world are continuing to increase at rates which some regard as being unsustainable; income disparities are rising; and many natural resources are being depleted beyond natural replacement rates. Technology development continues to outstrip the pace of administrative, legal and human systems to manage change. As a result, civic society might be expected to show increasing signs of stress. Already there have been many examples of this, such as the urban riots in a part of Los Angeles in the early 1990s, and more recently groups of protesters disrupting international business forums–such as those sponsored by the World Economic Forum (WEF) – demonstrating against the negative impact of globalization and the role of trans-national corporations.

Rapid change and a quest for greater competitiveness, combined with a growing global awareness over the need to achieve improved sustainability of development, have presented policy makers and businesses in all parts of the world with a paradox (Naisbitt 1994). Regional economies now more than ever face the challenge of how to balance becoming more competitive while at the same time needing to respond to pressure from local communities to maintain identity and to conserve and protect elements and values of cultural and natural heritage from exploitation. Those issues certainly present a formidable challenge to the sustainability of the so-called 'new' economic order, and have been seen by organizations such as the World Bank as particularly pressing issues in the mega city regions of the developing world (Cohen 1995).

The paradox of the so-called 'new economy' that has emerged as a result of the structural transitions that have been occurring since the 1970s, and which have gathered pace during the contemporary era of globalization, is that rapid change is making it very difficult to plan for the future with any degree of confidence, certainty or reliability. The dynamics of the processes of growth and development of the 'new' economy era challenges the very basis of planning—that is, being able to set desired outcomes and to manage processes to achieve them. In the past, analysts could anticipate certain aspects of the future with greater reliability, as business environments and societies were much more stable and often controlled. Governments and business developed infrastructure and services with the reasonable expectation that targeted economic outcomes would be achieved. But it is no longer the case that economic development may be planned with such confidence. The architecture that shaped the 'old' economic order and its relative certainty and security has been undergoing a metamorphosis to be replaced by more open, flexible and virtual structures, where the rules of engagement are in a state of constant flux and rapid change.

According to Florida (1995) the post-industrial modern economy

...is entering a new age of knowledge creation and continuous learning and that the territorial content plays an important part, it follows that regions are becoming focal points for knowledge creation and learning (p. 528).

He goes on to say:

....the shift to knowledge-intensive capitalism goes beyond the particular business and management strategies of individual firms. It involves the development of new interests and a broader infrastructure of the regional level on which individual firms and production complexes of firms can draw. The nature of this economic transformation makes regions key economic units in the global economy (p. 531).

Thus, during the last decade or so, there has emerged an increasing interest in the concept of the *learning region* and its implications for territorial production systems. The notion is that "maintaining competitive advantage follows from the creation of non-spatial resources which was contributed through learning processes" (Maillat and Kibir 2001, p. 255). According to Florida (1995), learning regions

...function as collectors and repositories of knowledge and ideas, and provide an underlying environment of infrastructure which facilitates the flow of knowledge, ideas and learning. Learning regions are increasingly important sources of innovation and economic growth, and are vehicles for globalization. (p. 528)

In this new-age society of knowledge-based, information-intensive economic activities, regional economic development will be influenced more significantly not only by exogenous factors but also by endogenous factors as well. Globally we are moving into an age were organizations and business need to learn to anticipate and manage in a flexible manner their economic futures rather than try to determine or control those future outcomes. Economic outcomes in the future will be managed through alliances and partnerships that combine values and knowledge rather than be based on grand plans and interventionist policies. Desired economic outcomes in the future will be won more on arguments, knowledge and persuasion. Thus, regional economic development analysts, planners and decision makers face increasing uncertainty concerning economic outcomes, and there is a growing debate and conflict over desired economic goals and values. This poses a great challenge to established systems of governance.

Many of the assumptions and approaches that regional scientists, planners and economists concerned with modelling and analysing economic growth and development have used in the past to analyze and plan for the development of regions may no longer be valid. While some groups in society argue that there are growing limitations on natural resources to service expanding populations and the demands of citizens for improvements in the quality of life for all, others argue that this is not so much a problem because of rapid technological innovation and faster take-up, and the substitution principle. But whoever is right or wrong, our thinking about competitiveness and its relationship to the development of regions must change. In Chap. 1 it was discussed how recent approaches to competitiveness are focused on *competitive advantage*. Such an approach is prone to overlook the need for development to be sustainable and to ensure greater equity for communities from the benefits of economic development. The challenge for policy makers in future will be to ensure more sustainable regional economic development outcomes.

The search to enhance regional competitiveness has been resulting in increased collaboration between firms, and between public, private and community sectors (Brown and Eisenhardt 1998; Collis and Montgomery 1995; Sohal et al. 1998). This is being driven by the need of business and government to increase efficiencies in the face of competition and often declining resources. Emerging from this push is a paradigm which places an emphasis on *collaborative competition*. Collaborative competition has created a new kind of business architecture which is formed by powerful networks, alliances, franchises and joint ventures—many of which are *virtual*. The emphasis on the architecture used to sell the comparative advantage of built infrastructure and generous public incentives to buy investment and jobs in regions will be less important policy considerations in this future state.

This emerging new paradigm for regional economic development favours regions that have the human capital and technologies necessary to generate new knowledge and information—and to competently reinvent it—that creates the strategic architecture to facilitate a services dominated economy. The new paradigm will force regions to closely examine:

(a) what factors give them competitiveness? and
(b) how they may combine resources, competencies, assets, human capital and knowledge to become more efficient to create environments favourable for new economic development opportunities?

This paradigm assumes added significance when linked to another new paradigm being driven from an agenda that relates to ecological and resource capacity to support human populations and greater demands for economic equity. This is one of the great issues facing governments and communities throughout the world, including regional economies in the future. How we ensure economic development is more sustainable and equitable without forfeiting competitive advantage is uncertain and represents a significant challenge in formulating regional development strategies and for implementing plans.

10.2 Future Directions of Regional Economic Strategy

During the 1990s, thinking on the new paradigm based on the concept of *collaborative advantage* (Huxham 1996) has advanced considerably, based on the premise that *networks, alliances,* and *partnerships* are the replacing the more interventionalist strategies of the past, and these changes in strategic thinking (Hamel 1996; Mintzberg and Quinn 1992) are becoming more important and are being more fully incorporated in regional economic development strategy under contemporary conditions of globalization, fast and flexible change.

Strategy itself is also undergoing a major transformation, with a clearer separation of planning from strategy (Mintzberg 1994). As discussed in Chap. 5, strategy is now more the framework that provides a path for development and a direction for planning; it precedes planning. *Planning* is now being seen more as the instruments providing the mechanism for strategy implementation. Future regional economic development strategy will likely have three important functions. These are to:

(a) Identify key *elements of capacity building* to support economic development— the key elements of capacity building are what termed the strategic architecture of a region.

(b) Define *strategic intent*, in terms of the direction, destiny and discovery of opportunities for economic development in a region.

(c) Define the main thrusts of strategies for achieving strategic intent for managing economic development processes and for capacity building in a region—setting a *future path.*

The role of planning is to provide the details of initiatives, actions, resources, management, timing and delivery of resources, infrastructure, competencies and other supporting structures to execute strategy. Strategy is thus continuous and dynamic, while planning is more methodical and applied to the achievement of specific projects and outcomes.

10.2.1 The Importance of Strategic Architecture in Supporting Strategic Intent

In Chaps. 5 and 8 it was argued how it is necessary if a region is to be successful and competitive in the future for regional economic development strategy to be directed towards achieving continuous capacity building to deliver clearly defined strategic outcomes. Capacity building for regional economic development involves the progressive construction of a kind of *strategic architecture* created by ongoing leveraging, stretching and development of competencies, resources and assets that are critical to supporting economic development processes. Strategic outcomes are shaped by something we have called *strategic intent*, a term that goes beyond a statement of goals and objectives used in most strategic and other planning processes. Strategic intent describes future *strategic directions* for a regional economy or development process at the local level. It conveys a *sense of destiny* about long-term markets and competitive positions. It also sets possibilities by providing a *sense of discovery* to explore new competitive and collaborative territories. Thus, strategic intent provides opportunities for choice on the *pathway to the future*. That statement of intent needs to set clear priorities for the development of the strategic architecture to support the development of regions.

Following the arguments put forward by Hamel and Prahalad (1994), *strategic architecture* might be seen as the 'game plan' or 'blue print' for realizing strategic intent on businesses strategy and development introduces diversity by capitalizing on these advantages by stretching and leveraging resources, competencies, and strategic infrastructure within and between industry sectors and clusters.

The concept of strategic architecture may be demonstrated by using the analogy of the Pompidou Centre in Paris. The centre and the streets and buildings that surround it create an environment enabling and encouraging a myriad of formal and informal events and activities to occur in that architecture that defines the place. Few of these activities are planned. The function of the place changes between day and night, and between the seasons. It is one of the most popular places in Paris as a major centre for entertainment, culture and businesses. However, it is not so much the physical architecture of the centre that makes its popular. The district is a living and learning environment for many people, and there is a high level of interaction and exchange between people who live and work in the same area. These activities - the opportunities to network and interact with other people in the centre - create a social hub that is its strategic architecture. This 'architecture' is largely invisible, but well known to those who use the place.

Much of the strategic architecture that can support the economic development of regions in the future similarly will be invisible or intangible. In the past, the strategic architecture of regional economies where their development was driven by comparative advantage, were strong in physical, financial and geographic terms. However, the strategic architecture required for regions to be successful in the 'new' economy of the contemporary global era is more in the form of technology, knowledge base and the virtual. Much greater knowledge is required about how to create that strategic architecture. This is where new tools for strategic planning—like *multi-sectoral analysis* (MSA) discussed in Chap. 7—can play a useful role in helping to develop strategic architecture by identifying:

(a) what competencies a region needs to build or maintain;
(b) what sector markets a region needs to develop or maintain;
(c) what strategic infrastructure to develop;
(d) what endowed resources to conserve and manage; and
(e) what approach to take to developing marketing intelligence.

10.2.2 The Future Thrust of Strategy for Regions Learning to Anticipate Futures

In Chap. 5 it was argued that one of the primary purposes of strategic intent is to set *paths* for the future. In formulating regional economic development strategies in support of strategic intent, a strategy needs to specify:

(a) that it creates an anticipated picture about the future that communities, business and organizations responsible for economic development can understand;
(b) what a region intends to do about shaping its economic future; and
(c) how the process of shaping and managing future economic development will be resourced and managed.

Many elements of the future are difficult to accurately predict, and sometimes they cannot even be envisaged. Predicting regional and national economic futures through methods of extrapolation need to be used with caution and with full appreciation of their limitations—or perhaps even avoided. It is becoming very difficult to accurately extrapolate futures, given the unpredictable nature of variables that impact upon regional economies from globalization, technology, legislative and societal change; but this is not to suggest that the regional development strategy should not attempt to do so. Communities, businesses and regional development organizations perhaps need to learn to anticipate a wide range of change and to visualize the multiple potential transformation impacts on their local economy. Anticipating futures involves investigating and creating avenues of economic opportunity in the environments in which a region has some influence over which it may have control. Anticipating futures involves developing measures to respond to change that may impact adversely upon communities. Anticipating futures thus involves a process of elevating communities to an improved state of awareness and to a position of greater preparedness to take advantage and manage all impacts of change. Thus, scenario generation and testing and futures simulation approaches become important tools to utilize in strategy planning for regional economic development.

In preparing strategies for regional economic development in future, it is important to identify and describe those *change agents* who are anticipated to influence both the development and future management of regions. These include considering processes of change relating to economic, technology, societal, environmental, and institutional (including legislative and governance) considerations. And in Chap. 8 we saw how leadership may be a crucial catalytic factor that enhances the ability and capacity of a region to be dynamic, entrepreneurial and successful. Analytical techniques and models discussed in earlier chapters can be of considerable value in helping regional organizations and business to evaluate current regional performance and competitiveness as well, as in developing a picture of possible futures outcomes that regional economic development strategy addresses. The purpose of strategy will be to move to guide stakeholders—including formal economic development organizations—to help them take advantage of outcomes as they evolve, and to prepare contingencies for addressing potential or actual adverse outcomes. It is impossible to cover all economic outcomes; thus development strategy should focus on future outcomes that build opportunities, enhance capacity and maintain the competitive strengths of a region.

Concurrent with a proliferation of strategies focused on the economic development of a region is the need for a second set of strategies for *maintenance*. These may be thought of as *defensive* strategies, as opposed to development strategies that are *offensive* or penetrating. Maintenance strategies should address:

(a) regional risk;
(b) maintain the efficiency of existing economic capacity; and
(c) leakage from the region.

Basically, maintenance strategies need to help a region to build immunity from undesirable externalities, both exogenous and endogenous that might weaken competitive advantage.

Since the early 1990s there have emerging four approaches to strategy (Mintzberg 1994) that are applicable to the development of regions. These are:

(a) strategy as convergence;
(b) strategy for endogenous growth;
(c) strategy as divergence; and
(d) strategy as both convergence and divergence, or dynamic strategy.

These are discussed in the sections that follow.

10.2.3 Strategy as Convergence

Some have argued that globalization is leading approaches to regional economic development strategy towards *convergence*. For a long time there has been growing evidence of agglomeration and specialization occurring in world cities (Hall 1966; Sassen 1994), which is leading to convergence of capital markets, research and development concentration, education, manufacturing and producer services (Daniels 1993; Giersch 1995; Hall 1995).

Building on the work of Porter (1991), in Chap. 6 it was shown how the identification and growth of *industry clusters* are now recognized as important elements of economic growth in regional economies. The advantage of focusing analysis on industry clusters is that they begin to develop a critical mass of competencies and smart infrastructure at a much more rapid rate than industry segregated approaches to capacity building. Clustering accelerates rates of innovation and leads to the generation of efficiencies that create competitive advantage in regions. Strategies for building industry clusters seem to be most successful when focused upon developing regional competencies in industries that already have some competitive advantage in a region. Weak competencies may be strengthened and built through leveraging and strategic alliances within and between industries, and across regions through virtual structures. The role of regional economic development organizations in cluster building is no longer just to facilitate process, but in addition to 'coach' industries to 'work as a team', respecting the skills and strengths of the individual players. Industry clusters need to work together to define common ground and direction for the development of a cluster. In Chap. 8 it was demonstrated how *leadership, networks* and *alliances* may play a key role in cementing and focusing the direction of cluster development, by focusing research and development and marketing on the frontiers of new discoveries and new markets, and by building value chains.

10.2.4 Strategy for Endogenous Growth

For a long time, regional economic development theory has told us that a stage comes in the development of regions where there is sufficient critical mass of population, technology, expertise and infrastructure to support the expansion of endogenous local industries. Many regions compete successfully with import industries and can develop into export base industries in time. Several references have been made throughout this book to *endogenous growth,* which can be stimulated in regional economic development strategies designed to enhance the existing and build new capacity. These include: developing social capital; building institutional cohesion; and ensuring strategic leadership.

Developing Social Capital. In Chap. 8 it was shown how social capital is now recognized as possibly a critical element that might enhance regional or local economic development. While there is significant debate about what we mean by the term 'social capital', it is generally agreed that it refers to the accumulative knowledge and expertise, networks of association and social cohesiveness that exists in communities that can be applied to support desired economic and social development outcomes for a region or community. Strategies for social capital development will involve support for continuous life-long learning, initiatives to improve community awareness of issues important social and economic well-being, encouraging the development of community and professional organizations and building trust amongst different groups that comprise a local community.

Building Institutional Cohesion. We saw in the discussion in Chap. 8 how this involves improving the efficiency and effectiveness and quality of services by public organizations that have a key role in facilitating endogenous or local economic development. Institutional cohesion-building strategies are those that will be directed to achieving multi-sectoral co-ordination of budgets and programs of service delivery by public organizations, the removal of institutional barriers, introduce more flexible delivery of services and place management. Place management might involve central governments allocating resources to place authorities that act on behalf of consumers in an area by selecting or buying inappropriate services from competing providers. These are all key elements of strategy for enhancing institution cohesion.

Strategic Leadership. In Chap. 8 it was shown how strategic leadership is very important for the building of social capital and institutional cohesion. It involves mentoring, training and educating a wide range of people in society to take on different leadership roles. These roles vary, but may involve entrepreneurs, political leaders and people with skills in stabilizing communities experiencing considerable disruption as the result of restructuring or loss of major industries. Identifying strategic leadership requirements involves careful analysis of the economic role played by different institutions and business, and how the leadership needs of these can be enhanced through a range of programs that will build leadership capacity.

10.2.5 Strategy as Divergence

Much strategy is increasingly focused on *divergence* or *diversification*. Diversification for many economic strategists means branching out—even in directions in which a region has no prior expertise or competence. In the past this has worked well for some regions, while for others such an approach has lead to a loss of competitiveness, with resources being spread too thinly as be of value in supporting new industry developments. The 'old' industrial economy developed and prospered in an age when national governments were very successful in stimulating and developing new industries in regions, and when businesses were often exclusively orientated towards servicing consumer demands in nationally protected economies (Dicken 1992). The 'new' economy of the post-industrial era is making it increasingly difficult for governments to plan and support those industries that belong to the services information sectors. Subsequently, many regions are on a path of becoming more specialized (Ohmae 1995; Saxenian 1994) while struggling for ideas to diversify. However the paradox is that for many of the most successful larger metropolitan regions, their economies are also diversified even though they have strong concentrations of particular services, and information and technology sector industries.

There are three basic strategies which policy makers might pursue to help diversify the economic base of regions. Each of these carries different levels of risk. The strategies involve elements of stretch and/or leverage developed by Hamel and Prahalad (1994). In Chap. 5 it was shown how:

(a) industry *stretch* involves expanding forward and backward linkages to add to the regional value chain;
(b) industry *leverage* involves cross industry collaboration with other industry sectors where there are strong possible synergies for business realizing economic development potential in white space; and
(c) industry *stretch* and *leverage* involves combinations one or more industries in value-adding and white space development.

10.2.6 Strategy as Both Convergence and Divergence: Dynamic Strategy

The concept of strategy being a continuous *interplay* between *convergence* and *divergence* was first developed by Mintzberg (1990). It involves careful time management of new initiatives to concentrate and diversify economic effort in response to endogenous and exogenous events that impact an economy. Strategies of convergence may be adopted for a period of time to build economic capacity only to expand later through initiatives involving stretch and leverage. Continuous expansion of a regional economy through leverage can sometimes rundown available resources and bring about the need for convergence to rebuild capacity. The strategic directions for economic development under strategies of continuous convergence and divergence will change at certain key points on the part of the future. Events which cause economies to substantially change direction and converge may result from actions such as the removal of tariffs and the closure of military bases, which require regional development strategies to redirect the development of a region along new paths. Divergence will occur when opportunities for new investment or natural asset development fit neatly with opportunities to stretch or leverage economic capacity to support new and innovative types of industry in a region to build new clusters or enhance existing ones. In developing strategies for convergence and divergence, constant monitoring of both the *endogenous* and *exogenous* environment is required to seek out new *opportunities* to identify and anticipate risk.

Strategies for the future need to create a stronger and richer milieu of critical ingredients that build upon and change successful formulae and which in turn strengthen *competitive advantage*. As discussed in Chap. 5, at the same time strategy needs to encourage experimentation to create or capture the *white spaces* of economic development opportunities that exists in the fertile ground at the frontiers and interface of strong regional industries. Unless regions have wealth and/or advantage to pursue strategies for economic development outside the thresholds of regional competencies, core business or natural advantage, there is a grave risk that they will meet the same demise that many highly diversified corporations in the late 1980s and early 1990s that strayed well beyond their capabilities and expertise.

10.3 Managing Risk

Risk is a matter that few firms or regions can afford to disregard if they are to survive in the competitive world of the 21st century. The types of risks that affect business, governments, regions and their communities take many forms. In businesses, the main focus of risk is financial and human capital. In government organizations, political and environmental risks are high priority concerns. In regions, the risks which have the potential to impact negatively upon an economy are much broader. Risks such as uncertainty about the future levels of regional investment, interruptions to regional communications and transport services, loss of regional markets through competition or substitution, the occurrence of natural and man-made disasters, the activities of pressure groups and political uncertainty–are events or activities which have the potential to affect, in varying degrees of severity, the economic competitiveness and overall performance of business and organizations in regions. The cumulative effect of those risks as perceived by business, investors, trading partners, governments and visitors is what we call broadly refer to as *regional risk* in this book. In Chap. 7 it was demonstrated how regional risk may be systematically assessed using the MSA methodology and how important it is for regional development strategy plans to incorporate procedures and mechanisms to ameliorate the potential impacts of those risks incorporated into regional development strategies and plans.

Regional risk will, of course, vary greatly among regions and that affects the type, duration and profitability of investment and the costs of capital as reflected in interest rates (Mason and Harrison 1995). Most organizations understand the need to develop strategies to insure or hedge against potential risk loss. However, the concept of *risk management* when applied to a regional economy is generally rather poorly understood, and is often disregarded by regional development organizations, governments, business and other decision-makers until a single or series of events occur that cause serious and often long-term irreparable damage to a region's economy. It is often not until the occurrence of events–such as a natural disaster, the sudden loss of markets, significant social disruption, or the closure of a major regional industry–that the full impact of risk to a region is realized. At the time the risk impacts, often it is too late to do anything preventative or and ameliorate the impact.

An important thrust of regional economic development strategy in the future will almost inevitably concern *regional risk management*. Just as larger corporations use a wide range of strategies to hedge against or reduce risk, business and organizations in regions will need to learn to work collectively to reduce the level of exposure of regional assets, investment, trade, social and human capital to risks. Failure to address regional risk in a contemporary global economy that is becoming more open, more competitive and increasingly dynamic will affect the credit worthiness of regions. For a long time it has been evident that there exist wide variations between nations, states and cities at any one point in time, and for a particular place over time, in the ratings of governments (including their administration) given by credit rating agencies. Thus it is important for regions to consider innovative ways to address and overcome actual and perceived levels of risk. This is a new dimension for the management for regions.

10.3.1 A Framework for Risk Management Plans

The framework that builds on the approaches to risk management analysis discussed in Chap. 7 is proposed in Fig. 10. 1. It develops a mechanism for selecting strategies and measures for risk management plans which can be applied at the regional and industry sector levels. The framework) provides a basis for industry

sectors to determine the types of risks to manage; the principles underpinning a
risk management plan; basic strategies and options which can be adopted for addressing the likelihood and impact of risks; and responsibilities for risk management.

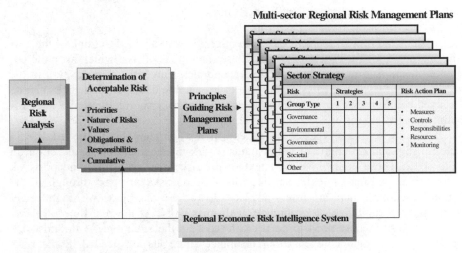

Fig. 10.1. Risk management planning framework

There are five elements to the risk management plan framework proposed in
Fig. 10.1:

(a)   analysis;
(b)   determination of acceptable risk;
(c)   principles guiding risk management plans;
(d)   multi-sector risk management plans; and
(e)   a regional economic risk intelligence system.

With the evolution of management and best practice standards, as we discussed
in Chap. 7 with respect to Australia and New Zealand Risk management Standards, the AS/NZS 4360:1999 suggests six basic strategies for risk management
that involve avoiding or reducing the likelihood and consequences, and transferring and retaining risks. These standards provide a useful basis from which to begin to develop a risk management framework and mechanisms to minimize and
manage regional economic risks. But a problem that emerged with Risk Management 4360:1999 was that risks were treated as threats. This has subsequently been
redressed in the Risk Management 4360:2004. Most of the strategies proposed by
the 1999 Standards aim to minimize or lessen the impact of specific risks. However, risks can create opportunities for business, and they can also hone competitiveness skills, which are important in order to maintain competitive advantage as
they spur on regional innovation, adaptation, responsiveness, entrepreneurship and
inventiveness. It is necessary, therefore, that risk management plans incorporate

strategies that capture opportunities for competitive advantage while, at the same time, ensuring that exposure or the consequences of risk events are minimized as far as possible.

10.3.2 Conceptualizing Risks as Cycles

To understand how to manage risks, we might be well advised to think of them as having a 'life cycle.' There are many causes and types of risk, which have the potential to impact on regional and local economies. Few risks ever develop to the level where they cause catastrophic loss or irreparable damage. Many risks can be allayed before they eventuate. Even after an event has occurred there is a need for strategies to help restore economic systems.

There are six basic types of strategies which we might apply to the risk lifecycle for regional economic or sector industry risk management plans, as set out in Fig. 10.2. These apply in the context of pre-risk time, real time, and post-risk event time, and the strategies act like filters. Some risks will pass through all the filters, assuming there are filters in place. At the pre-risk phase we might apply strategies to capitalize on, minimize or dismiss risks. At the real time of a risk event, measures may need to be taken to address the situation. At the post-risk stage, recovery and compensation/ insurance strategies are needed. Some risks will pass through all filters and the net result will be a loss situation.

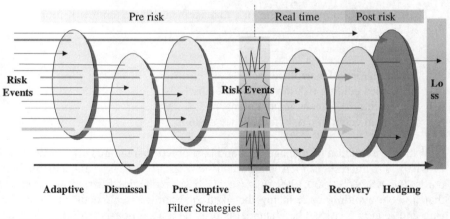

Fig. 10.2. The risk cycle strategy process

Adaptive risk strategies accept the premise that not all risks are bad or harmful occurrences. There is an assumption that certain types of risks occurring and impacting on regional economies are inevitable, but some of these are important in engendering change or innovation to replace existing systems or practices.

Dismissive strategy involves a deliberate decision to ignore or dismiss a risk. It assumes the possibility of certain types of risk is so small or large that it is not worth developing a strategy to address them.

Pre-emptive strategy involves laying out a path to minimize risks before they cause major harm. The aim is to confine risks within an acceptable range and path.

Responsive strategies are designed to reduce collateral damage. The strategy is to mount a carefully orchestrated defence against risks in order to stabilize situations as quickly as possible and to return systems to the status quo.

Recovery strategies involve developing measures to assist recovery after a major risk event or shock to an economy. One of the least-developed areas of risk management planning is post-risk event recovery.

Hedging strategies involve the use of insurance or other financial instruments in order to receive compensation or payment for loss. Hedging strategies are designed to transfer the responsibility for costs or liabilities that may be associated risk away from government, business and/or individuals.

Careful assessment is needed in developing and selecting a strategy to deal with a specific type of risk. In some cases more than one strategy may be applied depending on the nature, likelihood, impact, duration, frequency or opportunity costs.

10.4 The Evolution of Analytical Tools

Chapters 3 and 4 discussed a number of traditional basic and more sophisticated research tools that employ quantitative methods and use secondary data sources to measure and regional economic growth and development and evaluate regional performance. In Chap. 2 it was shown how a variety of both quantitative and qualitative methods have been used to develop methodologies to undertake a regional audit and to assess institutional capacity for regional and local development. In Chaps. 6 and 7 we discussed new emerging tools that involve quantitative, qualitative, and mixed or hybrid analytical techniques for analyzing regional competitiveness and assessing risk and opportunity. A wide variety of tools and methods to analyze, model and simulate possible future economic and social outcomes are available to assist in regional economic development strategy formulation and planning.

Quantitative techniques rely on the use of empirical data and tend to seek high levels of precision when used in the various applications described in Chaps. 2, 3 and 4. Quantitative techniques have been the foundation of analysis in regional science (see Isard 1956 1960). Over time they have evolved to a level where environmental, social and technological outcomes of economic processes may be modelled with a reasonably high level of precision. However, as all modelling researchers and empirical analysts are aware, models used to describe and simulate regional environments are only as reliable as the data used to run them.

Qualitative techniques applied to regional analysis have advanced significantly as the role of 'values' and 'choices' in decision-making have been increasingly recognized. Few decisions relating to economic development and infrastructure in regions will fail to take into consideration the views, attitudes and values of communities in which new investments are proposed. Nor can businesses and governments dismiss the 'attitudes', 'likes' and 'preferences' of consumers and markets in which they propose to sell or deliver products and services. Qualitative analysis has reached sophisticated levels in the area of modelling and interpretation of demands, preferences and choice assessment. It can also be used to good effect when combined with Delphi and other techniques to significantly influence decisions about future development outcomes. As demonstrated in Chap. 7, the new *Multi Sector Analysis* (MSA) tool developed by the authors has considerable potential as an informed qualitative approach to assess regional competitiveness and regional risk.

Mixed or hybrid techniques involve the utilization of both qualitative and quantitative techniques to provide explanations or direct possible outcomes from economic modelling and other analyses where data and information is imperfect or choices have to be made in the absence of information. As the world economy becomes more integrated and dynamic, the development and application of hybrid analytical tools will increase. The most significant areas of growth will be in the development of synthetic data to fill in gaps in data sets used for modelling. Synthetic data fills holes in imperfect data sets in this respect; it is not unlike creating synthetic parts to construct real life models of dinosaurs when only a few fossilized bones are recovered. Many of the synthetic parts of dinosaurs are probable or best estimates of true patterns or parts that made up the whole. The next step in the evolution and development of analytical tools for regional analysis will be the development of virtual models of regional economies. Such models will enable economic inputs to production and decision making systems to be modelled so that economic, social, environmental and development outcomes can be simulated and assessed. Those models will combine quantitative and qualitative data inputs and decision-making techniques to produce results that allow resources to be applied according to desired values and constraints (social, economic and environmental). While still in a nascent stage of development, models and analytical techniques which enable us to incorporate value systems as a path to achieving desired or probable economic outcomes will be in high demand by public policy and business analysts. There is a growing volume of research showing that it is value adding factors, the intangibles of knowledge, core competencies, and information that will drive the development of economies in future. Ways of modelling and analyzing those factors are yet in the early stages of development.

The marriage of qualitative and quantitative approaches of analysis to develop new kinds of hybrid analytical techniques to explain economic environments is on the horizon. It may never be possible to fully marry the two approaches. Adherents of quantitative and qualitative approaches often are heavily entrenched in pushing the frontiers of their respective analytical paradigms and many see little benefit in trying to combine the two approaches. But the continued imperfection of data suggests that there is much to be gained by pursuing a joint approach. The rich fields open for discovery and application for regional analysis are at the interface between qualitative and quantitative analysis. The development of mixed or hybrid tools that use combinations of quantitative and qualitative techniques, such as MSA described in Chap. 7, are tools providing useful outputs for policy analysts as well as for corporate business and community decision makers. The merging and development of quantitative and qualitative analytical tools will lead us to understand better and plan for more sustainable forms of regional economic development.

10.5 The Challenge of the Virtual Economy

The development of communication technologies which enable organizations, institutions, and individuals to operate and manage activities in virtual space and communicate in virtual reality is changing significantly the conduct of business, social and economic transactions. The impact of the virtual economy may not be as profound as some writers would have us believe. It will, however, likely necessitate rewriting the rulebook of economic development practice.

The virtual economy is the product of the information age. It is difficult to describe or define, but it is multi-dimensional and many of its elements are digital. There are many components of the virtual economy. These include:

(a) Organizations acting as brokers or catalysts for out-sourcing, coordinating, exchanging or disseminating goods and services using mainly telecommunications.
(b) The use of networks and alliances for business and other purposes.
(c) Virtual simulation of events or experiences for recreational, commercial or research purposes and face to face video conferencing between people for consultation, inquiry, education and learning.

Almost every part of society in the developed world is being touched by the virtual economy and increasingly will be in the near future, as will also become the case in the developing regions.

10.5.1 What Might Be the Impacts of the Virtual Economy?

Some of the challenges the virtual economy may have for the development of regions and communities in future are discussed below. The implications of these challenges will by no means be uniform.

Overcoming the Tyranny of Distance. Advances in transport and telecommunications will continue to shrink the dimensions and friction of distance and of time and access to information and knowledge. The ability of organizations to conduct business by minimizing transaction costs across space will increase greatly both customer and consumer choice regardless of geographic location. Thus distance may no longer be the same kind of barrier to business and markets it once was. Already its importance as a locational constraint is much lower and the significance of distance has been declining for a considerable time. The virtual economy makes the remotest parts of earth potentially accessible to individuals and corporations in a way that have never been possible before.

The virtual economy is leading to a second and more dynamic phase of globalization. The first (the commodity and merchandise phase) exposed many primary producing and manufacturing regions of the world to greater competition. The subsequent deregulation of many national economies during this phase fundamentally impacted the economic base of many regions and resulted in the loss of many traditional and nationally developed industries created during the period of post World War II reconstruction. The second phase of globalization (the service and information phase) began with the deregulation of international and national financial markets and the introduction of the Internet. This phase may have a far more profound impact on regions than the first as it can be expected to lead to greater competition and the nationalization and internationalization of regional services. For the first time in history, many industries comprising the non-basic economy of regions will be exposed to open competition from non-local providers offering universal and more specialized services via video interface systems. Distance/time costs once provided a protective 'tariff' for local service providers against external competition. The virtual economy will significantly reduce if not lift this protection. Further, the cost of distance factor has added significantly to the cost of business in regional areas.

Overcoming the cost of distance is a double-edged sword. Improved and cheaper access services will enhance the competitiveness of cities and regions; however competition will likely devastate many local IT and other service industries in less urbanized regions. Services likely to be substantially affected by the development of the virtual economy include retailing, information wholesaling, property real estate, insurance, finance, legal, IT, management services, education, health, community security and government services. The prospect that a significant proportion of once locally based services will be provided from remote locations is of increasing concern or should be for many regional organizations and businesses. In the sense that there might be a decline in jobs and even loss of some existing functions, rural regions, in particular, are seen as being 'threatened' by the virtual economy, as demonstrated by innovations in Tele-medicine, electronic banking, business, advisory and legal services and how they are sourced, priced, and scoped nationally and globally via advanced telecommunications technology. Many non-metropolitan regions that lack a critical scale of population, and many traditional services will become vulnerable to the impacts of the virtual economy. This is expected to lead to reduced demand for high level skills in some regions. With the exception of regions with high value adding production industries, many if not most non-metropolitan economies can expect to experience a decline in gross regional product per capita as the employment structure in such areas begins to reflect higher levels of low pay and routine process workers. Further, virtual services increasingly will be able to be substituted for local services that require analytical skills. This presents a significant challenge to regions already weakened by the first round effects of globalization. What we are suggesting here is that the virtual economy is likely to result in greater concentration of activity in core regions at the expense of peripheral regions in the context of this core/periphery theory of regional economic development.

The Impact of Virtual Economy will not Be Uniform. The impact of the virtual economy on regions and communities will thus not be uniform. The greatest impact of the virtual economy is likely to be in regions that have high levels of employment in industries involving business services and retailing. This will likely bring about a reduction in the demand for commercial office space - especially in non-metropolitan areas and regions where agglomeration economics are weak. That is already happening in many places, where substantial reductions in office space investments in regions that are still experiencing population growth have been observed. In other regions where there is strong clustering of high-level skills demand for office space and IT services continues to rise. In Australia and the United States, for example, where there is strong clustering of analytical skills in regions, there are heavier demands for commercial office and virtual technology services which creates demand for new and often different types of commercial office construction, including space in low density technology parks.

Regions and states which embark on supporting specialized activities that take advantage of virtual or e-commerce technology—as, for example, the competitive position created in Tele-marketing by the State of Nebraska in the United States—can benefit from the virtual economy. The virtual economy will be driven by the competency or human capital base of regions (Tapscott 1996). Thus regions that can develop strong core competencies that can be capitalized upon using telecommunications and other technologies stand to benefit most from the virtual economy.

Multi-national Organizations will Benefit Most. For multi-national corporations, e-commerce opens up new regional possibilities and markets. These organizations control over 50 percent of world trade (Korten 1995). In addition national and international franchise companies, of which there are over 1,600 in the United States, are proliferating, and are often highly suitable for e-commerce. For some service providers, no longer will it be necessary to invest heavily in capital costs to service locationally constrained and often marginally profitable markets 'on the ground' through local operations as it were. E-commerce will permit local closures of facilities, resulting in loss of jobs while customers anywhere can be supplied from centralized facilities. The other advantage for the multinationals and franchise operators is that e-commerce provides unique opportunities to acquire and on-sell the expertise of service industry specialists who choose, for life-style or other reasons, to live in non-metropolitan regions or remote regions. Many of these people can be positioned by large multi-national corporations in virtual space without frequently leaving their home to work. Likewise, individual operators can similarly locate to provide their outsourced functions to large companies and/or a range of firms. This also greatly reduces the costs to large and medium scale organizations that would normally have to relocate professionals for short periods of time from a home base to a remote location to work. Tele-conference interface systems are growing rapidly. They enable companies to operate in virtual space with employees, contractors, customers, distributors and suppliers of services, without having to travel long distances. The British Airways 'virtual centre' near Heathrow Airport in London enables staff to talk on video link to BA personnel anywhere in the airlines' global network (Parker 1999). The new virtual centre has led to a more seamless management structure, reduced operating costs and resulted in a higher level of employee satisfaction. Microsoft has been building the new Windows 2000 operating system using similar virtual reality technology, such as when Boeing built the Boeing 777 aircraft.

Cyber Communities as part of the new Virtual Infrastructure. The challenge facing local regional development organizations is how to build the infrastructure to capture opportunities created by the emergence of the virtual economy. The rapid growth in commercial and social transactions using the Internet is transforming the concept of community as a local spatial entity to an extended community in cyber space. Those cyber communities are built around ideas, beliefs, knowledge, problem issues and specialized activities and are not place specific (Tapscott 1996). The only spatial manifestation of cyber communities is a network controller's address and conference and seminars that bring like members of these communities together at specific locations.

The Internet is fostering communities of interest and associations across vast a-spatial networks that transcend national, regional, economic, social and cultural boundaries, bringing with it the recognition that technology will ultimately close the boundaries of language. The importance of networks challenges the idea that local communities can significantly influence economic activities. They can, but increasingly they are becoming dependent on cyber and other networks that enable global integration and sharing of ideas, knowledge and contact between people in regions that have similar interests. As regions become more internationalized, it will be the networks developed across cyber space that will have an important role in fostering business and social exchange. Accessing and using these networks to inform, guide and implement strategy will be a fundamental component of regional economic development. What appears to be emerging as we move towards a more virtual economy are strong local businesses (catalysts) that combine infrastructure, resources and core competencies with industries elsewhere to develop networked industries. The cyber space of community networks and alliances forms an important part of what can be referred to as virtual infrastructure. Virtual infrastructure gives business and investors in regions access to services providers that are not physically located in the region needing the services. Virtual infrastructure often involves major corporations working in alliances to provide a package of services to potential investors and developers in the regions. For local and regional services, overcoming competition created by the virtual infrastructure providers will be a challenge. Regional organizations will need to respond quickly to this challenge by developing strategic alliances and networks (including cyber space communities of business interest and expertise) that create the virtual infrastructure to match or even pose a threat to external service providers. This will necessitate regional business organizations and industry clusters working collaboratively to build virtual elements of 'smart' infrastructure to service networks of local communities. For small regions it will be important to build networks or link up with other regions that have similar sized economies or industry structures. Unless smaller regions develop a more collaborative approach to protect local industries from competition created by the virtual economy, large metropolitan service providers will decimate the local service capacity of weaker regions. There is little public sector policy can do to prevent this.

Quality Assurance. Ultimately it is the quality of services offered by individual businesses, organizations or clusters that will become the benchmark of competitive advantage in the virtual economy of the future. The major problem emerging in the virtual economy is that the quality of services is not assured or guaranteed. How does a business take back poor quality services when these can be destroyed with the press of a button. How does a customer judge the quality of service with a competitor? Much of the information disseminated in the virtual economy using the Internet is not certified and much of it contains errors. Thus, reputations of organizations and regions will be earned and lost quickly in the virtual economy depending on how well customers are perceived to be satisfied by the quality of services offered by service providers.

The Rewriting of Economic and Marketing Theory? The virtual economy will challenge many economic models and theories which have been supported by actions for more than a century. For many service industries capable of using virtual technologies, economies of scale will no longer be of concern to production costs. Information services can be duplicated many hundreds of times over, so that marginal costs of inputs are almost constant; for example, the marginal cost of producing a Microsoft software package to 100 customers is no different to that for 1,000 customers. The transaction costs associated with producing the product are the same regardless of volume or destination. The virtual economy will all but eliminate transaction time distance costs for many services, especially media, banking and insurance services. Another challenge to economic theory is that of barriers to entry to regional markets. Historically, service organizations seeking to expand trade and develop new markets in regions were faced with high establishment costs including offices and staff. The virtual economy no longer makes it necessary for organizations to open branch offices to tap new markets, thus eliminating barriers of entry into regional markets for services.

10.5.2 The Digital Divide

Widespread policy discussion is occurring about the so-called *digital divide*. The digital divide is the concept that some are benefiting enormously from the computer and information technology age while a disproportionate many are benefiting much less or are falling behind materially because they do not or no longer have the skills to find employment and an organizational niche. The truth surrounding the debate over this issue is confused because for every new story of a technology entrepreneur making a fortune on an initial public offering (IPO) there are thousands of stories of how individuals have used the Internet to break out of poverty and constrained living situations. For example, on Sunday February 13, 2000, the front page of the influential United States newspaper, the *Washington Post*, described how villagers in remote parts of China were able to find positions through the Internet in coastal areas that enabled them to dramatically improve their circumstances. Despite such anecdotes, there is significant evident that many are not participating in the digital age at levels that will improve or even maintain living circumstances. The article also tells stories of ex-cyber workers in the pre-eminent Silicon Valley region, who cannot find work and are living with the aid of public and community support.

So at the individual level there are winners and losers and they sort out, for the most part, according to education and skill. However, there is also a sorting out in geographic space with the technology regions the winners. Thus, it is no surprise that one of the most fashionable economic development strategies is focused on technology led economic development. Many regions might want to become the next Silicon Valley, but very few will.

There is another context in which the geographic digital divide emerges; namely, the relationship between a successful technology region and its hinterland, such as in the case of the Washington DC National Capital Region in the United States, where the digital benefits are concentrated in one region with minimal geographic spill-over effects. Such geographic digital divide situations are occurring in many parts of the world and beg for policy interventions to help lagging regions link to the more robust digital complete neighbouring regions.

10.5.3 The Importance of Human Interaction

There is a growing concern that the rate of economic and technology change is outstripping the ability of societies and individuals to keep pace. Income disparity in all nations and many regions is widening; so is the disparity in access to advanced technology and information. A long time ago James Burke (1978), in a television series on technology and change, warned:

...It seems inevitable that unless changes are made in the way information is disseminated, we will soon become a society consisting of two classes: the informed elite, and the rest. The dangers inherent in such a development are obvious (p. 294).

Human beings are social animals who value contact with their fellow humans. Technology has enabled the human race to advance its knowledge and skills to a level beyond the dreams of our forefathers. As our access to knowledge and information becomes more virtual, it is ironic that people seek more the experience of reality. Eco tourism is an example of an experience that can be exciting in virtual reality, but most people prefer the real thing. While the virtual economy will touch and change the lives on many people in future, it will never replace the need for people to meet and exchange ideas, goods and services in places.

Regional economic development involves much more than generating jobs, and improving the wealth and purchasing power of communities. The virtual economy breaks down one of the last barriers of competition in regions. The danger is that the virtual economy could create further polarization in society between the techno- competent and the techno-ignorant, the informed elite and the uninformed poor. These are dangers which we can ill afford to dismiss in the headlong rush to embrace the virtual economy. History teaches us many lessons about societies that have allowed gross disparities to occur in society. The virtual economy will need to be managed carefully to ensure greater equity of access to information across society. The virtual economy is an important tool that will enable many more people to gain access to services and exchange information and ideas. However, it will never replace the need for people to interact with each other in real time and space, thus ensuring the continued importance of agglomeration while simultaneous facilitating dispersal of activity.

10.6 Sustainable Development

Sustainable development is a concept now widely accepted as a desirable objective to guide urban and regional development and planning and it is also going to become more important as a principle to consider in regional economic development strategy. As discussed in Chaps. 1 and 2, it is a concept that has come increasingly to the fore, particularly since the late 1980s. It was shown how there exists a diverse range of ideas, concepts and schools of thought on what *sustainability* is and how it might be achieved.

Despite the many protocols, policies, initiatives and best practice examples to support sustainable development, many global indicators of sustainability continue to trend in a negative direction. It is the case that the Rio Earth Summit Agenda 21of 1992 (UN Department of Economic and Social Affairs 2004), the Johannesburg Summit of 2002 (UN Department of Economic and Social Affairs 2005) and the Millennium Goals (2000) (UN 2005) have demonstrated intense worldwide interest in sustainability issues. Green issues, and subsequently the 'brown' environmental issues of the urban areas, have been received increasing attention in government policies and regulations. Yet the approach in these protocols has had a limited impact on the core business activities of many public institutions and the business sector. However, many governments continue to give lip service or have refused to sign agreements such as the Kyoto Protocol on Greenhouse Gas emissions (UNFCCC 2004), which could substantially improve the sustainability of development and quality of life for all. And numerous corporations have used green imaging for promotion, but their practices are far from sustainable. Recent World Trade Organization (WTO), Global Economic Forum and World Bank meetings do demonstrate that many environmentalists and social activists perceive a less than satisfactory incorporation of their concerns in discussions on trade and industrial policy.

If sustainability is to be a meaningful societal goal, then it probably needs to have provable benefits at a personal, local, regional, and a corporate level, as well as at a national level. It is, however, at a personal and local level, not at the national or even regional level, where changes in attitudes and approaches to development will be most effectively made.

All of this might reflect a growing realization that if sustainable development is perceived as a peripheral and not a central concern, then there is little hope for enlightened development practice in the future. As a result, sustainable development may need to be increasingly focusing on the following:

(a) Value adding through increasing the economic, social and environmental benefits of projects and programs.
(b) Viewing the social and environmental consequences of economic activity as part of the solution, not as costs or problems.
(c) Encouraging organizations to understand how to optimize value-adding multipliers and value change.
(d) Ensuring investments become more strategic in order to gain from the benefits of cumulative causation.
(e) Persuading organizational management to become increasingly concerned with the management of externalities and balancing this with the internal local values and capacity to manage systems.
(f) Promoting multi-sector approaches to problem solving and decision making which will involve building completely new systems of governance.

The different values held in different societies, and the way within a given society those values might change over time with respect to the use of social capital, environmental capital, and economic capital, is depicted on the axes in Fig. 5.11 in Sect. 5.4.4. At the regional level, community interest has a strong influence on these values. The current approach to regional development might be represented by the triangle with a strong economic bias. The equilibrium or convergence approach to sustainable development suggests we must seek greater uniformity of social, environmental and economic outcomes (light grey centre triangle). If we seek to converge on solutions for sustainable development, this may require a reduction in economic expectations in order to achieve greater environmental and social outcomes.

In reality, few organizations or governments would accept a loss of economic position by reducing the standard of living and consumption to support desired social and environmental outcomes, unless forced to do so. This is one of the principal reasons why sustainable development has failed to gain support from many businesses and organizations involved in maximizing economic outcomes. Many organizations continue to see investment in human and social capital and the environment as a cost to business and not an investment in an asset.

If it is possible to use environmental and social multipliers to enhance economic multipliers, the threshold triangle of sustainability can be enhanced. This is shown in Fig. 5.11 where value adding occurs along the three axes of the sustainable development triangle. By encouraging greater leveraging of the three forms of capital, it is possible to increase the multiplier effect and expand the sustainable development triangle along the three value systems.

10.6.1 Using Sustainability as a Tool for Collaborative Competition

As we discussed in Chap. 2, an approach to sustainable development which aims to increase the multiplier and net positive cumulative causation effect of development in a region is more likely to win acceptance from investors, developers, economic development agencies and economists, provided it can be demonstrated in practice. A significant problem with using sustainability as a tool to develop competitive advantage for a region is the tendency for public institutions to close down opportunities for innovative solutions to development problems. Because of uncertainty and the perceived risks associated with innovative ideas and approaches to development, too often many good development solutions are rejected unless someone else has tried it and it may be demonstrated to have succeeded. Our thinking on sustainability will need to change if we are to make the concept work to support regional economic development.

How to develop and maintain robust competitive economies which promote sustainable development is something regions need to learn to do. However, regional administration and government institutions are often placed in the difficult position under decentralization of making tough choices in the trade-off between encouraging the development of competitive regional economies and addressing significant sustainability issues. Many see competitiveness and sustainability as opposing goals. This is because our understanding of competitiveness and sustainability has been somewhat clouded by conventional thinking. Nature teaches us that competition is important to sustainability as it enables natural systems to respond to change and to adapt. The challenge is for us to better understand how competitiveness can be used to induce positive change and innovation as part of the sustainability process.

Much contemporary thinking on regional competitiveness is postulated by economic development models and the concepts discussed in Chap. 1. The focus typically is on encouraging competitive economies to grow and gain greater market share. This introduces a notion that there will be 'winners' and 'losers'. Industry attraction theory and growth poles subscribe to this model of competitiveness, and in the long term are not sustainable. Growth pole models, especially if they are supply driven, result in labour and investment loss in poorer regions. The most successful growth poles are those which are demand driven and capitalize on their competitive advantages.

The emerging thinking on sustainability and regional development involves the *collaborative competition,* a concept we have referred to on numerous occasions throughout this book. Under this concept, regional competitiveness should not be concerned about 'winning' or 'losing' over who gets what share of the development pie; rather the focus should be on using collaboration to innovate and create a bigger and different pie for all to benefit, while recognizing that some regions/firms will benefit more from collaboration than others. The concept of collaborative competition encourages local competition to improve business and institutional efficiencies and cooperation between firms and institutions to create the economies of scale necessary to compete against larger corporations. In introduces the idea of collaborative advantage where there is an emphasis on inter and intra-regional collaboration and trade between firms, governments and institutions involving partnerships, strategic alliances and clustering for mutual advantage (Huxham 1996; Bradshaw and Blakely 1999).

The strategic intent of collaborative competition is to stretch and leverage resources (Hamel and Prahalad 1994). Its aim is to minimize resource inputs into production systems, reducing consumption and the development and restoration of total capital stock (that is, natural capital, social capital and environmental capital). Collaborative competition spawns myriads of possibilities, offering opportunities for increasing returns which underlies much of the thinking of endogenous growth theory. It provides opportunities to add value to production, distribution and marketing chains, thus stimulating innovation and product multipliers. In revisiting old ideas, technologies, products and processes it also provides opportunities to improve efficiencies. The concept of collaborative competition is thus entirely compatible with the goals of sustainability.

Common sense tells us that regions should be making better use of resources, increasing productivity, and ensuring the sustainability of development. The conventional thinking about sustainability probably need to be challenged; and innovative ideas and solutions to problems, which will ensure sustainable regional development outcomes, need to be encouraged. The evidence from the experience of highly developed regional clusters and networks would tend to indicate that regions (especially smaller regions) that have moved towards a focus on collaborative advantage and collaborative competition have been enjoying long-term sustainable growth which has had high multiplier effects on the development of natural and social capital. The evidence available from regions such as Silicon Valley and northern Italy, which have had a strong focus on collaborative advantage, demonstrates the importance of supporting best practices which encourage innovation and are underpinned by regional development principles.

10.6.2 The Urban Metabolism Model and Livability

A significant and as yet unresolved challenge is how to measure sustainable urban and regional development and how to operationalize the concept. Progress towards this end might be assisted through consideration of the urban metabolism model because of the particular framework it provides assisting us to consider issues to do with livability and quality of life of human settlements. The urban metabolism model incorporates the ecological, economic and socio-cultural approaches inherent in the integrated approach to sustainable development discussed above.

The urban metabolism approach may be seen as enhancing our understanding of the functionality of a city, its evolution, growth and performance, and as providing a conceptual basis for deriving quality of life measures. Figure 10.3 represents the urban *metabolism* metaphor (as conceptualized in Kuroda et al. 1993).

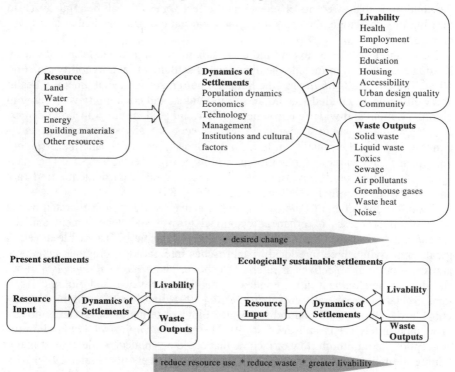

Fig. 10.3. The extended metabolism model of human settlements
Source: Newman et al. 1996

The term metabolism as applied to a city or region stems from the work of Wolman (1965) and it refers to "the total flow of material into and out of the [urban] system" (Boyden et al. 1981, p. 11). What takes place in a city in terms of population movements, the activities of households, firms and institutions and their use of resources and energy, is seen to transform a city through the process of urban growth, and in particular suburbanization.

Cities may be described as a 'metabolic process' in which:

(a) there are inputs of people, services and energy;
(b) these inputs are transformed into a distinctive quality of life; and
(c) people, activities, products and services generate wastes as outputs.

Obviously cities or regions vary in the efficiency of this metabolic process, some achieving a high quality of life for their inhabitants by utilizing relatively little energy and generating minimal waste, while producing high value-added activities. Others achieve a low quality of life by exploiting innate quantities of energy and omitting copious waste products, while producing low value-added products (Kuroda et al. 1993). Thus a challenge for contemporary society is how to measure systematically the character of this metabolic process and to identify the conditions that affect its efficiency and effectiveness. It can be argued that in meeting this challenge, the capacity to plan and build more sustainable cities will be realized.

Newman et al. (1996) extend the basic urban metabolism model proposed by Kuroda et al. (1993) to include the dynamics of settlements and livability (see Fig. 10.3). They define *livability* as "the human requirement for social amenity, health and well-being and include the notions of individual and community well-being in both the human and wider environment" (Newman et al. 1996, p.34). *Settlements* are seen as places where "communities evolve economically and socially in an attempt to improve their quality of life. The products of economic and industrial developments have value for livability as well as environmental impact...The management of the built environment and the natural environment need not necessarily be in conflict" (Newman et al. 1996, p. 34).

Central to the urban metabolism metaphor is the *quality of life* generated by the metabolism processes of urban and regional growth and development and the functionality of a city or region. The objective is to achieve sustainable development, embodying three interlocking approaches integrating ecological, economic and socio-cultural objectives as referred to above. The *ecological approach* views sustainable development in the context of stability of physical and biological systems crucial to the overall ecosystem (Perring 1991); but it is contentious that sustainable development is contradictory in ecological terms as ecosystems do not grow indefinitely (Daly and Cobb 1989). The *economic approach* is based on the concept of the minimum flow of income that could be created, while at least maintaining the renewable stock or assets that yield those benefits (Maler 1990). Underlying this is the notion of optimality and economic efficiency in the use of scarce resources. The *socio cultural approach* aspires to maintain the stability of social and cultural systems, and central to this is a recognition that people are at the centre of concerns for sustainable development, being entitled to a healthy and productive life in harmony with the environment.

This holistic view of sustainability in urban development presents considerable challenges concerning the complexity of relationships, with institutions being central to the social processes both as to how we measure the parameters of sustainable development and how society develops strategies and implements actions to pursue sustainability. Thus, the sustainable development of a city or region has as much to do with changing institutional behaviour, culture and value systems as it has to do with the development of new approaches to urban planning. Newton (1997) notes how the policy debate on sustainable development in Australia has been focused on the sustainability of continuing urban growth and consumption (especially of energy), on the management of infrastructure, and urban form and structure, issues to which we return later.

It is our contention that principles of sustainability will become more important factors to build into the strategic intent of regional economic development planning, and that there will be a greater need to interface regional economic development planning with broader regional planning strategy frameworks. Kemp et al. (1997) provide a broader discussion of how the urban metabolism framework can be applied and the type of regional data systems that are needed to implement the concept and to measure regional performance against sustainability criteria. The type of spatial decision support systems discussed in Chap. 9 are useful in this context.

10.7 Being Fast and Flexible in Search of Sustainable Development

It is for certain that we live in a rapidly changing and an increasing uncertain and competitive world; a world where competitive advantage is being transformed by new paradigms of collaborative advantage and sustainable development. This book we has introduced the reader to a wide range of new concepts and ideas for economic regional analysis and strategy which are useful for measuring and evaluating this transition and for formulating development approaches, policies and programs for planning how to position a region to be successful in the future. To illustrate a large number case studies and examples from around the world have been used. While much of the material contained in the book is related to developed countries, many of the tools and strategies outlined will have wide application in developing and newly industrialized nations. Over the past 50 years the focus of regional economic development has moved beyond the need to generate more employment, industries, infrastructure and housing. Increasingly, economic development is focused on improving the total well-being of people in communities and the environment in which they live. Many people and even whole communities are beginning to question the value of economic development if societies and communities are not safe and healthy, enjoying a good quality of life, and being allowed to develop in areas such as culture, art and sport, that provide meaning, happiness and a sense of community. The search for more sustainable approaches to economic development has resulted in social and environmental values being embraced as an integral part of the desired outcomes for regional development and for regional economic development strategy to be oriented towards a region taking responsibility for its own destiny, as discussed in Chap. 5.

The increasing emphasis being placed on the role of endogenous forces in regional economic development raises significant issues for the role of policy and strategy and their meaning. Karlsson et al. (2001, pp. 5-6) pose these questions:

...What is the role for national policy? Is there a role for national policy? Are top-down policies capable of providing desired regional dynamics or facilitation the achievement of national goals? Are bottom-up policies capable of impacting national goals positively or negatively? Further, are there combinations of top-down, inside and bottom-up policies that offer the best guide to development regionally or nationally? Finally, as with all policy it is ultimately dependent upon culture and values, which vary significantly from region to region and from country to country.

Another important aspect of the role of policy in regional economic development is emphasized by Karlsson et al. (2001):

...In the old economy of the industrial age, policy from higher levels of government was used to drive development but with mixed results. In the new economy of the later 20th century, development is often driven most efficiently by endogenous self-organization and self-adjustment processes. Thus, there is an important question regarding the degree to which it is possible to create and implement policies that will efficiently and effectively induce self-organizing and adjustment processes. Information technology, capital in a global market context, flexible production systems, freeing of trade from restrictions, life-long learning and flexible employment, speed and agility in the value delivery chain, and smart infrastructure are hallmarks of the new economy. Institutionally and express-through governance, we are currently rather inept in building these further into our regional economic development policies and strategies (p. 6).

There is an implicit recognition in this book that we are in the process of constructing or transiting to a 'new' kind of economy. This is an economy that will be shaped by the proliferation and diffusion of information, knowledge, and technology, and the forces of societal change and of environmental values. No one can predict the exact nature of that future. Historically we have tried to use protective barriers to shape the future to suit our own ends. It has been evident since the 1970s that this is no longer a feasible approach. We need to learn to better anticipate and manage the future. The tools and strategies canvassed in this book are intended to be used by students and practitioners of regional economic development to help them both to better understand the present and to better anticipate and plan for, and to manage the future. However, considerable work still needs to be done to develop new analytical tools and to improve the processes for the development of strategies and plans to manage regional economic development. These will, of course, continue to evolve and it is inevitable that advances will be achieved over time. With each advance in planning strategy formulation and its implementation, hopefully we will be better able to equip regions with the wherewithal to be fast and flexible in their search for sustainable development, and for them to be competitive and collaborative in a rapidly changing world. Much remains to be done to address this exciting challenge.

Figures

Tables

References

Aacker D (1998) Strategic market management. 5th edn, John Wiley, New York

Abel T, Stough RR (1999) Social capital and regional economic development. Paper Presented at the Southern Regional Science Association Meeting, April 15-17, Richmond, VA

Abramson G (1998) Cluster power. CIO Entreprise Magazine 11: 48–57

Adams B, Parr J (1997) Boundary crossers: case studies of how ten American metropolitan regions work. Research Report. University of Maryland, Baltimore

Advisory Committee to the White House (1978) White House conference on balanced national growth and economic development. Department of Commerce, Washington, DC

Australian Housing and Urban Research Institute (AHURI) (1995) The internationalisation of the Far North Queensland regional economy: final report. Part 1: Regional trends analysis and part 2: Capabilities, capacities and potential for economic and trade development. Report to the Far North Queensland region economic development strategy. Australian Housing and Urban Research Institute, Brisbane

Alexander JW (1954) The basic-nonbasic concept of urban economic function. Economic Geography 30: 246–261

Ali AI, Lerme CS (1990) Determination of comparative advantage for the economy of states in the U.S. (Mimeograph), the University of Massachusetts, MA

Ali AI, Lerme CS, Nakosteen RA (1993) Assessment of Intergovernmental revenue transfers. Socio-Economic Planning Sciences 27 (2): 109–118

Amin A (1999) An institutionalist perspective on regional economic development. International Journal of Urban and Regional Research 23 (2): 365-78

Amin A, Goddard J (1986) Technological change, industrial restructuring and regional development. Allen and Unwin, London

Amin A, Housener J (eds) (1997) Beyond market and hierarchy: interactive governance and social complexity. Edward Elgar, Aldershot

Andrew BP (1997) Tourism and the economic development of tourism. Annals of Tourism Research 24 (3): 721–735

Andrews K, Swanson J (1995) Does public infrastructure affect regional performance. Growth and Change 26: 204–216

Anselin L, Bera A (1997) Exploratory spatial data analysis linking SpaceStat and ArcView. In: Fischer M, Getis A (eds) Recent developments in spatial analysis. Springer, Berlin Heidelberg New York, pp 35-59

Anselin L, Getis A (1992) Statistical analysis and Geographic Information Systems. Annals of Regional Science 26 (3): 9–33

Ansoff HI (1965) Corporate strategy: an analytic approach to business policy for growth and expansion. McGraw-Hill, New York

Arcelus FJ (1984) An extension of shift-share analysis. Growth and Change 13 (1): 3–8

Arena PM, Stough RR, Trice M (1996) The contribution of Newport news shipbuilding to the Virginia economy. Center for Regional Analysis, School of Public Policy, George Mason University, Fairfax, VA

Armstrong P (2001) Science, enterprise and profit: ideology in the knowledge-driven economy. Economy and Society 30 (4): 524-552

Arrow KJ (1962) The economic implications of learning by doing. Review of Economic Studies 29: 155-173

Arthur WB (1994) Increasing returns and path dependency in the economy. University of Michigan Press, Ann Arbor, MI

Aschauer D (1989) Is public expenditure productive? Journal of Monetary Economics 39 (41): 41-57

Ashby DL (1960) The shift-share analysis of regional growth: a reply. The Southern Economic Journal 33: 577–581.

Audretsch D, Feldman M (1996) Innovative clusters and the industry life cycle. Review of Industrial Organisation 11 (2): 253–273

Banker RD, Charnes A, Cooper WW (1984) Some Models for estimating technical and scale inefficiencies in Data Envelopment Analysis. Management Science 30 (9): 1078–1092

Barff RA (1987) Industrial Clustering and the organization of production: a point pattern analysis of manufacturing in Cincinnati, Ohio. Annals of the Association of American Geographers 77: 89–103

Barff RA, Knight PL III (1988) Dynamic shift-share analysis. Growth and Change 19 (2): 1–10

Barrand J, Guigou C (1984) Analyse Structurelle. University Thesis. Université Paris Dauphine, Paris

Barro RJ (1990) Government spending in a simple model of endogenous growth. Journal of Political Economy 98: S103–S125

Bartik TJ (1991) Who benefits from state and local economic development policies. W.E. Upjohn Institute for Employment Research, Kalamazoo, MI

Bartik TJ (1999) Growing state economies: how taxes and public services affect private-sector performance. In: Sawicky MB (ed) The end of welfare? consequences of federal devolution for the nation. Armonk, London, pp 95–126

Batty M , Xie Y (1994) From cells to cities. Environment and Planning 21: 531–548

Baumol WJ, Blackman SAB, Wolff EN (1989) Productivity and American leadership. MIT Press, Cambridge, MA

Beacon Council (1985) Dade county strategic plan. Beacon Community Council, Dade County, FL

Beckmann M (1968) Location theory. Random House, New York

Bell D (1976) The coming of the post-industrial society. Basic Books, New York

Bennis WG, Nanus B (1991) Leaders: strategies for taking charge. Harper and Row Publishers, New York

Bergman E (1981) Citizen guide to economic development in job loss communities. Center for Urban and Regional Studies, Chapel Hill, NC

Bergman EM, Feser EJ (1999a) Industrial and regional clusters: concepts and comparative applications. Web Book of Regional Science, RRI-West Virginia University, Morgantown, WV

Bergman EM, Feser EJ (1999b) Applying industry value-chain cluster concepts to regional development. In: Roelandt TJ.A, den Hertog P (eds) Cluster analysis and cluster-based policy. New perspectives and rationale in innovation policy. OECD, Paris

Bergman EM, Feser EJ (2000) National industry cluster templates: a framework for applied regional cluster analysis. Regional Studies 34 (1): 1–19

Bergman EM, Feser EJ (2001) Innovation system effects on technological diffusion in a regional value chain. European Planning Studies 9(5): 626-648

Bergman EM, Feser EJ, Sweeney S (1997) Targeting North Carolina manufacturing: understanding a state economy through national industrial cluster analysis. SRE Discussion Paper 55. Vienna University of Economics and Business Administration, Vienna

Bergman EM, Maier G, Lehner P (2001) Banning the Bahn: preferred transport options by firms of Austrian industrial clusters. Paper Presented at 40th Western Regional Science Association, February 2001, Palm Springs, CA

Berry BJL (1967) Geography of market centres and retail distribution. Prentice Hall, Englewood Cliffs, New Jersey

Berry BJL, Garrison WL (1958) Functional bases of the central place hierarchy. Economic Geography 34: 145–154

Berzeg K, Koran T (1978) The empirical content of shift-share analysis. Journal of Regional Science 18 (3): 463-469

Berzeg K, Koran T (1984) A note on statistical approaches to shift-share analysis. Journal of Regional Science 24 (2): 277-285

Bianchi G (1996) Galileo used to live here: Tuscany hi tech, the network and its poles. Research and Development Management 26 (3): 199–213

Bingham RD, Mier R (eds) (1993) Theories of local economic development: perspectives from across the disciplines. Sage Publications, Newbury Park, CA

Blakely EJ (1994) Planning local economic development: theory and practice. 2nd edn, Sage Publications, Thousand Oaks, CA

Blakely EJ, Bowman K (1986) Taking local development initiatives: a guide to economic and employment development for local government authorities. Discussion Paper 129, Australian Institute of Urban Studies, Canberra

Blakely EJ, Jensen R, Stimson RJ, Avery B, Robinson J (1991) An economic development strategy the Brisbane region. 12th meeting of Pacific Regional Science Conference, 7 July, Cairns, Australia

Bluestone B, Harrison B, Baker L (1981) Corporate flight: the causes and consequences of economic dislocation. Progressive Alliance Books, Washington, DC

Bolton R (1998) A critical examination of the concept of social capital. Paper presented to the 37th Western Regional Science Association Conference, 18-22 February, Monterey, CA

Bordecki M (1984) A Delphi approach. New York University, New York

Borman M, Taylor J, Williams H (1994) Common knowledge inter-organisational networking and local economic development. In: Cuadrado-Roura J, Nijkamp P, Salva P (eds) Moving the frontiers: economic restructuring, regional development and emerging networks. Avebury, England, pp 213–224

Bosscher R, Voytek K (1990) Local strategic planning: a primer for local area analysis. US Department of Commerce, Washington, DC

Boulton WR, Pecht M, Tucker W, Wenberg S (1997) Singaporean and Malaysian electronics industries. In Kelly M, Boulton WR (eds) Electronics manufacturing in the Pacific Rim. World Technology Evaluation Center, Baltimore, MD. Available at http://www.wtec.org/loyola/em/04_05.htm

Boyden S, Millar K, Newcombe K, O'Neill B (1981) The ecology of a city and its people: the case study of Hong Kong. Australian National University Press, Canberra

Boyle R (1988) Corporate headquarters as economic development targets. Economic Development Review 6 (1): 50-56

Bower L (1983) The two faces of management. Mentor Books, New York

Braczyk HJ, Cooke P, Heidenreich M (eds) (1998) Regional innovation systems: the role of governances in a globalised world. UCL Press, London

Bradshaw TK, Blakely EJ (1999) What are "third-wave" state economic development efforts? from incentives to industrial policy. Economic Development Quarterly 13 (3): 229-244

Bronfenbrenner M (1985) Japanese productivity experience. In: Baumol WJ, McLennan K (eds) Productivity growth and U.S. competitiveness. Oxford University Press, New York

Brown R, Robinson J, Gannon J, Jensen R (1994) Prefeasibility study of the gateway ports industrial development project. Queensland Premiers Department, Office of the Coordinator General, Brisbane

Brown S, Eisenhardt K (1998) Competing on the edge: strategy as structured chaos. Harvard Business School Press, Boston, MA

Brownowski J (1973) The ascent of man. British Broadcasting Corporation, London

Bruntland Commission (1987) Our common future. World Commission on Environment and Development, New York

Brusco S (1995) From Marshallian industrial districts to 21st century local production systems. OECD International Seminar on Local Systems of Small Firms and Job Creation. Paris

Bryson JM (1988) Strategic planning for public and non-profit organisations: a guide to strengthening and sustaining organisational achievement. Jossey Bass, San Francisco

Bryson JM, Crosby BC (1992) Leadership for the common good and tackling public problems in a shared-power world. Jossey Bass Publishers, San Francisco, CA

Bryson JM, Einsweiler RC (eds) (1988) Strategic planning: threats and opportunities for planners. American Planning Association, Planners Press, Washington, DC

Bunch C (1987) Passionate politics. Harper Collins, New York

Burke J (1978) Connections. McMillian, London

Burns JM (1978) Leadership. Harper and Row, Publishers, New York

Burrough P, McDonnell R (1998) Principles of Geographical Information Systems. Oxford University , Oxford

Button K, Costa A (1997) Economic efficiency gains from urban public transport regulatory reform: two case studies of changes in Europe. Annals of Regional Science 33 (4): 425-438

Byars L (1987) Strategic management - planning and implementation. HarperCollins, new York

Calavita N, Caves R (1994) Planners' attitudes toward growth: a comparative case study. Journal of the American Planning Association 60 (4): 483–500

Campagni RP (1995) The concept of innovative milieu and its relevance for public policies in European Lagging Regions. Papers in Regional Science 74 (4): 317–340

Campbell H, Stough RR (1994) Forecasting the future of Northern Virginia. In: Stough RR (ed) Proceedings of the Second Annual Conference on the Future of the Northern Virginia Economy, Fairfax, VA. Center for Regional Analysis, The Institute of Public Policy, George Mason University

Carnoy M (1993) Multinationals in a changing world economy: whither the nation-state. In: Carnoy M, Castells M, Cohen S, Cardoso F (eds) The new global economy in the information age: reflections on our changing world. The Pennsylvania State University Press, Pennyslvania, pp 45–96

Carnoy M, Castells M, Cohen S, Cardoso F (eds) (1993)The new global economy in the information age: reflections on our changing world. The Pennsylvania State University Press, Pennyslvania

Carson R (1962) The Silent spring. Fawcett Crest, New York

Castells M (1989) The information city, economic restructuring and urban and regional process. Basil Blackwell, Oxford

Castells M, Hall P (1994): Technopoles of the world: the making of 21st century industrial complexes. Routledge, London.

Chang P-L, Hwang S-N, Cheng W-Y (1995) Using Data Envelopment Analysis to measure the achievement and change of regional development in Taiwan. Journal of Environmental Management 43: 49–66

Charnes A, Cooper WW, Rhodes EL (1978) Measuring the efficiency of decision making units. European Journal of Operational Research 2 (6): 429–444

Charnes A, Clark CT, Cooper WW, Golany B (1985) A development study of data envelopment analysis in measuring the efficiency of maintenance units in U.S. air forces. Annals of Operation Research 2: 95–112

Charnes A, Cooper WW, Li S (1989) Using DEA to evaluate relative efficiencies in the economic performance of Chinese cities. Socio-Economic Planning Sciences 23 (6): 325–344

Chou YH (1997) Exploring spatial analysis in Geographical Information Systems. Onward Press, Santa Fe, New Mexico

Chrisman N (1997) Exploring Geographic Information Systems. John Wiley and Sons, New York

Christaller W (1933) Die zentralen Orte in Süddeutschland. Fischer, Jena

Church RL, Murray AT (1999) GIS, business site selection, and locational analysis (Unpublished manuscript)

City of Brisbane (1991) The Brisbane Plan: A City Strategy. The City of Brisbane

Clarke M (1990) Geographical Information Systems and Model based analysis: towards effective decision support systems. In: Scholten HJ, Stillwell JCM (eds) Geographic Information Systems in urban and regional planning. Kluwer, Dordrecht, pp 165–175

Clarke KC, Hoppen S, Gaydos L (1997) A self-modifying cellular automaton of historical urbanization in the San Francisco bay area. Environment and Planning B 24: 247–261

Clingermayer JC, Feiock RC (2001) Institutional constraints and policy choice: an exploration of local government. State University of New York Press, New York

Clones D, Keeney L, Anderson B, Barbarin A, Manley D (1988) Development report card for the states. 12th edn, Corporation for Enterprise Development, Washington, DC

Cochrane R (1992) Understanding Local Economic Development in a Comparative Context. In: Wolman H, Stoker G (eds): Economic Development Quarterly 6 (5): 415

Cohen MA (1995) Cities and the prospects of nations. In: Cities and the new global economy. Proceedings of the International Conference Presented by the OECD and the Australian Government, 1: 165–172

Cohen S, Fields G (1999) Social capital and capital gains in Silicon Valley. California Management Review 41 (2): 108-130

Coleman J (1988) Social capital in the creation of social capital. American Journal of Sociology 94: 95-120

Collis D, Montgomery C (1995) Competing on resources: strategy in the 1990s. Harvard Business Committee for Urban Economic Development, Washington, DC

Commoner B (1972) The closing circle. Confronting the environmental crisis. Jonathan Cape, London

Constanza R (1994) The value of the world's ecosystem services and natural capital. Nature 387: 253-260

Conway RS Jr. (1990) The Washington projection and simulation model: a regional inter-industry econometric model. International Regional Science Review 13: 141-166

Cook LM, Munnell AH, Bodie Z, (eds) (1991) How does public infrastructure affect regional economic performance? Is there a shortfall in public capital investment? Federal Reserve Bank of Boston Conference Series No. 34. Boston, MA

Cooke P (1996) Reinventing the region: firms, clusters and networks in economic development. In Daniels P, Lever W (eds) The global economy in transition. Longman, Harlow

Cooke P, Morgan K (1993) The network paradigm. Environment and Planning D 11: 543–564

Cooke P, Morgan K (1994) Growth regions under duress: renewal strategies in Baden-Württemberg and Emilia Romagna. In: Amin A, Thrift N (eds) Globalization, institutions and regional development in Europe. Oxford University Press, Oxford, pp 91–117

Cooke P, Morgan K (1998) The associational economy: firms, regions, and innovation. Oxford University Press, Oxford

Cooper WW, Li S, Tirupati D (1995) Technology choice with stochastic demands and dynamic allocations: a two product analysis. Journal of Operations and Production Management, Special Issue on the Economics of Operations Management 15(9): 70-88

Council for Urban Economic Development (CUED) (1992) Census of economic development organisations. Washington, DC

Crainer S (1997) The ultimate book of business quotations. Capstone, Oxford

Crook C (1996) Strategic planning in the contemporary world: nonlinearity, complexity, and incredible technological change. American Planner 26 (1): 24–35

Cunnington B (1991) The strategic alliance: a clever tool for a clever country. The Australian Entrepreneur 2(1): 35–47

Curry L (1967) Central places in the random spatial economy. Journal of Regional Science 7: 217–238

Cyert RM, March JG (1963) A behavioural theory of the firm. Prentice Hall, Inc., Englewood, New Jersey

Czamanski S (1964) A model of urban growth. Papers of the Regional Science Association 13: 177–200

Czamanski S (1965) Industrial location and urban growth. Town Planning Review 36: 165–180

Czamanski S (1976): Study of spatial industrial complexes. Institute of Public Affairs, Halifax, Nova Scotia

Daly H, Cobb J (1989) For the common good. Beacon Press, Boston

Daly M, Roberts B (1998) The application of porters diamond model to evaluate the competitiveness of tradable services in a regional economy: a case study of tropical North Queensland. Paper presented to the 37th Annual Conference of Western Regional Science Association, 18-22 February, Monterey, CA

Daniels P (1993) Service industries in the world economy. Blackwell, Oxford

Davidson B (1998) Historic analogy: a way through the institutional maze. Australian and New Zealand Regional Science Association Annual Conference, September, Tanunda, South Australia

Dawson J (1982) Shift-share analysis: a bibliographic review of technique and applications. Vance Bibliographies, Monticello, IL

De Waele M (1998) The management of policy making and implementation : conceptualising development. Human Systems Management 17:1- 14

Degreene KB (1993a) Evolutionary structure in the informational environmental field of large scale human systems. Journal of Social & Evolutionary Systems 16 (2): 215–230

Degreene KB (1993b) The growth of exhaustion. European Journal of Operational Research 69: 14–25

de Groot H, Nijkamp P, Stough RR (eds) (2004) Entrepreneurship and regional economic development: a spatial perspective. Edward Elgar Publishing, Cheltenham Camberley Northampton

Delbecq AL, Van de Ven AH, Gustafson DH (1975) Group techniques for program planning, a guide to nominal group technique and Delphi processes. Scott Foresman, Glenview, IL

Densham P (1994) Integrating GIS and spatial modelling: visual interactive modelling and location selection. Geographical Systems 1: 203–219

Department of Foreign Affairs and Trade (1994) Country economic brief. Briefs for China, Korea, Hong Kong, Indonesia, Singapore, New Zealand, the Philippines, PNG and Thailand. Department of Foreign Affairs, Canberra, Australia

Department of Housing and Urban Development (1996) America's new economy and the challenge of cities. HUD Report on Metropolitan Economic Strategy, Department of Housing and Urban Development, Washington DC

DeSantis M (1993) Leadership, resource endowments, and regional economic development. Ph.D. Dissertation, Fairfax, VA, George Mason University

DeSantis M, Stough RR (1999) Fast adjusting urban regions, leadership and regional economic development. Region et Développement 10: 37-56

Dicken P (1992) Global Shift: The internationalisation of economic activity. 2nd edn, Guilford Press, New York

Dinc M, Haynes KE (1998a) International Trade and shift-share analysis: a specification note. Economic Development Quarterly 12 (4): 337-343

Dinc M, Haynes KE (1998b) International trade and productivity in the European Union and its impact on employment change. Working paper, the School of Public Policy, George Mason University, Fairfax, VA

Dinc M, Haynes KE (1999) Sources of regional inefficiency: an integrated shift-share, data envelopment analysis and input–output approach. Annals of Regional Science 33: 469-489

Dinc M, Stough RR (1997) Intertemporal DEA: Using DEA to examine local government efficiency in Fairfax, County, Virginia, U.S.A. Working Paper, the Institute of Public Policy, George Mason University

Dinc M, Haynes KE, Qiangsheng L (1998) A comparative evaluation of shift-share models and their extensions. Australasian Journal of Regional Studies 4: 275-302

Dinc M, Haynes KE, Tarimcilar M (2001) Integrating decision support models for regional development policy making: an exploratory step. (Unpublished Manuscript), School of Public Policy, George Mason University, Fairfax, VA

Doeringer P, Terkla D (1995) Business strategy and cross-industry clusters. Economic Development Quarterly 9 (3): 225-237

Doig J, Hargrove E (1987) Leadership and innovation: a biographical perspective on entrepreneurs in government. John Hopkins University Press, Baltimore, MD

Downs A (1992) Stuck in Traffic. The Brookings Institution, Washington, DC

Drucker, P (1991) The changed world economy. In: Scott Fosler R (ed) International City. Management Association, Washington, DC

Dunn ES (1960) A statistical and analytical technique for regional analysis. Papers of the Regional Science Association 6: 97–112

Durlauf SN, YH Peyton (eds) (1945) Social Dynamics. Brookings Institution Press, Washington, D.C. [(2001),MIT Press, Cambridge MA]

Elkington J (1997) Cannibals with forks: the triple bottom line of 21st century business.. Capstone, Oxford

Emergy F, Trist E (1965) The casual texture of organisational environments. Human Relations February: 21–31

Emmerson R, Ramanthan R, Ramm R (1975) On the analysis of regional growth patterns. Journal of Regional Science 15: 17–28

Enache M (1994) EMSPlan. EMI Consulting, Bethesda, MD

Enache M(1995) Quick tour of decision software: decision for Windows version 1.3, computer program. George Mason University, Institute of Public Policy, Fairfax, VA

Enright M (1995) Organisation and coordination in geographically concentrated industries. In: Lamoreaux NR, Raff DMG (eds) Coordination and information: historical coordination and information: historical perspectives on the organization of enterprise. National Bureau of Economic Research Conference Report Series, University of Chicago Press, Chicago, pp 103–142

Enright M, Scott E, Dodwell D (1997) The Hong Kong advantage. Oxford University Press, Hong Kong

Enright M, Scott E, West J (1996) Hong Kong's competitiveness. Vision 2047. Workshop. March, Hong Kong

Erickson RA (1994) Technology, industrial restructuring and regional development. Growth and Change 25: 353–379

Erickson RA, Leinbach T (1979) Characteristics of branch plants attracted to non metropolitan areas. In: Lonsdale R, Seyter HL (eds) Non metropolitan industrialisation. Winston, Washington, DC

Erneste H, Meier V (eds) (1992) Regional development and contemporary industrial response: extending flexible specialisation. Belhaven Press, London

Esteban-Marquillas JM (1972) Shift-Share Analysis Revisited. Regional and Urban Economics 2 (3): 249–261

Fafchamps M (2004) Market institutions in sub-Saharan Africa: theory and evidence. MIT Press, Cambridge MA

Fainstein S (1983) Restructuring the city: the political economy of urban redevelopment. Longman, New York

Fairholm GW (1994) Leadership and the culture of trust. Praeger, Westport, CO

Far Eastern Economic Review (1995) Asia 1995 yearbook: a review of the events of 1994, Hong Kong, p 56

Fare R, Grosskopf S (1996) Intertemporal production frontiers: with dynamic DEA. Kluwer Academic Publishers, Boston, MA

Farrel MJ (1957) The measurement of productive efficiency. Journal of the Royal Statistical Society Series A 120 (3): 253–290

Federal Committee on Standard Metropolitan Statistical Areas (1979) The metropolitan statistical area classification. December Statistical Reporter, Washington, DC

Ferguson RO, Sargent LF (1958) Linear programming. McGraw Hill, New York

Feser EJ (1998) Enterprises, external economics and economic development. Journal of Planning Literature 12: 282–302

Feser EJ (2001) Agglomeration, enterprise size and productivity. In: Johansson B, Karlsson C, Stough RR (eds) Theories of endogenous growth. Springer, Berlin Heidelberg New York, pp 231–251

Fischer MM, Nijkamp P (1992) Geographic Information Systems and spatial analysis. Annals of Regional Science 26 (3): 3–17

Florida R (1995) Toward the learning region Futures 27 (5): 527-537

Florida R (2002) The rise of the Creative Class: and how it's transforming work, leisure, and everyday life. Basic Books, New York

Florida R (2005) The flight of the Creative Class. The new global competition for talent. HarperBusiness, HarperCollins, New York

Fosler RS (1992) State economic development policy: the emerging paradigm. Economic Development Quarterly 6(3): 22–41

Fothergill S, Gudgin G (1979) In defense of shift-share. Urban Studies 16 (3): 309–319

Friedland R, Bielby WT (1981) The power of business in the city. Urban Policy Analysis, Sage Publications, Beverly Hills, CA

Fritz OM, Sonis M, Hewings GJD (1998) A Miyazawa analysis of interactions between polluting and non-polluting sectors. Structural Change and Economic Dynamics 9: 289-305

Fukuyama F (1995) Trust: the social virtues and creation of prosperity. Free Press, New York

Galster G (1998a) A stock/flow model of defining racially integrated neighbourhoods, Journal of Urban Affairs 20: 43-51

Galster G (1998b) An econometric model of the metropolitan opportunity structure; cumulative causation among city markets, social problems and underserved areas. Fannie Mae Foundation and Urban Institute, Washington, DC

Gardner JW (1990) On leadership. The Free Press, New York

Garlick S, Taylor M. Plummer P (2006 forthcoming): An enterprising approach to regional growth: the role of VET in regional development. NCVER (National Council for Vocational Education Research), Australian National Training Authority, Adelaide

Gertler MS (1988) The limits to flexibility: comments on the post-fordist vision of production and its geography. Transactions Institute of British Geographers 13: 419–432

Giarrantani F (1991) A note on public-private partnerships and the restructuring of a regional economy. Working Paper, Department of Economics, University of Pittsburgh, Pittsburgh, PA

Gibson R (1996) Rethinking business. In: Gibson R (ed) Rethinking the future: rethinking business, principles, competition, control, leadership, markets and the world. Nicholas Brealey, London, pp 1–14

Giersch H (1995) Urban agglomeration and economic growth. Springer Berlin Heidelberg New York

Giloth R, Meier R (1989) Spatial change and social justice: alternative economic development in Chicago. In: Beauregard R (ed) Restructuring and Political Response. Sage, Newbury Park, CA

Glaser EL (1994) Cities, information, and economic growth. Cityscape 1: 9–47

Godet M (1990) Integration of scenarios and strategic management: using relevant, consistent and likely scenarios. Futures 22 (7): 730–739

Godet M (1991) From anticipation to action: a handbook of strategic prospective. UNESCO, Paris

Godet M, Roubelat F (1996) Creating the future: the use and misuse of scenarios. Long-Range-Planning 29 (2): 164–171

Golany B, Roll Y (1989) An application procedure for DEA. Omega 17 (3): 237–250

Government of Singapore (2006) Government assistance. Accessed on 12 March 2006. Available at http://www.business.gov.sg/EN/Government/GovernmentAssistance/index.htm

Grabowski R (1994) The state and economic development. Studies in Comparative International Development 29 (1): 3–17

Granovetter M (1985) Economic action and social structure: the problems of embeddedness. American Journal of Sociology 91: 481-510

Gray B (1989) Collaborating: finding common ground for multiparty problems. Jossey-Bass Publishers, San Francisco, CA

Greenhut ML (1956) Plant location in theory and in practice: the economics of space. University of North Carolina Press. Chapel Hill, NC

Grossman GM, Helpman E (1991) Innovation and growth in the global economy. MIT Press, Cambridge, MA

Grove D, Huszar L (1964) The towns of Ghana. Ghana University Press, Accra

Gunnersson J (1977) Production systems and hierarchies of centres. Martinus Nijoff, Leiden

Haddad EA, Hewings GJD (1999)The short-run regional effects of new investments and technological upgrade in the Brazilian automobile industry: an interregional computable general equilibrium analysis. Oxford Development Studies 27(3): 359-383

Haddad EA, Hewings GJD (2000) Linkages and interdependence in the Brazilian economy: an evaluation of the inter-regional input-output system. Revista Economica do Nordeste, Fortaleza, 31 (3): 330-367

Haddad EA, Hewings GJD, Sonis M (1999) Trade and interdependence in the economic growth process: a multiplier analysis for Latin America. Economia Aplicada 3 (2): 205-237

Hagerstrand T (1966) Aspects of the spatial structure of social communication and the diffusion of information. Papers, Regional Science Association 15: 27–42

Hall P (1966) The world cities. McGraw-Hill, New York

Hall P (1988) The geography of the Fifth Kondratieff. In: Kirjassa MD, Allen J (eds) Uneven re-development - cities and regions in transition. Hodder and Stoughton, London

Hall P (1995) The European city: past and future. Paper presented to The European City: Sustaining Urban Quality Conference, April, Copenhagen

Hall P (1995) The roots of urban innovation: culture, technology and urban order. In: Cities and the New Global Economy, Proceedings of an International Conference presented by the OECD and the Australian Government. Australian Government Publishing Service, Canberra, pp 275–293

Hall P, Markusen A (eds) (1985) Silicon landscapes. Allen and Unwin, Boston, MA

Hamel G (1996) Reinventing the basis for competition. In: Gibson R (ed) Rethinking the future: rethinking business, principles, competition, control, leadership, markets and the world. Nicholas Brealey, London, pp 77–92

Hamel G (1996) Strategy as revolution. Harvard Business Review, July/August: 69–71

Hamel G, Prahalad C (1993) Strategy as stretch and leverage. Harvard Business Review, March/April: 75-84

Hamel G, Prahalad C (1993) Competing for the future: breakthrough strategies for seizing control of your industry and creating the markets of tomorrow. Harvard University Press, New York

Hammer M (1997) Beyond the end of management. In: Gibson R (ed) Rethinking the future. Nicholas Brealey Publishing, London, pp 94–105

Hansen N (1992) Competition, trust and reciprocity in the development of innovative regional milleux. Papers in Regional Science 71: 95–105

Harrison B, Kluver J (1989) Reassessing the 'Massachusetts miracle': reindustrialization and balanced growth, or convergence to 'Manhattanization'? Environment and Planning A 21: 771–801

Harrison WJ, Pearson KR (1994) Computing solutions for large general equilibrium models using GEMPACK. Preliminary Working Paper No. IP-64, IMPACT Project, Monash University, Clayton

Harrison WJ, Pearson KR (1996) An introduction to GEMPACK. GEMPACK User Documentation GPD-1, 3rd edn, IMPACT Project and KPSOFT, Monash University, Clayton, Victoria, Australia

Hatry P, Fall M, Singer TO, Liner EB (1990) Monitoring the outcomes of economic development programs: a manual. The Urban Institute Press, Washington, DC

Haynes KE, Machunda ZB (1987) Considerations in extending shift-share analysis: a note. Growth and Change 18: 69–78

Haynes KE, Machunda ZB, Stough RR (1990a) New extensions of shift-share for assessing regional economic change. Paper prepared for First PRSCO Workshop, July, Bandung, Indonesia

Haynes KE, Stough RR, Shroff HFE (1990b) New methodology in context: Data Envelopment Analysis. Computers, Environment and Urban Systems 14 (2): 85–88

Haynes KE, Dinc M (1997) Productivity change in manufacturing regions: a multifactor shift share approach. Growth and Change 28: 201-221

Heenan D, Bennis W G (1999) Co-leaders: the power of great partnerships. New York: John Wiley

Held J (1996) Clusters as an economic development tool: beyond the pitfalls. Economic Development Quarterly 10: 249–261

Henton D (1995) Reinventing Silicon Valley: creating a total quality community. In: Cities and the new global economy. Proceedings of an international conference presented by the OECD and the Australian Government. Australian Government Publishing Service, Canberra, pp 306–326

Hewings GJD (1985) Regional input-output analysis. Sage Publications, Beverly Hills

Hewings GJD, Sonis M, Guo J, Israilevich PR, Schindler GR (1998) The hollowing-out process in the Chicago economy, 1975-2011. Geographical Analysis 30 (3): 217-233

Higgins B, Savoie D (1988) Regional economic development essays in honour of François Perroux. Unwin Hyman, Boston, MA

Hill M (1973) Planning for multiple objectives: an approach to the evaluation of transportation plans. Regional Science Research Institute, Philadelphia

Hill M (1973) The Goals Achievement Matrix (GAM) method of evaluation. University North Carolina, Chapel Hill, NC

Hirschman AO (1958) The strategy of economic development. Yale University Press, New Haven, CT

Hitomi K, Okuyama Y, Hewings GJD, Sonis M (2000) The role of interregional trade in generating change in the regional economies of Japan, 1980-1990. Economic Systems Research 12 (4): 515-537

Hodgetts R (1982) Management: theory, process and practice. CBS College Publishing, New York

Hodgson G (1998) The approach of institutional economics. Journal of Economic Literature, 36 (1): 166-92

Hofstede G (ed) (1997) Cultures and organizations: software of the mind. McGraw-Hill, New York

Hoover EM (1948) The location of economic activity. McGraw-Hill, New York

Hoover EM (1971) An introduction to regional economies. Alfred A. Knopf, New York

Hughes RL, Ginnett RC, Curphy GJ (1998) Contingency theories of leadership. In: Hickman GR (ed) Leading organizations: perspectives for a new era. Sage Publications, Thousand Oaks, California, pp 141-157

Humphrey J (1995) Introduction. World Development 23: 1–7

Humphrey J, Schmitiz H (1995) Principles for promoting cluster and networks of SMEs. Discussion Paper No. 1, Small and Medium Enterprise Programme. United Nations Industrial Development Organization, Vienna

Huxham C (1996) Collaborative advantage. Sage, Thousand Oaks, London

Illeris S (1993) An inductive theory of regional development. Papers in Regional Science 72: 113–134

Imbroscio D (1995) An alternative approach to urban economic development: exploring the dimensions and prospects of a 'self-reliance' strategy. Urban Affairs 30: 840-867

IMPLAN (1993) USDA forest service and Minnesota IMPLAN group. Lake Elmo, MN

Information Design Associates and ICF/Kaiser International Inc. (1997) Cluster-based economic development: a key to regional competitiveness. Report to the US Department of Commerce, Washington, DC

International Management Development–World Economic Forum (IMD–WEF) (1995) World competitiveness report. International Management Development, Lausanne, Switzerland

Irwin M, Tolbert C, Lyson T (1999) There is no place like home: nonmigration and civic management. Environment and Planning A 31 (12): 2223–2238

Isard W (1956) Location and space economy. The MT Press, Cambridge, MA

Isard W (1960) Methods of regional analysis. John Wiley, New York

Ishikawa Y (1992) The 1970s migration turnaround in Japan revisited: a shift-share approach. Papers in Regional Science 71 (2): 153–173

Israilevich PR, Hewings GJD, Sonis M, Schindler GR (1997) Forecasting structural change with a regional econometric input-output model. Journal of Regional Science 37 (4): 565-590

Jacobs D, de Man AP (1996) Clusters, industrial policy and firm strategy: a menu approach. Technology Analysis and Strategic Management 8: 425–437

Jacobs D, de-Jong MW (1992) Industrial clusters and the competitiveness of the Netherlands: empirical results and conceptual issues. Economist-Leiden 140 (2): 233–252

Jacobs J (1969) The economy of cities. Random House, New York

Jacobs M (1991) The green economy: environment, sustainable development, and the politics of the future. Pluto Press, London

Jarillo JC (1988) On strategic networks. Strategic Management Journal 9 (1): 31–42

Jensen RC, Hewings GJD, Sonis M, West GR (1988) On a taxonomy of economies. Australian Journal of Regional Studies 2: 3-24

Jessop B (1998) The narrative of enterprise and the enterprise of narrative: place marketing and the entrepreneurial city. In: Hall T, Hubbard P (eds) The entrepreneurial city: geographies of politics, regime, and representation. John Wiley & Sons, Chichester, pp 77-99

Jessop B, Sum N-L (2000) An entrepreneurial city in action: Hong Kong's emerging strategies in and for (inter)urban competition. Urban Studies, Vol. 37 (12): 2287-2313

Jin DJ, Stough RR (1996) Agile cities: the role of intelligent transportation systems in building the learning infrastructure for metropolitan development. In Proceedings of the International Symposium on Technology and society: technical expertise and public decisions, Princeton University, Princeton, NJ, pp 448-456

Jin DJ, Stough RR (1998) Learning and learning capability in the Fordist and post-Fordist age: an integrative framework. Environment and Planning A 30: 1255-1278

Johansson B, Karlsson C (2001) Geographic transition costs and specialisation opportunities of small and medium size regions: scale economies and market extension. In: Johansson B, Karlsson C, Stough RR (eds) Theories of endogenous regional growth. Springer, Heidelberg Berlin New York, pp 150–180

Johansson B, Karlsson C, Stough RR (eds) (2001a) Theories of endogenous regional growth. Springer Heidelberg Berlin New York

Johansson B, Karlsson C, Stough RR (2001b) Theories of endogenous growth. In Johansson B, Karlsson C, Stough RR (eds) Theories of endogenous regional growth. Springer, Heidelberg Berlin New York, pp 406-414

Jorgenson DW (1995) Productivity, volume 1: postwar U.S. economic growth. The MIT Press, Cambridge, MA

Judd D, Parkinson M (1990) Leadership and urban regeneration: cities in North America and Europe. Sage Publications, London

Kanter R (1995) World class: thriving locally in the global economy. Simon and Schuster, New York

Karlsson C, Johansson B, Stough RR (2001) Introduction: endogenous regional growth and policy. In: Johansson B, Karlsson C, Stough RR (eds) Theories of endogenous regional growth. Springer, Heidelberg Berlin New York, pp 3–13

Karlsson C, Johansson B, Stough RR (eds) (2005) Industrial clusters and inter-firm networks. Edward Elgar, Cheltenham, UK

Kash DE (1989) Perpetual innovation: the new world of competition. Basic Books, New York

Kasper W, Bennett J, Jackson S, Markowski S (1992) The international attractiveness of regions: a case study of the Gladstone Fitzroy region in Central Queensland. Technical Report. Centre for Management Logistics, Canberra

Keane J, Allison J (2001) The intersection of the learning region and local and regional economic development: Analysing the role of higher education. Regional Studies 33: 896-902

Kemp D, Manicaros M, Mullins P, Simpson R, Stimson R, Western J (1997) Urban metabolism: a framework for evaluating the viability, livability and sustainability of south-east Queensland. Research Monograph No 2, Queensland University of Technology, for the Australian Housing and Urban Research Institute

Kendrick JW (1973) Postwar productivity trends in the United States, 1948–1969. National Bureau of Economic Research, New York

Kendrick JW (1983) Interindustry differences in productivity growth. American Enterprise Institute, Washington, DC.

Kendrick JW (1984) Improving company productivity. The Johns Hopkins University Press, Baltimore, MD

Kessidaies C (1993) The contribution of infrastructure to economic development: a review of the experience and policy implications. World Bank Discussion Paper 213. World Bank, Washington, DC

Kidd PT (1994) Agile manufacturing. Fording new frontiers. Addison-Wesley, New York

Kilpatrick HE (1998) Empirical complexity: a study of dynamic increasing returns in the semiconductor industry and its policy implications. Ph.D. Dissertation, George Mason University, Fairfax, VA

King LJ (1984) Central place theory. Sage Publications, Beverly Hills, CA

Kirzner IM (1973) Competition and Entrepreneurship. University of Chicago Press, Chicago

Klein H (2001) The Montgomery county advanced transportation management system. In: Stough RR (ed), Intelligent transportation systems: cases and policies. Edward Elgar, Cheltenham, England, pp 110-130

Knight RV, Gappert G (eds) (1989) Cities in a global society. Sage Publications, Newbury Park

Knudsen DC, Barff R (1991) Shift-share analysis as a linear model. Environment and Planning A 23: 421–431

Kondratiev N (1935) The long waves in economic life. Review of Economic Statistics 17 (Part 2): 105–115

Korten D (1995) When corporations rule the world. Kumarian Press, West Hartford, CT

Kouzes JM, Posner BZ (1987) The leadership challenge: how to get extraordinary things done in organisations. Jossey-Bass, San Francisco, CA

Kozlowski J, Hill G (eds) (1993) Towards planning for sustainable development: a guide for the UET method. Aldershot, England

Krugman P (1981) Trade, accumulation and uneven development. Journal of Development Economics 8: 149–161

Krugman P (1990) Rethinking international trade. MIT Press, Cambridge, MA

Krugman P (1991) Geography and trade. MIT Press, Cambridge, MA

Krugman P (1995) Development, geography and economic theory. MIT Press, Boston

Kuhn TS (1996) The structure of scientific revolutions. 3rd edn, University of Chicago Press, Chicago, IL

Kuroda T, Berk R, Koizumi A, Marans R, Mizobashi I, Ness G, Niino K, Rickwell R, Thomas S (1993) An international symposium on urban metabolism: a background paper. University of Michigan Global Change Project, Ann Arbor, MI

Kurre JK, Weller BR (1989) Forecasting the local economy, using time-series and shift-share techniques. Environment and Planning A 21: 753–770

Lakshmanan TR, Johansson B (1985) Consequences of energy developments: an approach to assessment and management. In: Lakshmanan TR, Johansson B (eds): Large-scale energy projects: assessment of regional consequences. an international comparison of experiences with models and methods. Studies in Regional Science and Urban Economics Series vol. 12, North-Holland, Amsterdam, pp 1-24

Laumann E, Pappi F (1976) Networks of collective action: a perspective on community influence systems. Academic Press, New York

Laurini R, Thompson D (1992) Fundamentals of spatial information systems. Academic Press, London

Ledebur LC, Moomaw RL (1983) A shift-share analysis of regional labor productivity in manufacturing. Growth and Change 14 (1): 2–9

Lefebvre J (1982) L'analyse structurelle, méthodes et développements. Université de Paris Dauphine, Paris

Leipzieger DM (ed) (1997) Lessons from East Asia. The University of Michigan Press, Ann Arbor, MI

Leontief W (1941) The structure of the American economy. Oxford University Press, Oxford

Leontief W (1953) Studies in the structure of the American economy. Oxford University Press, Oxford

Lewis G (1993) Australia's competitiveness: a strategic - management approach. In: Lewis G, Morkel A, Hubbard G (eds) Australian strategic management. Prentice Hall Australia, Sydney, pp 345–359

Lichfield N (1996) Community impact evaluation. UCL Press, London

Likert R (1961) New patterns of management. McGraw Hill, New York

Lindblom CE (1965) The intelligence of democracy: decision making through mutual adjustment. Collier- McMillan, London

Lindfield M (1998) Multi-Sectoral investment planning in Ho Chi Minh City: industry clusters and strategic infrastructure. United Nations Development Program, Ho Chi Minh, VietNam

Liyanage S (1995) Breeding innovation clusters through collaborative research networks. Technovation 15 (9): 553–567

Lösch A (1940) Die räumliche Ordung der Wirtschaft. Jena, Fischer. Translated by Woglom WH, Stolper WF (1954).The Economics of location. Yale University Press, New Haven

Lucas RE (1988) On the mechanics of economic development. Journal of Monetary Economics 22: 3–42

Lui J (1996) A comparison of a Delphi forecast of Hawaii tourism with actual performance. Paper presented at the Annual Western Regional Science Conference, February, Napa, CA

Luke J (1988) Managing interconnectedness: the challenge of shared power. In: Bryson JM, Einsweiler RC, (eds) Shared power: what is it? How does it work? How can we make it work better? University Press of America, Lanham, MD

Maier G (2001) History, spatial structure, and regional growth: lessons for policy making. In: Johansson B, Karlsson C, Stough RR (eds) Theories of endogenous regional growth. Springer, Heidelberg Berlin New York, pp 111–134

Maillat D (1995) Territorial dynamic, innovative milieu and regional policy. Entrepreneurship and Regional Development 7: 157-65

Maillat D, Kebir L (2001) The learning region and territorial production systems. In: Johansson B, Karlsson C, Stough RR (eds)Theories of endogenous regional growth. Springer, Heidelberg Berlin New York, pp 255–277

Malecki E (1991) Technology and economic development: the dynamics of local, regional and national competitiveness. Longman Scientific and Technical, Harlow

Malecki E (1998) How development occurs: local knowledge, social capital and institutional embeddedness. Paper presented at the Annual Meeting of Southern Regional Science Association, Savannah, GA

Maler KG (1990) Economic theory and environmental degradation: a survey of some problems. Revista de Analisis Economico 5 (2): 7–17

Markusen A (1985) Profit cycles, oligopoly and regional development. MIT Press, Cambridge, MA

Markusen A, Nooponen H, Driessen K (1991) international trade, productivity, and us job growth: a shift-share interpretation. International Regional Science Review 14 (1): 15–39

Marshall A (1920) Principles of economics. 8th edn, Macmillan, London

Martin R, Sunley P (1996) Paul Krugman's geographical economics and its implications for regional development theory: A critical assessment. Economic Geography 72 (3): 259-292

Maskell P, Eskelinen H, Hannibalsson I, Malmberg A, Vatne E (1998) Competitiveness, localised learning and regional development: specialisation and prosperity in small open economies. Routledge, London

Maslow A (1954) Motivation and personality. Harper, New York.

Mason CM, Harrison R (1995) Closing the regional equity capital gap—the role of informal venture capital. Small Business Economics 7 (2): 153–172

Matsuhashi K (1997) Application of multi-criteria analysis to land-use planning. International Institute for Applied Systems Analysis Working Papers ir97091, Laxenburg, Austria

Maunsell Pty Ltd, Australian Housing and Urban Research Institute (AHURI) (1998) Brisbane gateway ports area strategy. State of Queensland Department of Economic Development and Trade, Brisbane

McGee TG (1995) System of cities and networked landscapes: new cultural formations and urban built environments in the Asia-Pacific region. Paper presented at the Pacific Rim Council on Urban Development Conference, 17-19 October, Brisbane

McGovern A (1997) Modelling the South East Queensland economy. Pacific Regional Science Conference Organization and Regional Science Association International Australia and New Zealand Annual Conference, December, Wellington, New Zealand

McGuirk PM, Winchester HPM, Dunn KM (1998) On losing the local in responding to urban decline: the Honeysuckle redevelopment, New South Wales. In: Hubbard P, Hall PH (eds) The entrepreneurial city. Geographies of politics, regime and representation. John Wiley & Sons, Chichester, pp 107-128

McKinsey & Company (1994) Lead local compete global: unlocking the growth potential of Australian regions. Office of Regional Development, Department of Housing and Regional Development, Canberra

Meeker DO (1976) 1976 Report on national growth and development: the changing issues for national growth. Department of Housing and Urban Development, Washington, DC

Merced County (1991) Merced county economic development strategy. Office of Economic and Strategic Development, Merced County, CA

Miller RE, Blair PD (1985) Input-output analysis: foundations and extensions. Prentice Hall, Englewood Cliffs, CA

Mills ES (1972): An aggregate model of resource allocation in a metropolitan area. In: Edel M, Rothenberg MJ (eds) Readings in urban economics. Macmillan, New York

Minkin B (1995) Future in sight: 100 most important trends, implications and predictions that will most impact businesses and the world economy into the 21st century. McMillan, New York

Mintzberg H (1990) Strategy formation: schools of thought. In: Fredrickson J (ed) Perspectives on strategic management. Harper Business, New York, pp 105–235

Mintzberg H (1994) The rise and fall of strategic planning. Harvard Business Review, Jan/Feb: 107–114

Mintzberg H, Quinn J (1992) The strategy process: concepts and context. Prentice Hall, New Jersey

Mitchell WC (1927) Business cycles. The problem and its setting. National Bureau of Economic Research, New York

Moore JE (1996) The death of competition: leadership and strategy in the age of business ecosystems. Harper Press New York

Munasinghe M (1993) Environmental economics and sustainable development. World Bank Environment Paper No 3, The World Bank, Washington, DC

Munasinghe M (1994) The economist's approach to sustainable development. In: Serageldin I, Steer A (eds) Making development sustainable, environmentally sustainable development. Occasional Paper Series, vol. 2, World Bank, Washington, DC

Munnell AL (ed) (1991) Is there a shortfall in public capital investment? Federal Reserve Bank of Boston, Boston, MA

Myrdal G (1957) Economic theory and underdeveloped regions. Duckworth Press, London

Naisbitt J (1994) Global paradox: the bigger the world economy, the more powerful its smallest players. Allen and Unwin, St. Leonards, NSW, Australia

Nelson RR, Winters SG (1982) An evolutionary theory of economic change. Harvard University Press, Cambridge, MA

Neustadt RE, May ER (1990) Thinking in time: the uses of history for decision makers. Free Press, New York

Newman P, Birrell B, Holmes D, Mathers C, Newton P, Oakley G, O'Connor A, Walker B, Spessa A, Tait D (1996) Human settlements, Australia: state of the environment, 1996. CSIRO Publishing, Collingwood

Newton P (ed) (1997) Re-shaping cities for a more sustainable future: exploring the link between urban form, air quality, energy and greenhouse gas emissions. Australian Housing and Urban Research Institute Research Monograph No 6, Queensland University of Technology, Brisbane

Nijkamp P, Scholten HJ (1993) Spatial information systems: design, modelling, and use in planning. International Journal of Geographical Systems 7(1): 85–96

Nijkamp P, Rietveld P, Voogd M (1990) Multi-criteria evaluation in physical planning. North Holland, Amsterdam

Nooponen H, Markusen A, Driessen K (1997) Trade and American cities: who has the comparative advantage? Economic Development Quarterly 11 (1): 67-87

North D (1990) Institutions, institutional change and economic performance. Cambridge University Press, Cambridge

Norton RD, Rees J (1979) The product cycle and the decentralization of north American manufacturing. Regional Studies 13: 141–151

O'Connor K (1996) The Australian capital city report. Centre for Population and Urban Research, Monash University and the Australian Housing and Urban Research Institute, Melbourne

Organization for Economic Co-operation and Development (OECD) (1986) The revitalization of urban economies. OECD, Paris

Organization for Economic Co-operation and Development (OECD) (2000) Learning regions and cities: knowledge, learning and regional innovation systems. OECD, Paris

Office of Management and Budget (1990) Revised standards for defining metropolitan areas in the 1990s. Notice, Federal Register, March 30, Washington, DC

Ohmae K (1995) The end of the nation state: the rise of regional economics. Free Press, New York

Oppenshaw S (1990) Spatial analysis and geographical information systems: a review of progress and possibilities. In: Scholten HJ, Stillwell JCM (eds) Geographic information systems in urban and regional planning. Kluwer, Dordrecht, pp 153–163

Osborne D (1988) Laboratories of democracy. Harvard Business School Press, Cambridge, MA

Ouchi WG (1981) Theory Z: American business can meet the Japanese challenge. Addison Wesley, Reading, MA

Park SO (1995) Networks and competitive advantages of new industrial districts. Paper presented at Pacific Regional Science Conference Organisation, 14th Biennial Conference, July, Taipei

Parker S (1999) The future has landed. Business Review Weekly. September 28: 88–89

Parkinson M (1990) Leadership and regeneration in Liverpool: confusion, confrontation, or coalition. In Judd D, Parkinson M (eds) Leadership and urban regeneration Newbury Park: Sage Publications, pp 241-257

Patten C (1991) The competitiveness of small firms. Cambridge University Press, Cambridge

Pearce D (1997) The MIT dictionary of modern economics. MIT Press, Boston, MA

Pearce D, Markandya A, Barbier EB (1989) Blueprint for a green economy. Earthscan, London

Perloff HS, Dunn ES, Lampard EE, Muth RF (1960) Regions, resources and economic growth. University of Nebraska Press, Lincoln, NE

Perring C (1991) Ecological sustainability and environmental research. Centre for Resource and Environmental Studies, Australian National University, Canberra

Perrott B (1996) Managing strategic issues in the public service. Long Range Planning 29 (3): 337–345

Perroux F (1950) Economic space: theory and application. Journal of Economics 4: 90–97

Plummer P, Taylor M (2001) Theories of Local Economic Growth: Concepts, models and measurement, Environment and Planning A 33: 385-399

Porter ME (1980) Competitive strategy: techniques for analyzing industries and competitors. New York Free Press, New York

Porter ME (1985) Competitive advantage: creating and sustaining superior performance. New York Free Press, New York

Porter ME (1986) Competition in global industries. Cambridge, Harvard Business School Press, Boston, MA

Porter ME (1990) The competitive advantage of nations. MacMillan, New York

Porter ME (1991a) The competitiveness advantage of Massachusetts. Harvard Business School, Boston, MA

Porter ME (1991b) Towards a dynamic theory of strategy. Strategic Management Journal 12: 95–117

Porter ME (1996) What is strategy? Harvard-Business-Review 74 (6): 61–78

Porter ME (1998) Location, clusters and the "new" micro-economics of competition. Business-Economics 33 (1): 7–13

Porter ME (1998) On competition. Harvard Business School Press, Boston

Porter ME (2000) Locations, clusters and company strategy. In: Clark G, Feldman M, Gertler M (eds) The Oxford handbook of economic geography. Oxford University Press, Oxford, pp 253-74

Prahalad C (1996) Strategies for growth. In: Gibson R (ed) the Rethinking the future: rethinking business: principles, competition, control, leadership, markets and world. Nicholas Brealey, London, pp 49–61

Prud-homme R (1995) On the economic role of cities, in cities and the new global economy. In: Proceedings of International Conference presented by the OECD and the Australian Government. Australian Government Publishing Service, Canberra, pp 728–742

Putnam R (1993) The prosperous community: social capital and public life. The American Prospect Spring: 35–42

Putnam RD, Leonardi R, Nanetti R (1993) Making democracy work: civic traditions in modern Italy. Princeton University Press, Princeton, NJ

Qiangsheng L (1997) Regional dynamics and growth advantages of the Washington metropolitan economy: an extended shift-share approach. PhD Dissertation, the Institute of Public Policy, George Mason University, Fairfax, VA

Rebelo S (1991) Long run policy analysis and long run growth. Journal of Political Economy 98: S71–S102

Rees J (1979) State technology programs and industry experience in the USA. Review of Urban and Regional Development Studies 3: 39–59

Rees J (2001) Technology and regional development: theory revisited. In: Johansson B, Karlsson C, Stough RR (eds) Theories of endogenous regional growth. Springer, Heidelberg Berlin New York, pp 94–110

Rees J, Hewings GJ, Stafford H (1981) Industrial locations and regional systems. Bergin, New York

Reich RB (1991) The work of nations: preparing ourselves for 21st century capitalism. 2nd edn, Vintage Books, New York

Richardson HW (1973) Regional growth theory. MacMillan, London

Richardson HW (1978) The state of regional economies: a review article. International Regional Science Review 31: 1–48

Rigby DL (1992) The impact of output and productivity changes on manufacturing employment. Growth and Change 23: 405–427

Rigby DL, Anderson WP (1993) Employment change, growth and productivity in Canadian manufacturing: an extension and application of shift-share analysis. Canadian Journal of Regional Science XVI (1): 69–88

Robbins G (1990) Scenario planning: a strategic alternative. Public Management 77 (3): 4–8

Roberts BH (1997) Inventing a future for regions: new frameworks for regional economic development planning. Paper presented to the Annual Conference of Western Regional Science Association, February, Hawaii

Roberts BH (1999) Anticipating regional risk. Paper presented at the Australian and New Zealand Regional Science Association Annual Conference, September, Newcastle

Roberts BH (2000) Benchmarking the competitiveness of the Far North Queensland regional economy. Queensland University of Technology, Brisbane

Roberts BH (2003) Regional risk management and economic development. Australasian Journal of Regional Sciences 9(1): 67-96

Roberts BH, Dean J (2001) An economic development strategy for Cairns. In: Williams J, Stimson R (eds) *International urban planning settings: lessons of success*. International Review of Comparative Public Policy. Elsevier, London

Roberts BH, Lindfield M (2000) Managing the provision infrastructure in support of industry cluster development: the Case of Ho Chi Minh City. Journal of Public Affairs Management (1): 115 -147

Roberts BH, Stimson RJ (1998) Multi-sectoral qualitative analysis: a tool for assessing the competitiveness of regions and developing strategies for economic development. Annals of Regional Science 32 (4): 459–467

Robertson R (1992) Globalisation: social theory and global culture. Sage Publications, Beverly Hills, CA

Robinson CJ (1989) Municipal approaches to economic development. Journal of the American Planning Association 55: 283-295

Roelandt T, den Hertog P, van Sinderen J, Vollaard B (1997) Cluster analysis and cluster policy in the Netherlands. Paper Presented at an OECD Workshop, October, Amsterdam

Romer PM (1986) Increasing returns and long run growth. Journal of Political Economy 94: 1002–1037

Romer PM (1990) Endogenous technological change. Journal of Political Economy 98: S71–S102

Rosebury N (1994) Exploring the black box: technology, economics and history. Cambridge University Press, Cambridge

Rosenfeld S (1997) Bringing business clusters into the mainstream of economic development. European Planning Studies 5: 3–23

Rost JC (1991) Leadership for the twenty-first century. Praeger, New York

Rostow WW (1960) The process of economic growth. 2nd edn, Clarendon Press, Oxford

Rowen SH (ed) (1998) Behind East Asian growth: the political and social foundations of prosperity. Routledge, New York, London

Rugman AM (1991) Diamond in the rough. Business Quarterly 55: 61–63

Saaty TL (1980) The analytic hierarchy process. McGraw-Hill, New York

Sable CF (1989) Flexible specialization and the re-emergence of regional economics. In: Hirst P, Zeitlin J (eds) Reversing industrial decline. Berg Publishers, Oxford

Saint-Paul R, Teniere-Buchot P (1974) Innovation et évaluation techniques, selection des projects, méthodes et prévisions. Enterprise Moderne d'Editions, Paris

Salazar M, Stough RR (2006) Sovereignty and economic development with some examples from the Atlantic community. In: Eaton D (ed) The end of sovereignty? A transatlantic perspective. Transatlantic Public Policy Series, Vol. 2, Muenster: LIT Verlag, pp 287-308

Sarafoglou N, Haynes KE (1990) Regional efficiencies of building sector research in Sweden: an introduction. Computers, Environment and Urban Systems 14 (2): 117–132

Sassen S (1994) Cities in a global economy. Pine Forge Press, Thousand Oaks.

Saxenian A (1994) Regional advantage: culture and competition in Silicon Valley and Route 128. Harvard University Press, Cambridge, MA

Schoemaker P (1995) Scenario planning: a tool for strategic thinking. Sloan Management Review 36: 25–40

Schriner J (1995) Picking your neighborhood. Industry-Week 244: 71

Schumpeter JA (1934) The theory of economic development. Harvard University Press, Cambridge, MA

Scott AJ (1986) Industrial organization and location: division of labor, the firm, and spatial process. Economic Geography 63: 215–231

Scott AJ (1988) New industrial spaces: flexible production organization and regional development in North America and Western Europe. Pion, London

Scott AJ (1992) The collaborative order of flexible production agglomerations: lessons for local economic development policy and strategic choice. Economic Geography 68: 219–233

Scott AJ, Storper M (1992) Industrialization and regional development. In: Scott AJ, Storper M (eds) Pathways to industrialization and regional development. Routledge, New York, pp 3–17

Segil L (1996) Intelligent business alliances: how to profit using today's most important strategic tool. Times Books, London

Segil L (1998) Strategic business alliances. Random House, New York

Seiford LM, Thrall RM (1990) Recent developments in DEA; the mathematical programming approach to frontier analysis. Journal of Econometrics 46 (1): 7–38

Sengupta JK (1995) Dynamics of Data Envelopment Analysis. Kluwer Academic Publishers, Boston

Shaffer R (1989) Community economies. Iowa State University Press, Ames, IA

Sheppard E, Leitner H (1998) Economic uncertainty, inter-urban competition and the efficacy of entrepreneurialism. In: Hall T, Hubbard P (eds) The entrepreneurial city: geographies of politics, regime, and representation. John Wiley & Sons, Chichester, pp 285-307

Sihag BS, McDonough CC (1989) Shift-share analysis: the international dimension. Growth and Change 20(3): 80–88

Sihag BS, McDonough CC (1991) The incorporation of multiple bases into shift-share analysis. Growth and Change 22:1-9

Silicon Valley Joint Venture Network (1995): The joint venture way: lessons for regional rejuvenation. Joint Venture Silicon Valley Network

Simmie J (1997) Introduction. In: Simmie J (ed) Innovation, networks and learning regions? Jessica Kingsley, London, pp 1–23

Simon JL, Kahn H (eds) (1984) The resourceful earth: A response to Global 2000. Blackwell, Oxford

Skott P, Auerbach P (1995) Cumulative causation and the "new" theories of economic growth. Journal of Post Keynesian Economics 17(3): 381-402

Smilor RW, Wakelin M (1990) Smart infrastructure and economic development: the role of technology and global networks. In: Kosmetsky G, Smilor RW (eds) The technopolis phenomenon. IC2 Institute, University of Texas, Austin, TX, pp 53–75

Smith DM (1971) Industrial location: an economic geographical analysis. John Wiley and Sons, New York

Smith SE (1991) Shift share analysis of change in occupational sex composition. Social Science Research 20: 437–453

Sohal A, Perry M, Pratt T (1998) Developing partnerships and networks: learning from practices in Australia. Technovation 18 (4): 245–251

Solow RM (1956) A contribution to the theory of economic growth. Quarterly Journal of Economics 70: 65–94

Solow RM (2000) Growth theory: an exposition. Oxford University Press, New York

Sonis M, Hewings GJD (1998) Temporal Leontief inverse. Macroeconomic Dynamics 2: 89-114

Sonis M, Hewings GJD (1999) Economic landscapes: multiplier product matrix analysis for multiregional input-output systems. Hitotsubashi Journal of Economics 40 (1): 59-74

Sonis M, Hewings GJD, Lee JK (1994) Interpreting Spatial economic structure and spatial multipliers: three perspectives. Geographical Analysis 26 (2): 122-151

Sonis M, Guilhoto JJM, Hewings GJD, Martins EB (1995) Linkages, key sectors, and structural change: some new perspectives. The Developing Economies 33(3): 233-270

Sonis M, Hewings GJD, Guo J (2000) A new image of classical key sector analysis: minimum information decomposition of the Leontief inverse. Economic Systems Research 12 (3): 401-423

Sorkin DL, Ferris NB, Hudak J (undated) Strategies for cities and countries: a strategic planning guide. US Department of Housing and Urban Development, Office of Policy Development and Research, and Public Technology Inc., Washington, DC

Speigel H W (1991) The growth of economic thought. Duke University Press, London

Starrett D (1978) Market allocations of local choice in a model with free mobility. Journal of Economic Theory 17: 21–37

Steiner M (2001) Clustering and economic change: new policy orientations—the case of Styria. In: Johansson B, Karlsson C, Stough RR (eds) Theories of endogenous regional growth. Springer Heidelberg Berlin New York, pp 278-298

Steinfels P (1999) A social ethicist carries the language of economics into the world of religion, and, though history's tide opposes him, finds benefits for both. New York Times: 7

Sternburg E (1991) The sectoral clusters in economic development policy; lessons from Rochester and Buffalo. Economic Development Quarterly 4: 342–356

Stevens BH, Moore CL (1980) A critical review of the literature on shift-share as a forecasting technique. Journal of Regional Science 20 (4): 419–437

Stigler GJ (1951) The division of labour is limited by the extent of the market. Journal of Political Economy 59: 185–193

Stimson RJ (1991a) A report on strategies for the development, growth and management of Brisbane City. Brisbane City Council, Brisbane

Stimson, RJ (1991b): Brisbane magnet city, strategies for the development, growth and management of Brisbane. Brisbane City Council, Brisbane

Stimson RJ, Stough RR, Roberts BH (eds) (2002) Regional economic development. Springer, Heidelberg Berlin New York

Stimson RJ, Robson A, Shyy T-K (2004) Shift share analysis and modelling growth across Queensland's regions. Paper presented to North American Regional Science Association International, November, Seattle, WA

Stimson RJ, Stough RR, Salazar M (2005) Leadership and institutional factors in endogenous regional economic development. Investigaciones Regionales, 7 (October): 23-52

Storper M, Scott A J (eds) (1992) Pathways to industrialization and regional development. Routledge, London

Stough RR (1990) Potentially irreversible global trends and changes: local and regional strategies for survival. American Association for the Advancement of Science Annual Conference, 17 -20 February, New Orleans

Stough RR (1995) Industry sector analysis of the Northern Virginia Economy. In: Stough RR (ed) Proceedings of the Second Annual Conference on the Future of the Northern Virginia Economy. Center for Regional Analysis, The Institute of Public Policy and the Northern Virginia Business Roundtable, George Mason University, Fairfax, VA, pp 3–37

Stough RR (1997) Linking technology and traditional sectors of the Northern Virginia economy. In: Stough RR (ed) Proceedings of the Fifth Annual Conference on the Future of the Northern Virginia Economy, May 28, Center for Regional Analysis, School of Public Policy, George Mason University, Fairfax, VA, pp 72-106

Stough RR (1998) Endogenous growth in a regional context. Annals of Regional Science 32: 1-5

Stough RR (2001) Endogenous growth theory and the role of institutions in regional economic development. In: Johansson B, Karlsson C, Stough RR (eds) Theories of endogenous regional growth. Springer, Heidelberg Berlin New York, pp 17–48

Stough RR, Maggio ME (1994) Evaluating IVHS/ITS transportation infrastructure in a metropolitan area. In: Proceedings of the Korea-USA Symposium on IVHS and GIS-T. U.S. and Korean National Science Foundations, Seoul, Korea

Stough RR, Popino J, Campbell HC (1995) Technology in the Greater Washington region. Technical Report, Center for Regional Analysis, Institute of Public Policy, George Mason University, Fairfax, VA

Stough RR, Haynes KE, Campbell HS Jr. (1997) Small business entrepreneurship in the high technology services sector: an assessment for the edge cities of the U.S. national capital region. Small Business Economics 9: 1–14

Stough RR, Kulkarni R, Riggle J, Haynes KE (2000) Technology and industrial cluster analysis: some new methods. Paper presented at the 1st South Africa Regional Science Meeting, January, Port Elizabeth, South Africa

Stough RR, Kulkarni R, Riggle J (2000) Technology in Virginia's regions. Virginia's Center for Innovative Technology, Herndon, VA

Stough RR, DeSantis MF, Stimson RJ, Roberts BH (2001) Leadership in regional economic development strategic planning. In: Kumssa A, McGee, TG (eds) New regional development paradigms. Edington DW, Fernandez AL, Hoshino C (eds) Volume 2: New regions – concepts, issues and development practices. Greenwood Press in cooperation with the United Nations and the United Nations Centre for Regional Development, Westport, CN, pp 175-191

Stough RR, Kulkarni R, Paelinck J (2002) ICT and knowledge challenges for entrepreneurs in regional economic development. In: Acs ZJ, Groot HLF, Nijkamp P (eds), The emergence of the knowledge economy. Springer, Heidelberg Berlin New York, pp. 195-214

Stough RR, Maggio ME, Jin D (2001) Methodological and technical challenges in regional evaluation of ITS: induced and direct effects. In: Stough RR (ed) Intelligent transportation systems: cases and policies. Edward Elgar, Cheltenham, pp 13-46

Streeck W (1992) Social Institutions and Economic Performance: Studies in Industrial Relations in Advanced Capitalist Economies. Sage Publications, Newburg Park, CA

Sui DZ (1995) Spatial economic impacts of new town development in Hong Kong: a GIS-based shift shift-share analysis. Socio-Economic Planning Science 29 (3): 227–243

Sun Tzu (600 BC) (1988) The art of war. T. Cleary, Trans. Random House, Boston, MA

Sweeney SH, Feser EJ (1998) Plant size and clustering of manufacturing activity. Geographical Analysis 31 (1): 45–64

Tapscott D (1996) The digital economy: promise and peril in the age of networked intelligence. McGraw Hill, New York

Taylor M (2003) Drivers of local growth: ideologies, ambiguities and policies. Paper presented to the Annual Conference of Australia and New Zealand Regional Science Association International, 28 September- 1 October, Freemantle, WA

Teitz M, Blakely EJ (1985) Unpublished course materials. University of California, Berkeley, CA

Theil H, Gosh R (1980) A comparison of shift-share and the RAS adjustment. Regional Science and Urban Economics 10: 175–180

Thomas MD (1975) Growth pole theory, technological change and regional economic growth. Papers of the Regional Science Association 34: 3–25

Thrift N (2001) It's the romance, not the finance that makes the business worth pursuing: disclosing a new market culture. Economy and Society 30 (4): 412-432

Tiebout C (1962) The community economic base study. Committee for Economic Development, New York

Toffler A (1980) The third wave. William Morrow, New York

Toft GS, Stough RR (1986) Transportation employment as a source of regional economic growth: a shift-share approach. Transportation Research Board, National Research Council, Washington, DC

Turner RK (1993) Sustainability: principles and practices. In: Turner RK (ed) Sustainable environmental economics and management. Belhaven Press, London

United Nations (2005) UN Millenium development goals 2000. Available at http://www.un.org/millenniumgoals

United Nations Department of Economic and Social Affairs (2004) Agenda 21. Published on 15 December 2004. Available at http://www.un.org/esa/sustdev/documents/agenda21/

United Nations Department of Economic and Social Affairs (2005) Earth summit 2002: Johannesburg plan of implementation. Published on 11 August 2005. Available at http://www.un.org/esa/sustdev/documents/WSSD_POI_PD/English/POIToc.htm

United Nations Framework Convention on Climate Change (2004) Kyoto protocol: framework on climate change. Published on 02 August 2004. Available at http://unfccc.int/resource/docs/convkp/kpeng.html

U.S. Bureau of the Census (1969) Census of county business patterns. U.S. Department of Commerce, Washington, DC

U.S. Bureau of the Census (1982) Census of county business patterns. U.S. Department of Commerce, Washington, DC

U.S. Bureau of the Census (1995) County and city data book. U.S. Department of Commerce, Washington, DC

U.S. Department of Housing and Urban Development (HUD) (1996) America's new economy and the challenge of cities. HUD Report on Metropolitan Economic Strategy, Washington, DC

Ullman EL, Dacey MF (1960) The minimum requirements approach to the urban economic base. Papers of the Regional Science Association 6: 175–194

Vazquez-Barquero A (2002) Endogenous development. Networking, innovation, institutions and cities. Routledge, London

van der Linden JA, Oosterhaven J, Cuello FA, Hewings GJD, Sonis M (2000) Fields of influence of productivity change in EU intercountry input-output tables, 1970-1980. Environment and Planning A 32: 1287-1305 van Lier HN (1994) Land use planning in perspective of sustainability: an introduction. In: de Buck AJ, Jaarsma CF, Jurgens CR, van Lier HN (eds) Sustainable land use planning, ISOMUL developments in landscape management and urban planning. Elsevier, Amsterdam, pp 1–12

Van Neuman J, Morgenson O (1943) Theory of games and economic behaviour. Princeton University Press, Princeton, NJ

Vernez-Moudon A, Hubner M (eds) (2000) Monitoring land supply with geographic information systems. John Wiley and Sons, New York

Vining DR, Dobronyi JB, Otness MA, Schwinn JE (1982) A Principal Axis Shift in the American Spatial Economy. Professional Geographer 34 (3): 270–278

von Thünen KH (1826) Der isolierte Staat in Beziehung auf Landwirtschaft und National-ökonomie. Perthes, Hamburg. (Wartemberg CM (1966) Von Thünen's isolated State. English translation, Pergamon, Oxford)

Waites M (1995) Economic development: building and economic future. State Government News 38, Phoenix, AZ

Waites M (2000) The added value of the industry cluster approach to economic analysis, strategy development, and service delivery. Economic Development Quarterly 14 (1): 35-50

Ward D, Stimson R, Murray A (2000) A spatial decision support system model for planning real time optimal allocation of regional growth: a case study of the Gold Coast Sub-region in South East Queensland. Paper presented at the 6th World Congress of the Regional Science Association International, May, Lugarno, Switzerland

Warren ER (1980) States and urban strategies. National Academy of Public Administration, and Department of Housing and Urban Developments, Washington, DC

Webber MJ (1984) Industrial location. Sage Publications, Beverly Hills, CA

Weber D (1990) A spatial decision support system for bank location: a case study. Technical Report 90–9. National Center for Geographic Information and Analysis, Santa Barbara, CA

West G (1998) Structural change and regional and urban economics. Paper presented at the Annual Conference of the Australian and New Zealand Regional Science Association, September, Tanunda, South Australia

White RW (1977) Dynamic central place theory: results of a simulation approach. Geographical Analysis 9: 226–243

Whitt JA, Lammers JC (1991) The art of growth: ties between development organisations and performing arts. Urban Affairs Quarterly 26: 376–93

Whyte W (1968) The last landscape. Doubleday, New York

Wilde P (1996) Heuristic evaluation. Available at http://www.dcs.qmw.ac.uk/courses/ISD.local/students/eval/Heuristic_Evaluation.html

Williams I (1996) Networks, clusters and export development. Firm connections. Trade NZ 4 (1): 10-12

Williams O (1975) Markets and hierarchies. The Free Press, New York

Williamson OE (1994) Institutions and economic organization – the governance perspective. The World Bank, Washington, DC

Wolman H (1987) U.S. urban economic performance: what accounts for success and failure? Journal of Urban Affairs 9: 1-17

Wolpert J (1988) The geography of generosity: metropolitan disparities in donations and support for amenities. Annals of American Geographers 78: 665–679

Womack JP, Jones DT, Roos D (1990) The machine that changed the world. Rawson Associates, New York

Worbys M (1995) GIS: a computing perspective. Taylor and Francis, London

Yang G, Stough RR (2005) A preliminary analysis of functional spatial clustering: the case of the Baltimore metropolitan region. In: Karlsson C, Johansson B, Stough RR (eds) Industrial clusters and inter-firm networks. Edward Elgar, Cheltenham, UK, pp 303-320

Yves L, Hamel G (1998) Alliance advantage: the art of creating value through partnering. Harvard Business School, Boston, MA

Index